Spatial Augmente

Spatial Augmented Reality

Merging Real and Virtual Worlds

Oliver Bimber
Bauhaus-University, Weimar

Ramesh Raskar
Mitsubishi Electric Research Laboratory,
Cambridge, MA

A K Peters
Wellesley, Massachusetts

Editorial, Sales, and Customer Service Office

A K Peters, Ltd.
888 Worcester Street, Suite 230
Wellesley, MA 02482
www.akpeters.com

Library of Congress Cataloging-in-Publication Data

Bimber, Oliver, 1973–
Spatial augmented reality : merging real and virtual worlds / Oliver Bimber, Ramesh Raskar.
p. cm.
Includes bibliographical references and index.
ISBN 1-56881-230-2
1. Computer graphics- 2. Virtual reality. I. Raskar, Ramesh II. Title.

T385.B5533 2004
006.8–dc22

2005043110

Printed in the United States of America
09 08 07 06 05 10 9 8 7 6 5 4 3 2 1

To Mel

—O. B.

To my parents

—R. R.

Contents

Preface xi

1 A Brief Introduction to Augmented Reality 1
- 1.1 What is Augmented Reality 1
- 1.2 Today's Challenges 4
- 1.3 Spatial Augmented Reality 7
- 1.4 Outline of the Book 8

2 Fundamentals: From Photons to Pixels 13
- 2.1 Light in a Nutshell 14
- 2.2 Geometric Optics 17
- 2.3 Visual Depth Perception 30
- 2.4 Rendering Pipelines 44
- 2.5 Summary and Discussion 66

3 Augmented Reality Displays 71
- 3.1 Head-Attached Displays 72
- 3.2 Hand-Held Displays 79
- 3.3 Spatial Displays 83
- 3.4 Summary and Discussion 90

4 Geometric Projection Concepts 93
- 4.1 Geometric Model 93
- 4.2 Rendering Framework 98
- 4.3 Calibration Goals 104
- 4.4 Display Environments and Applications 108
- 4.5 Summary and Discussion 108

5 Creating Images with Spatial Projection Displays 111
5.1 Planar Displays 111
5.2 Non-Planar Display 126
5.3 Projector Overlap Intensity Blending 129
5.4 Quadric Curved Displays 133
5.5 Illuminating Objects 142
5.6 Summary and Discussion 147

6 Generating Optical Overlays 149
6.1 Transparent Screens 150
6.2 Mirror Beam Combiners 152
6.3 Planar Mirror Beam Combiners 152
6.4 Screen Transformation and Curved Screens 163
6.5 Moving Components 166
6.6 Multi-Plane Beam Combiners 166
6.7 Curved Mirror Beam Combiners 174
6.8 Summary and Discussion 206

7 Projector-Based Illumination and Augmentation 213
7.1 Image-Based Illumination: Changing Surface Appearance . 214
7.2 Creating Consistent Occlusion 220
7.3 Creating Consistent Illumination 227
7.4 Augmenting Optical Holograms 244
7.5 Augmenting Flat and Textured Surfaces 255
7.6 Augmenting Geometrically Non-Trivial Textured Surfaces . 267
7.7 Summary and Discussion 276

8 Examples of Spatial AR Displays 279
8.1 Shader Lamps 280
8.2 Being There 286
8.3 iLamps: Mobile Projectors 288
8.4 The Extended Virtual Table 291
8.5 The Virtual Showcase 297
8.6 The HoloStation 302
8.7 Augmented Paintings 308
8.8 Smart Projectors 314
8.9 Summary and Discussion 320

9 The Future 321
9.1 Displays 321
9.2 Supporting Elements 326
9.3 Summary and Discussion 329

A Calibration of a Projector (or a Camera) 331
A.1 Source code for Calibration of a Projector (or a camera) . . 331
A.2 Quadric Image Transfer 334

B OpenGL's Transformation Pipeline Partially Re-Implemented 337
B.1 General Definitions 337
B.2 Projection Functions 337
B.3 Transformation Functions 338
B.4 Additional Functions 340

Bibliography 343

Index 363

Preface

Spatial Augmented Reality is a rapidly emerging field which concerns everyone working in digital art and media who uses any aspects of augmented reality and is interested in cutting-edge technology of display technologies and the impact of computer graphics. We believe that a rich pallet of different display technologies, mobile and non-mobile, must be considered and adapted to fit a given application so that one can choose the most efficient technology. While this broader view is common in the very established area of virtual reality, it is becoming more accepted in augmented reality which has been dominated by research involving mobile devices.

This book reflects our research efforts over several years and the material has been refined in several courses that we taught at the invitation of Eurographics and ACM SIGGRAPH.

Who Should Read This Book

In order for a broad spectrum of readers—system designers, programmers, artists, etc— to profit from the book, we require no particular programming experience or mathematical background. However, a general knowledge of basic computer graphics techniques, 3D tools, and optics will be useful.

The reader will learn about techniques involving both hardware and software to implement spatial augmented reality installations. Many Cg and OpenGL code fragments, together with algorithms, formulas, drawings, and photographs will guide the interested readers who want to experiment with their own spatial augmented reality installations.

By including a number of exemplary displays examples from different environments, such as museums, edutainment settings, research projects, and industrial settings, we want to stimulate our readers to imagine novel

AR installations and to implement them. Supplementary material can be found at http://www.spatialar.com/.

About the Cover

The images at the top of the front cover show a rainbow hologram of a dinosaur (Deinonychus) skull (found in North America). It has been augmented with reconstructed soft tissue and artificial shading and occlusion effects. The soft tissue data, provided by Lawrence M. Witmer of Ohio University, were rendered autostereoscopically. A replica of the skull was holographed by Tim Frieb at the Holowood holographic studio in Bamberg, Germany. The hologram was reconstructed by projected digital light that could be controlled and synchronized to the rendered graphics. This enabled a seamless integration of interactive graphical elements into optical holograms.

The image at the bottom show an example of Shader Lamps: an augmentation of a white wooden model of the Taj Mahal with two projectors. The wooden model was built by Linda Welch, George Spindler and Marty Spindler in the late 1970s. In 1999, at the University of North Carolina, Greg Welch spray painted the wooden model white and Kok-Lim Low scanned it with a robotic arm to create a 3D model. The wooden model is shown illuminated with images rendered with real time animation of a sunrise.

The art work on the back cover was created by Matthias Hanzlik. The sketches show early concepts of SAR prototypes. They have all been realized and are described in the book.

Acknowledgements

The material presented in this book would not have been realized without the help and dedication of many students and colleagues. We want to thank them first of all: Gordon Wetzstein, Anselm Grundhöfer, Sebastian Knödel, Franz Coriand, Alexander Kleppe, Erich Bruns, Stefanie Zollmann, Tobias Langlotz, Mathias Möhring, Christian Lessig, Sebastian Derkau, Tim Gollub, Andreas Emmerling, Thomas Klemmer, Uwe Hahne, Paul Föckler, Christian Nitschke, Brian Ng, Kok-Lim Low, Wei-Chao Chen, Michael Brown, Aditi Majumder, Matt Cutts, Deepak Bandyopadhyay, Thomas Willwacher, Srinivas Rao, Yao Wang, and Johnny Lee.

We also want to thank all colleagues in the field who provided additional image material: Vincent Lepetit, Pascal Fua, Simon Gibson, Kiyoshi Kiyokawa, Hong Hua, Susumu Tachi, Daniel Wagner, Dieter Schmalstieg, George Stetten, Tetsuro Ogi, Barney Kaelin, Manfred Bogen, Kok-Lim Low, Claudio Pinhanez, Masahiko Inami, Eric Foxlin, Michael Schnaider, Bernd Schwald, and Chad Dyner.

Special thanks go to those institutions and people who made this research possible: Bauhaus-University Weimar, Mitsubishi Electric Research Laboratories, Jeroen van Baar, Paul Beardsley, Paul Dietz, Cliff Forlines, Joe Marks, Darren Leigh, Bill Yerazunis, Remo Ziegler, Yoshihiro Ashizaki, Masatoshi Kameyama, Keiichi Shiotani, Fraunhofer Institute for Computer Graphics, José L. Encarnação, Bodo Urban, Erhard Berndt, University of North Carolina at Chapel Hill, Henry Fuchs, Greg Welch, Herman Towles, Fraunhofer Center for Research in Computer Graphics, Miguel Encarnação, Bert Herzog, David Zeltzer, Deutsche Forschungsgemeinschaft, and European Union.

We thank the Institution of Electronics & Electrical Engineers (IEEE), the Association for Computing Machinery (ACM), Elsevier, MIT Press, Springer-Verlag, and Blackwell Publishing for granting permission to republish some of the material used in this book.

Many thanks to Thomas Zeidler and to Alice and Klaus Peters for helping us manage the process of writing this book alongside our everyday duties.

Oliver Bimber
Weimar April 2005

Ramesh Raskar
Cambridge, MA April 2005

1

A Brief Introduction to Augmented Reality

Like *Virtual Reality* (VR), *Augmented Reality* (AR) is becoming an emerging edutainment platform for museums. Many artists have started using this technology in semi-permanent exhibitions. Industrial use of augmented reality is also on the rise. Some of these efforts are, however, limited to using off-the-shelf head-worn displays. New, application-specific alternative display approaches pave the way towards flexibility, higher efficiency, and new applications for augmented reality in many non-mobile application domains. Novel approaches have taken augmented reality beyond traditional eye-worn or hand-held displays, enabling new application areas for museums, edutainment, research, industry, and the art community. This book discusses *spatial augmented reality* (SAR) approaches that exploit large optical elements and video-projectors, as well as interactive rendering algorithms, calibration techniques, and display examples. It provides a comprehensive overview with detailed mathematics equations and formulas, code fragments, and implementation instructions that enable interested readers to realize spatial AR displays by themselves.

This chapter will give a brief and general introduction into augmented reality and its current research challenges. It also outlines the remaining chapters of the book.

1.1 What is Augmented Reality

The terms *virtual reality* and *cyberspace* have become very popular outside the research community within the last two decades. Science fiction movies, such as *Star Trek*, have not only brought this concept to the public, but have also influenced the research community more than they are willing to admit. Most of us associate these terms with the technological possibility to dive into a completely synthetic, computer-generated world—sometimes referred to as a *virtual environment*. In a virtual environment our senses, such as vision, hearing, haptics, smell, etc., are controlled by a computer while our actions influence the produced stimuli. *Star Trek*'s Holodeck is probably one of the most popular examples. Although some bits and pieces of the Holodeck have been realized today, most of it is still science fiction.

So what is augmented reality then? As is the case for virtual reality, several formal definitions and classifications for augmented reality exist (e.g., [109, 110]). Some define AR as a special case of VR; others argue that AR is a more general concept and see VR as a special case of AR. We do not want to make a formal definition here, but rather leave it to the reader to philosophize on their own. The fact is that in contrast to traditional VR, in AR the real environment is not completely suppressed; instead it plays a dominant role. Rather than immersing a person into a completely synthetic world, AR attempts to embed synthetic supplements into the real environment (or into a live video of the real environment). This leads to a fundamental problem: a real environment is much more difficult to control than a completely synthetic one. Figure 1.1 shows some examples of augmented reality applications.

As stated previously, augmented reality means to integrate synthetic information into the real environment. With this statement in mind, would a TV screen playing a cartoon movie, or a radio playing music, then be an AR display? Most of us would say no—but why not? Obviously, there is more to it. The augmented information has to have a much stronger link to the real environment. This link is mostly a spatial relation between the augmentations and the real environment. We call this link *registration*. R2-D2's spatial projection of Princess Leia in *Star Wars* would be a popular science fiction example for augmented reality. Some technological approaches that mimic a holographic-like spatial projection, like the Holodeck, do exist today. But once again, the technical implementation as shown in *Star Wars* still remains a Hollywood illusion.

Some say that Ivan Sutherland established the theoretical foundations of virtual reality in 1965, describing what in his opinion would be the *ultimate display* [182]:

> The ultimate display would, of course, be a room within which the computer can control the existence of matter. A chair displayed in such a room would be good enough to sit in. Hand-

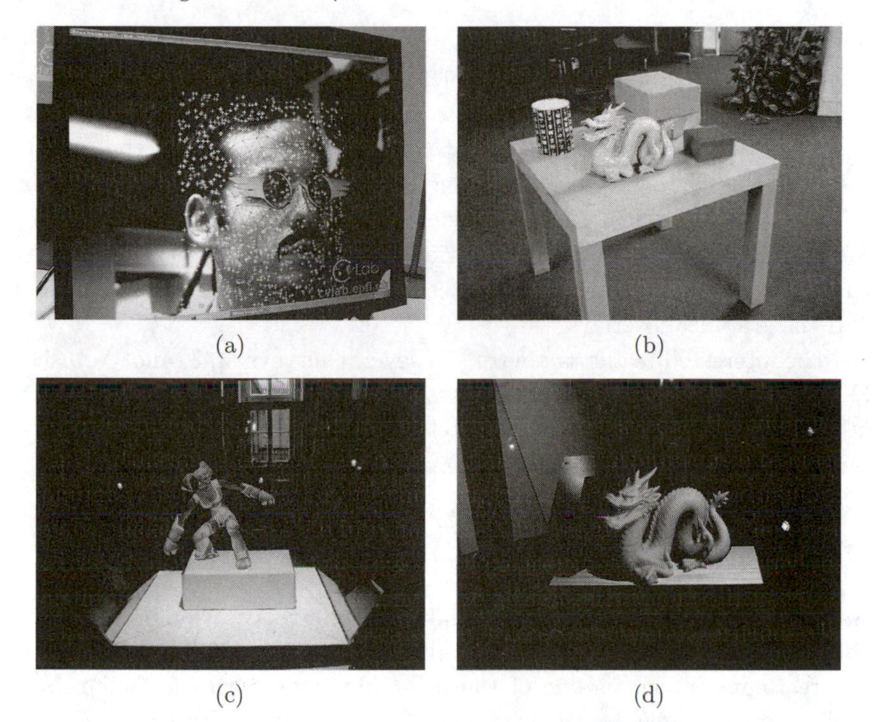

Figure 1.1. Example of augmented reality applications. The glasses, mustache, dragons, and fighter figure are synthetic: (a) and (b) augmenting a video of the real environment; (c) and (d) augmenting the real environment optically. (*Images: (a) courtesy of Vincent Lepetit, EPFL [87]; (b) courtesy of Simon Gibson [55], Advanced Interfaces Group © University of Manchester 2005; (c) and (d) prototypes implemented by the Barhaus-University Weimar.*)

> cuffs displayed in such a room would be confining, and a bullet displayed in such a room would be fatal. With appropriate programming, such a display could literally be the Wonderland into which Alice walked.

However, technical virtual reality display solutions were proposed much earlier. In the late 1950s, for instance, a young cinematographer named Mort Heilig invented the *Sensorama* simulator, which was a one-person demo unit that combined 3D movies, stereo sound, mechanical vibrations, fan-blown air, and aromas. Stereoscopy even dates back to 1832 when Charles Wheatstone invented the *stereoscopic viewer*.

Then why did Sutherland's suggestions lay the foundation for virtual reality? In contrast to existing systems, he stressed that the user of such an ultimate display should be able to interact with the virtual environment. This led him to the development of the first functioning *Head-Mounted Display* (HMD) [183], which was also the birth of augmented reality. He used half-silvered mirrors as optical combiners that allowed the user to see both the computer-generated images reflected from *cathode ray tubes* (CRTs) and objects in the room, simultaneously. In addition, he used mechanical and ultrasonic head position sensors to measure the position of the user's head. This ensured a correct registration of the real environment and the graphical overlays.

The interested reader is referred to several surveys [4, 5] and Web sites [3, 193] of augmented reality projects and achievements. Section 1.2 gives a brief overview of today's technical challenges for augmented reality. It is beyond the scope of this book to discuss these challenges in great detail.

1.2 Today's Challenges

As mentioned previously, a correct and consistent registration between synthetic augmentations (usually three-dimensional graphical elements) and the real environment is one of the most important tasks for augmented reality. For example, to achieve this for a moving user requires the system to continuously determine the user's position within the environment.

Thus the *tracking* and *registration* problem is one of the most fundamental challenges in AR research today. The precise, fast, and robust tracking of the observer, as well as the real and virtual objects within the environment, is critical for convincing AR applications. In general, we can differentiate between *outside-in* and *inside-out tracking* if absolute tracking within a global coordinate system has to be achieved. The first type, outside-in, refers to systems that apply fixed sensors within the environment that track emitters on moving targets. The second type, inside-out, uses sensors that are attached to moving targets. These sensors are able to determine their positions relative to fixed mounted emitters in the environment. Usually these two tracking types are employed to classify camera-based approaches only—but they are well suited to describe other tracking technologies as well.

After *mechanical* and *electromagnetic tracking*, *optical tracking* became very popular. While *infrared* solutions can achieve a high precision and a high tracking speed, *marker-based tracking*, using conventional cameras, represent a low-cost option. Tracking solutions that do not require artificial markers, called *markerless tracking*, remains the most challenging, and at the same time, the most promising tracking solution for future augmented reality applications. Figure 1.1(a) shows an example of a markerless face tracker.

Much research effort is spent to improve performance, precision, robustness, and affordability of tracking systems. High-quality tracking within large environments, such as the outdoors, is still very difficult to achieve even with today's technology, such as a *Global Positioning System* (GPS) in combination with relative measuring devices like *gyroscopes* and *accelerometers*. A general survey on different tracking technology [164] can be used for additional reading.

Besides tracking, *display technology* is another basic building block for augmented reality. As mentioned previously, head-mounted displays are the dominant display technology for AR applications today. However, they still suffer from optical (e.g., limited field of view and fixed focus), technical (e.g., limited resolution and unstable image registration relative to eyes) and human-factor (e.g., weight and size) limitations. The reason for this dominance might be the long time unique possibility of HMDs to support *mobile AR applications*. The increasing technological capabilities of *cell phones* and *Personal Digital Assistants* (PDAs), however, clear the way to more promising display platforms in the near future. In addition, not all AR applications require mobility. In these cases, *spatial display configurations* are much more efficient.

The third basic element for augmented reality is *real-time rendering*. Since AR mainly concentrates on superimposing the real environment with graphical elements, fast and realistic rendering methods play an important role. An ultimate goal could be to integrate graphical objects into the real environment in such a way that the observer can no longer distinguish between real and virtual. Note that not all AR applications really make this requirement. But if so, then besides perfect tracking and display technologies, *photo-realistic real-time rendering* would be another requisite. Graphical objects, even if rendered in a high visual quality, would have to be integrated into the real environment in a consistent way. For instance, they have to follow a consistent *occlusion*, *shadow-casting*, and *inter-reflection* behavior, as Figure 1.1 demonstrates.

Realistic, non-real-time capable *global illumination* techniques, such as ray-tracing or *radiosity*, can be used if no interactive frame rates are required, . But for interactive applications, faster image generation methods have to be used to avoid a large *system lag* and a resultant *misregistration* after fast user motions. The improving *hardware acceleration* of today's

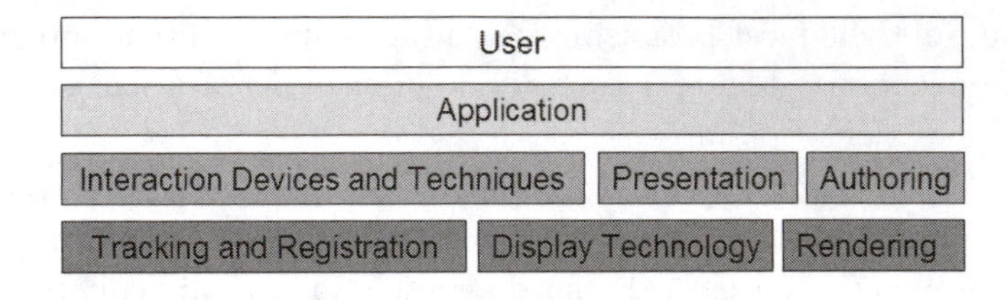

Figure 1.2. Building blocks for augmented reality.

graphics cards make a lot possible, as is shown throughout the following chapters. The ongoing paradigm shift of the computer graphics community from the old *fixed function rendering pipelines* to *programmable pipelines* strongly influences the rendering capabilities of AR applications. Examples of consistent rendering techniques for augmented reality have been discussed in different scientific publications and in the course of several conference tutorials [176, 17].

Figure 1.2 illustrates some general building blocks for augmented reality. As we can see, the previously discussed challenges (tracking and registration, display technology and rendering) represent fundamental components. On top of this base level, more advanced modules can be found: *interaction devices* and *techniques*, *presentation*, and *authoring*. If we take a comparative look at virtual reality again, we can see that the base technology of today's VR is much more mature. In contrast to VR where a large portion of research is now being shifted to the second layer, the AR community still has to tackle substantial problems on the base level.

Ideas and early implementations of presentation techniques, authoring tools, and interaction devices/techniques for AR applications are just emerging. Some of them are derived from the existing counterparts in related areas such as virtual reality, multimedia, or digital storytelling. Others are new and adapted more to the problem domain of augmented reality. However, it is yet too early to spot matured concepts and philosophies at this level.

The third layer, the *application*, is finally the interface to the user. Using augmented reality, our overall goal is to implement applications that are tools which allow us to solve problems more effectively. Consequently, augmented reality is no more than a human-computer interface which has the potential to be more efficient for some applications than others. Although many ideas for possible applications of this interface exist, not many have actually become applicable today. One reason for this is the immature base layer. With a stable core technology, augmented reality does have the potential to address many application areas more effectively. Some virtual reality applications, for instance, have already managed to become real tools outside the research community. It is also clear, that broader base levels will lead to a broader application spectrum.

Some *software frameworks* (e.g., [165]) are being realized that comprise several of these parts. Good software engineering will be important for the efficient handling of an increasing pallet of new tools and techniques. Finally, user studies have to be carried out to provide measures of how effective augmented reality really is.

1.3 Spatial Augmented Reality

The roots of virtual reality and augmented reality are not that far apart. After almost forty years of research and development, however, they do not follow the same technological paths anymore. In the early 1990s, *projection-based surround screen displays* became popular. One of the most well-known is the *CAVE* [35]—a multi-sided, immersive projection room. But there are other examples of *semi-immersive* wall-like and table-like displays or *immersive* cylindrical and spherical *spatial displays*. In general, spatial displays detach the display technology from the user and integrate it into the environment. Compared to head- or body-attached displays, spatial displays offer many advantages and solve several problems that are related to visual quality (e.g., resolution, field-of-view, focus, etc.), technical issues (e.g., tracking, lighting, etc.), and human factors (e.g., cumbersomeness, etc.), but they are limited to non-mobile applications..

The virtual reality community has oriented themselves away from head-mounted displays and towards spatial displays. Today, a large variety of spatial displays make up an estimated 90% of all VR displays. Head-mounted displays, however, are still the dominant displays for augmented reality. The reason for this might lie in the strong focus of mobile AR applications—requiring mobile displays.

Video see-through and optical see-through head-mounted displays have been the traditional output technologies for augmented reality applications for almost forty years. However, they still suffer from several technological and ergonomic drawbacks which prevent them from being used effectively in many application areas. In an all-purpose context, HMDs are used in many non-mobile AR applications. This affects the efficiency of these applications and does not currently allow them to expand beyond laboratory demonstrations. In the future, other mobile devices, such as cell phones or PDAs might replace HMDs in many mobile areas. Head-mounted displays will also be enhanced by future technology, leading to a variety of new and different possibilities for mobile AR.

Furthermore, we believe that for non-mobile applications a rich pallet of different spatial display configurations can be as beneficial for augmented reality, as they have been for virtual reality. Novel approaches have taken augmented reality beyond traditional eye-worn or hand-held displays enabling additional application areas. New display paradigms exploit large spatially-aligned optical elements, such as mirror beam combiners, transparent screens, or holograms, as well as video projectors. Thus, we call this technological variation *spatial augmented reality* (SAR). In many situations, SAR displays are able to overcome technological and ergonomic limitations of conventional AR systems. Due to the decrease in cost and availability of projection technology, personal computers, and graphics hardware, there has been a considerable interest in exploiting SAR systems in universities, research laboratories, museums, industry, and the art community. Parallels to the development of virtual environments from head-attached displays to spatial projection screens can be clearly drawn. We believe that an analog evolution of augmented reality has the potential to yield a similar successful factor in many application domains. Thereby, SAR and body-attached AR are not competitive, but complementary.

1.4 Outline of the Book

This book provides survey and implementation details of modern techniques for spatial augmented reality systems and aims to enable the interested reader to realize such systems on his or her own. This is supported by more than 200 illustrations and many concrete code fragments.

After laying foundations in optics, interactive rendering, and perspective geometry, we discuss conventional mobile AR displays and present spatial augmented reality approaches that are overcoming some of their limitations. We present state-of-the-art concepts, details about hardware and software implementations, and current areas of application in domains such as museums, edutainment, research, and industrial areas. We draw parallels between display techniques used for virtual reality and augmented reality and stimulate thinking about the alternative approaches for AR.

One potential goal of AR is to create a high level of consistency between real and virtual environments. This book describes techniques for the optical combination of virtual and real environments using mirror beam combiners, transparent screens, and holograms. It presents projector-based augmentation of geometrically complex and textured display surfaces, which along with optical combiners achieve consistent illumination and occlusion effects. We present many spatial display examples, such as Shader Lamps, Virtual Showcases, Extended Virtual Tables, interactive holograms, apparent motion, Augmented Paintings, and Smart Projectors.

Finally, we discuss the current problems, future possibilities, and enabling technologies of spatial augmented reality.

Chapter 2 lays the foundation for the topics discussed in this book. It starts with a discussion on light. The atomic view on light will give some hints on how it is generated—knowing about its properties allows us to realize how it travels through space.

From electromagnetic waves, a more geometric view on light can be abstracted. This is beneficial for describing how simple optical elements, such as mirrors and lenses, work. This chapter will describe how images are formed by bundling real and virtual light rays in a single spatial spot or area in three-dimensional space. Furthermore, it is explained that the structure and functionality of the human eye (as the final destination of visible light produced by a display) is as complex as an optical system itself, and that the binocular interplay of two eyes leads to visual depth perception. The depth perception can be tricked by viewing flat stereo images on a stereoscopic display. The principles of stereoscopic vision and presentation, as well as a classification of stereoscopic and autostereoscopic displays, will be discussed in this chapter as well. We will illustrate how images that are presented on stereoscopic displays are computed. Basic rendering concepts, such as components of traditional fixed function rendering pipelines and techniques like multi-pass rendering, but also modern programmable rendering pipelines will be described.

Chapter 3 classifies current augmented reality displays into head-attached, hand-held, and spatial displays. It gives examples of particular displays that are representative for each class and discusses their advantages and disadvantages. Retinal displays, video see-through and optical see-through head-mounted displays, and head-mounted projectors are presented first in the context of head-attached displays. For hand-held displays, personal digital assistants, cell phones, hand-held projectors, and several optical see-through variations are outlined. Finally, spatial AR displays are presented. First examples include screen-based video see-through displays, spatial optical see-through displays, and projector-based spatial displays. The following chapters will explain how to realize such displays, both from a hardware and software point of view.

Chapter 4 reviews the fundamental geometric concepts in using a projector for displaying images. A projector can be treated as a dual of a camera. The geometric relationship between the two-dimensional pixels in the projector frame buffer and the three-dimensional points in the world illuminated by those pixels is described using perspective projection. The chapter introduces a general framework to express the link between geometric components involved in a display system. The framework leads to a simpler rendering technique and a better understanding of the calibration goals. We describe the procedure to calibrate projectors and render images for planar as well non-planar displays along with issues in creating seamless images using multiple projectors.

Chapter 5 expands on the geometric framework introduced in the previous chapter and describes the concrete issues in calibration and rendering for various types of display surfaces. The techniques use parametric as well as non-parametric approaches. For planar surface, a procedure based on homography transformation is effective. For arbitrary non-planar displays, we outline a two-pass scheme, and for curved displays, we describe a scheme based on quadric image transfer. Finally, the chapter discusses the specific projector-based augmentation problem where images are projected not on display screens, but directly onto real-world objects.

Chapter 6 explains spatial optical see-through displays in detail. An essential component of an optical see-through display is the optical combiner—an optical element that mixes the light emitted by the illuminated real environment with the light produced with an image source that displays the rendered graphics. Creating graphical overlays with spatial optical see-through displays is similar to rendering images for spatial projection screens for some optical combiners. For others, however, it is more complex and requires additional steps before the rendered graphics are displayed and optically combined. While monitors, diffuse projection screens, or video projectors usually serve as light emitting image sources, two different types of optical combiners are normally used for such displays: transparent screens and half-silvered mirror beam combiners. Rendering techniques that support creating correct graphical overlays with both types of optical combiners and with different images sources will be discussed in this chapter. In particular, Chapter 6 will discuss rendering techniques for spatial optical see-through displays which apply transparent screens, as well as planar and curved mirror beam combiners, in many different configurations. We describe how optical effects, such as reflection and refraction, can be neutralized by hardware-accelerated rendering techniques and present building blocks that can easily be integrated into an existing software framework.

Chapter 7 outlines interactive rendering techniques for projector-based illumination and augmentation. It starts with an overview of methods that allow augmenting artificial surface appearance, such as shading, shadows, and highlights, of geometrically non-trivial objects. Calibrated projectors are used to create these effects on physical objects with uniformly white surface color. We then describe how to use calibrated projectors in combination with optical see-through configurations (described in Chapter 6) for the illumination of arbitrary real objects. This allows digitization of the illumination of the real environment and synchronization with the rendering process that generates the graphical overlays.

The final goal is to create consistent occlusion effects between real and virtual objects in any situation and to support single or multiple observers. The surface appearance of real objects with non-trivial surface color and texture can also be modified with a projector-based illumination. Such an approach, for instance, allows the creation of consistent global or local lighting situations between real and virtual objects. Appropriate rendering techniques are also described in this chapter. A projector-based illumination also makes it possible to integrate graphical augmentations into optical holograms. In this case, variations of the algorithms explained previously let us replace a physical environment by a high-quality optical hologram. Finally, this chapter presents real-time color-correction algorithms that, in combination with an appropriate geometric correction, allow an augmentation of arbitrary (colored/textured) three-dimensional surfaces with computer generated graphics.

Chapter 8 brings together the previous, more technical chapters in an application-oriented approach and describes several existing spatial AR display configurations. It first outlines examples that utilize the projector-based augmentation concept in both a small desktop approach (e.g., Shader Lamps) and a large immersive configuration (e.g., the *Being There* project). In addition, an interactive extension, called *iLamps*, that uses hand-held projectors is described. Furthermore, several spatial optical see-through variations that support single or multiple users, such as the *Extended Virtual Table* and the *Virtual Showcase* are explained. It is shown how they can be combined with projector-based illumination techniques to present real and virtual environments consistently. A scientific workstation, the *HoloStation*, is presented which combines optical hologram records of fossils with interactive computer simulations. Finally, two configurations (*Augmented Paintings* and *Smart Projectors*) are presented. They use real-time color correction and geometric warping to augment artistic paintings with multimedia presentations, as well as to make projector-based home-entertainment possible without artificial canvases.

Potential application areas for the display configurations described in this chapter are industrial design and visualization (e.g., Shader Lamps, iLamps, Extended Virtual Table), scientific simulations (e.g., HoloStation), inertial design and architecture (e.g., Being There), digital storytelling and next-generation edutainment tools for museums (e.g., Virtual Showcase and Augmented Paintings), and home-entertainment (e.g., Smart Projector). However, the interested reader can easily derive further application domains, such as those in an artistic context.

Another goal of this chapter is to show that spatial augmented reality display configurations can be applied successfully and efficiently outside research laboratories. The Virtual Showcase, for instance, has been presented to more than 120,000 visitors at more than 11 exhibitions in museums, trade shows, and conferences. Unattended running times of four months and more are an indicator for the fact that it is possible to make the technology (soft- and hardware) robust enough to be used by museums and other public places.

Chapter 9 postulates future directions in spatial augmented reality. Many new opportunities are based on emerging hardware components, and they are briefly reviewed. The chapter discusses innovative optics for displays, new materials such as light emitting polymers, promising developments in sensor networks including those using photosensors, and the excitement surrounding radio frequency identification tags. These technology developments will not only open new possibilities for SAR, but also for other AR display concepts, such as hand-held and head-attached displays.

2

Fundamentals: From Photons to Pixels

This chapter lays the foundation for the topics discussed in this book. It will not describe all aspects in detail but will introduce them on a level that is sufficient for understanding the material in the following chapters. We strongly encourage the reader to consult the secondary literature.

We start our journey at the most basic element that is relevant for all display technology—light. The atomic view on light will give us some hints on how it is generated. Knowing about its properties allows us to understand how it travels through space.

Starting from electromagnetic waves, we abstract our view on light to a more geometric concept that is beneficial for describing how simple optical elements, such as mirrors and lenses, work. We see how images are formed by bundling real and virtual light rays in a single spatial spot or area in three-dimensional space.

In addition, we learn that the structure and functionality of the human eye (as the final destination of visible light produced by a display) is as complex as an optical system itself, and that the binocular interplay of two eyes leads to visual depth perception.

Depth perception, however, can be tricked by viewing flat stereo images on a stereoscopic display. The principles of stereoscopic vision and presentation, as well as a classification of stereoscopic and autostereoscopic displays, will be discussed.

From stereoscopic displays, we will see how images presented on these displays are computed. Basic rendering concepts, such as components of traditional fixed function rendering pipelines and techniques like multi-pass rendering will be described. Modern programmable rendering pipelines will be discussed as well.

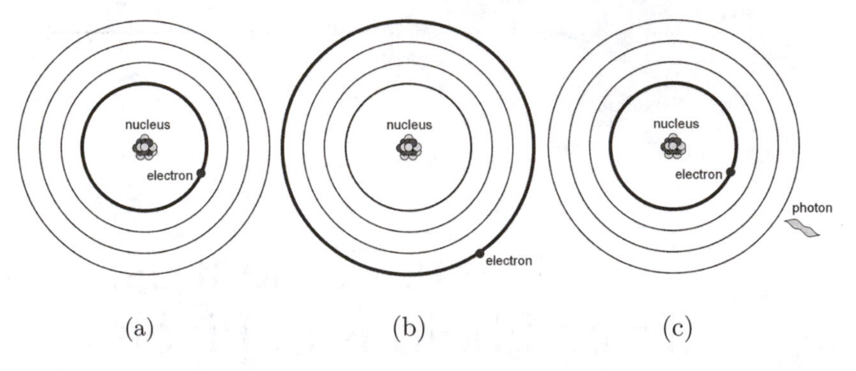

Figure 2.1. Planetary model of atom: (a) electron orbiting the nucleus in a non-excited state; (b) excited electron after quantum leap; (c) electron releasing a photon while dropping back into a non-excited state.

2.1 Light in a Nutshell

To explain what light is, we want to start at a very low level—at an atomic level. Illustrated by the well-known *planetary model* by Niels Bohr (1913), atoms consist of a *nucleus* and *electrons* that orbit the nucleus (Figure 2.1). They are held in the orbit by an electrical force. The nucleus itself consists of *protons* (having a positive electrical charge) and *neutrons* (having no electrical charge). The electrons have a negative electrical charge and can move from atom to atom. This flow is called *electricity.*

The level at which an electron orbits the nucleus (i.e., the distance of the electron to the nucleus) is called its *energy state*. By default, an electron exists at the lowest energy state—that is the orbit closest to the nucleus. If excited by external energy (e.g., heat) the electron can move from lower to higher energy states (i.e., further away from the nucleus). This shift from a lower to a higher energy state is called *quantum leap*. Since all energy is always preserved (it might be converted, but it is never lost), the electrons have to release energy when they drop back to lower energy states. They do so by releasing packages of energy. Albert Einstein called these packages *photons*.

Photons have a frequency that relates to the amount of energy they carry which, in turn, relates to the size of the drop from the higher state to the lower one. They behave like waves—they travel in waves with a specific *phase*, *frequency*, and *amplitude*, but they have no mass. These *electromagnetic waves* travel in a range of frequencies called electromagnetic (EM) spectrum, which was described by J. C. Maxwell (1864–1873).

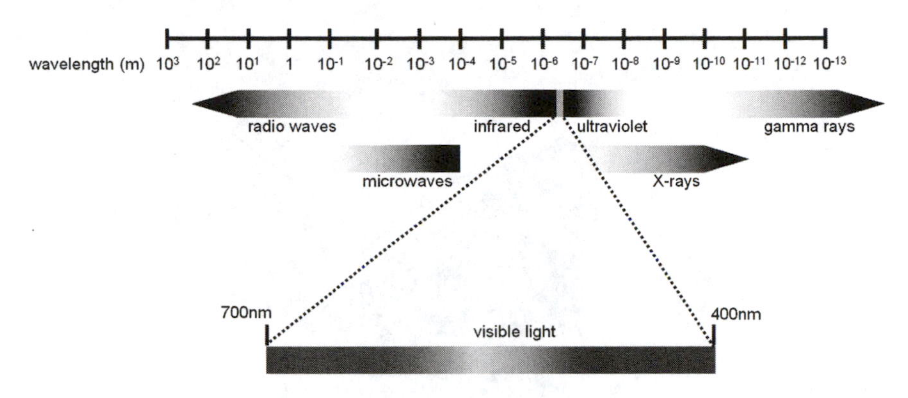

Figure 2.2. EM spectrum and spectrum of visible light. (*See Plate I.*)

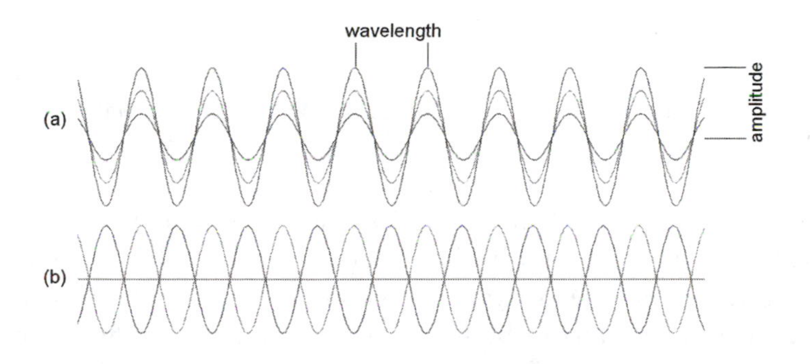

Figure 2.3. Interference of light waves: (a) amplification with zero phase shift, (b) cancellation with π phase shift.

A small part of this spectrum is the EM radiation that can be perceived as *visible light*.

Since light behaves like waves, it shares many properties of other waves. Thomas Young showed in the early 1800s that light waves can interfere with each other.

Depending on their phase, frequency, and amplitude, multiple light waves can amplify or cancel each other out (Figure 2.3). Light that consists of only one wavelength is called *monochromatic light*. Light waves that are in phase in both time and space are called *coherent*. Monochromaticity and low divergence are two properties of coherent light.

If a photon passes by an excited electron, the electron will release a photon with the same properties. This effect—called *stimulated emission*—was predicted by Einstein and is used today to produce coherent laser

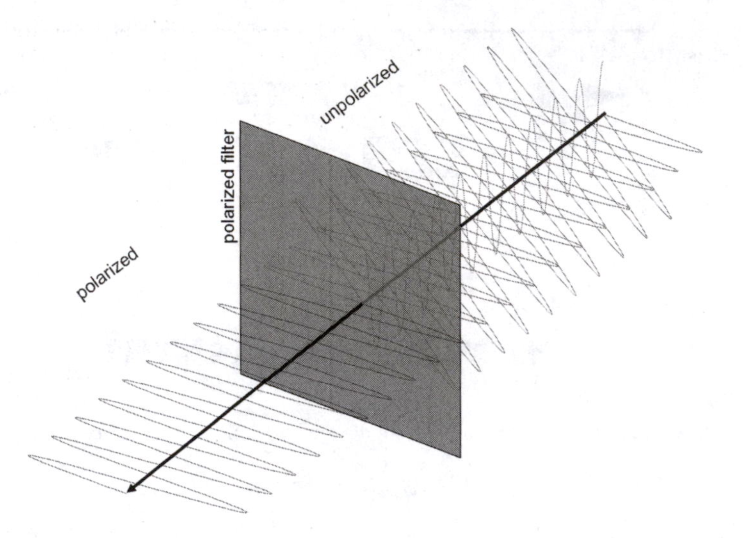

Figure 2.4. Polarization of light: only light waves with a specific orientation pass through the filter.

light. In general, "normal" light consists of an arbitrary superimposition of multiple incoherent light waves that create a complex *interference pattern*.

Light travels in a composition of waves with a variety of different orientations. It can be *polarized* by selecting waves with a specific orientation. The filtered portion is called *polarized light*. Augustine Fresnel explained this phenomenon in the 19th century.

Depending on the material properties, light can be *reflected*, *refracted*, *scattered*, or *light!absorbed* by matter. If reflected, light is bounced off a surface. Imperfections on the reflecting surface causes the light to be scattered (*diffused*) in different directions. Light can also be scattered when it collides with small particles (like molecules). The amount of scattering depends on the size of the particle with respect to the wavelength of the light. If the particle (e.g., a dust particle) is larger than the wavelength, light will be reflected. If the particle (e.g., a gas molecule) is smaller, then light will be absorbed. Such molecules will then radiate light at the frequency of the absorbed light in different directions. John Rayleigh explained this effect in the 1870s, thus this process is called *Rayleigh scattering*. Light can also be absorbed and its energy converted (e.g., into heat). Refraction occurs when light travels across the boundaries of two mediums. In a vacuum, light travels at 299.792 km/s (the speed of light). If travelling through a denser medium, it is slowed down which causes it to alter its direction.

The amount of refraction also depends on the wavelength of the light—as described by Isaac Newton who showed that white light splits into different angles depending on its wavelength.

2.2 Geometric Optics

In general, optics refers to all appearances that are perceived by the human visual system. The physical reason for these appearances, the light, was analyzed at an early time, and the basic principles of geometric optics and wave optics were outlined in the 19th century. In geometric optics, the light is represented by individual rays that, in ordinary media, are represented by straight lines. An ordinary media is *homogeneous* (the same at all points) and *isotropic* (the same for all directions). One of the basic hypotheses of geometric optics, the principle of P. de Fermat (1657), allows the representation of light rays within isotropic media that are independent of the light's wave nature. Today this hypothesis is known as the *principle of the optical path length*, and it states that the time that light travels on the path between two points is minimal.

2.2.1 Snell's Laws

The following laws were discovered by W. Snellius in 1621. They can also be derived from Fermat's principle. They describe the reflection and refraction behavior of straight light rays at the interfacing surface between two *homogeneous media*.

Figure 2.5 illustrates the interfacing surface that separates two homogeneous media with the two indices of refraction η_1 and η_2. A light ray r intersects the surface at i (with the normal vector n) and is refracted into r'.

The vectorized form of Snell's' law is given by $\eta_2 r' - \eta_1 r = an$, where r, r' and n are normalized and a is real.

Laws of refraction. We can derive the following *laws of refraction* from the vectorized form of Snell's law.

Theorem 2.2.1 (First refraction theorem.) *Since $r' = (\eta_1 r + an)/\eta_2$, the refracted ray r' lies on the plane that is spanned by r and n. This plane is called the* plane of incidence.

Theorem 2.2.2 (Second refraction theorem.) *If we compute the cross-product with n and the vectorized form of Snell's law, we obtain* $\eta_2(n \times r') = \eta_1(n \times r)$. *If we define the* angle of incidence α_i *and the* angle of refraction α_t, *we can substitute the cross-products and obtain Snell's law of refraction:* $\eta_1 \sin \alpha_i = \eta_2 \sin \alpha_t$.

Laws of reflection. Since the vector relation applies in general, we can assume r and r' to be located within the same medium with a refraction index η_1. Consequently, r is reflected at i into r' (Figure 2.6).

We can derive the following two theorems of reflection from the vectorized form of Snell's law.

Theorem 2.2.3 (First reflection theorem.) *Since* $r' = r + (a/\eta_1)n$, *the reflected ray* r' *lies on the plane of incidence.*

Theorem 2.2.4 (Second reflection theorem.) *If we compute the cross-product with n and the vectorized form of Snell's law, we obtain* $n \times r' = n \times r$. *If we reassign* α_t *to be the* angle of reflection, *we can substitute the cross-products and obtain Snell's law of reflection:* $-\sin \alpha_t = \sin \alpha_i$ *or* $-\alpha_t = \alpha_i$ *for* $-\pi/2 \leq \alpha_i \leq \pi/2$ *and* $-\pi/2 \leq \alpha_t \leq \pi/2$.

Note that the law of reflection is formally based on the assumption that $\eta_2 = -\eta_1$.

Critical angle and total internal reflection. Since $-1 \leq \sin \alpha_i \leq 1$, we can derive $-(\eta_1/\eta_2) \leq \sin \alpha_t \leq (\eta_1/\eta_2)$ from the second refraction theorem. It therefore holds that $-(\pi/2) \leq \alpha_i \leq (\pi/2)$ and $-\gamma \leq \alpha_t \leq \gamma$, whereby γ is called the *critical angle* and is defined by $\gamma = \sin^{-1}(\eta_1/\eta_2)$.

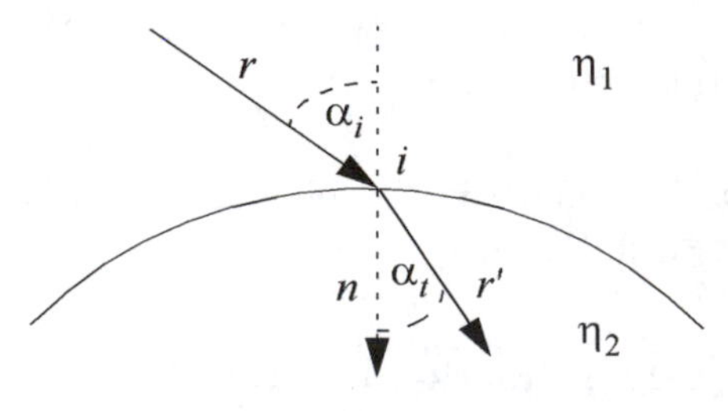

Figure 2.5. Snell's law of refraction.

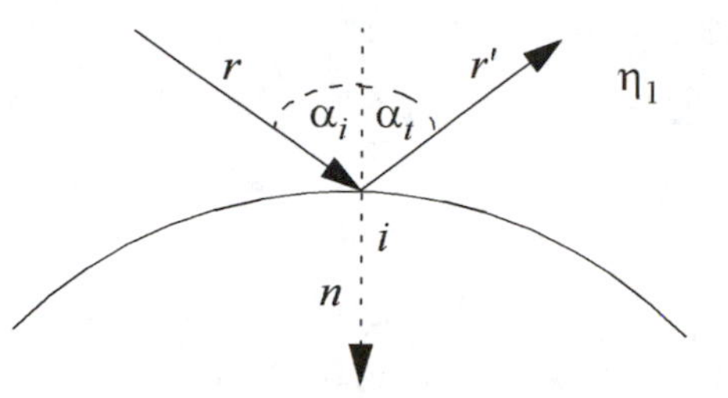

Figure 2.6. Snell's law of reflection.

If α_i becomes sufficiently large when entering an optically sparser medium, then α_t exceeds 90° and r is reflected from the interfacing surface, rather than being transmitted. This phenomenon is known as *total internal reflection*.

We can differentiate between two cases:

1. r enters an optically denser medium ($\eta_1 < \eta_2$): r is refracted for all angles of incidence α_i. If $\alpha_i = \pi/2$, then $\alpha_t = \sin^{-1}(\eta_1/\eta_2) = \gamma$.

2. r enters an optically sparser medium ($\eta_1 > \eta_2$): If $\alpha_i < \gamma = sin^{-1}(\eta_1/\eta_2)$, then r is refracted. Otherwise, r is reflected, due to total internal reflection.

2.2.2 The Formation of Point Images

Optical instruments can form *images* from a number of point-like light sources (so-called *objects*). Light rays that are emitted from an object can be reflected and refracted within the optical instrument and are finally perceived by a detector (e.g., the human eye or a photographic film). If all light rays that are emitted from the same object p_o travel through the optical system which bundles them within the same image p_i, then the points p_o and p_i are called a *stigmatic pair*. Consequently, this image-formation property is called *stigmatism*, and the optical system that supports stigmatism between all object-image pairs is called an *absolute optical system*.

The basic precondition for stigmatism can also be derived from Fermat's principle. It states that the optical path length for every light ray travelling from p_o to p_i is constant:

$$L(p_o \rightarrow p_i) = \eta_1(i_x - p_o) + L(i_x \rightarrow j_x) + \eta_2(p_i - j_x) = \text{const}$$

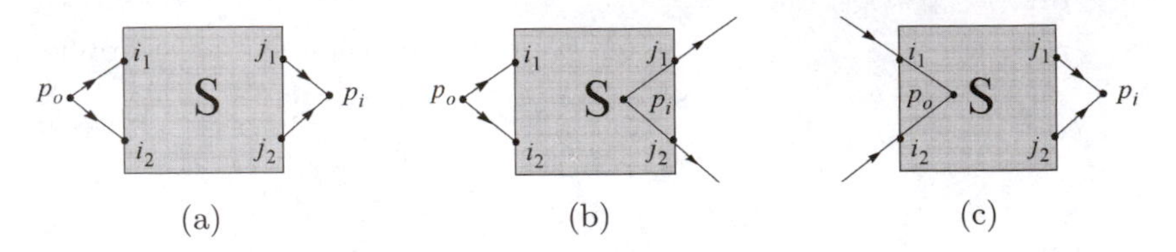

Figure 2.7. Stigmatic image formation. (a) real object, real image, (b) real object, virtual image, and (c) virtual object, real image.

where η_1 and η_2 are the refraction indices at the entrance and the exit of the optical system.

If points (objects or images) are formed by a direct intersection of light rays, then these points are called *real*. Figure 2.7(a) illustrates the *real object* p_o whose emitted light rays pass through an optical system (the filled square) and intersect at the *real image* p_i.

If light rays do not directly intersect at a point (e.g., if they diverge after exiting the optical instrument), they can form *virtual points*. Since human observers are only able to detect the directions of light rays, rays diverging from an optical system can appear to intersect within the system. These images are called *virtual images* (Figure 2.7(b)).

The location of virtual points can be determined by extending the exiting light rays in the negative direction. Consequently, this portion of the optical path is negative and must be subtracted from the total path length.

As illustrated in Figure 2.7(c), objects can also be virtual. In this case, the entering light rays have to be extended to find the location of the corresponding virtual object. Similar to the relationship of the optical path to a virtual image, the sub-path to a virtual object also has to be subtracted from the total path length.

The production of absolute optical systems is difficult, since the only surfaces that are easy to build and support stigmatism (some only for a single object-image pair) are planar or spherical surfaces. Therefore, most optical instruments only approximate stigmatic image formation. The introduced deviation from the ideal image is called *aberration*. Some examples of reflective and refractive optical systems are given in the following sections.

2.2.3 Reflective Optics

In the case of exclusively reflective optical systems (*mirrors*), the medium that light rays travel through is homogeneous, thus $\eta_1 = \eta_2 = \eta$ and

$i_x = j_x$. Consequently, the optical path length equation can be simplified:

$$L(p_o \to p_i) = \eta((i_x - p_o) + (p_i - i_x)) = \text{const}.$$

It can be further idealized that a mirror is surrounded by air, and that the medium air is approximately equivalent to the medium of a vacuum ($\eta = 1$), then two stigmatic points which are formed within air are defined by

$$L(p_o \to p_i) = (i_x - p_o) + (p_i - i_x) = \text{const}.$$

Planar mirrors. In the case of planar mirrors p_o is real while p_i is virtual (Figure 2.8 (a)), and all points i_x of the simplified optical path equation describe the surface of a rotation-hyperboloid with its two focal points in p_o and p_i. Planes represent a special variant of a rotation-hyperboloid, where $L(p_o \to p_i) = 0$. Planar mirrors are absolute optical systems that map each object p_o to exactly one image p_i. Since this mapping is bijective, invertible, and symmetrical for all points, it provides stigmatism between all objects and images. This means that images which are generated from multiple image points preserve the geometric properties of the reflected objects that are represented by the corresponding object points.

If we represent the mirror plane by its normalized plane equation within the three-dimensional space $f(x, y, z) = ax+by+cz+d = 0$, then the image for a corresponding object can be computed as follows: With respect to Figure 2.8(a), it can be seen that the distance from p_o to the mirror plane equals the distance from the mirror plane to p_i (i.e., $a = a'$). This can be derived from the simplified optical path equation with simple triangulation.

If we now define the ray $r_p = p_o + \lambda n$, where $n = (a, b, c)$ ($f(x, y, z)$ is normalized) is the normal vector perpendicular to the mirror plane and λ an arbitrary extension factor of n, we can insert the components of r_p into f and solve for λ:

$$f(r_p) = nr_p + d = n(p_o + \lambda n) + d = 0 \to \lambda = -\frac{1}{nn}(np_o + d).$$

Since $|n| = 1$, we can set $nn = 1$ and solve $\lambda = -(np_o + d) = a = a'$. Consequently the intersection of r_p with f is given by

$$i_p = p_o - (np_o + d)n.$$

Since $a = a'$, the image point p_i results from

$$p_i = p_o - 2(np_o + d)n. \tag{2.1}$$

In Chapter 6, we show how Equation (2.1) can be expressed as a homogeneous 4×4 transformation matrix that can be integrated into fixed function rendering pipelines to support optical combination with mirror beam combiners.

With respect to Snell's first reflection theorem, we can determine the reflection ray r' of the original ray r as

$$r' = r - 2n(nr). \tag{2.2}$$

In the case of planar mirrors, n is constant for all surface points i. However, this equation is also valid for non-planar mirrors with individual normal vectors at each surface point. In this case, the intersection i of r with the mirror surface and the normal n at i has to be inserted in Equations (2.1) and (2.2). Note that Equation (2.2) is a common equation used by ray-tracers to compute specular reflection rays. The curvilinear behavior of reflections at non-planar mirrors can be well expressed with ray-tracing techniques, since they are based on the optical foundations of light rays.

Non-planar mirrors. In contrast to planar mirrors, non-planar mirrors do not provide stigmatism between all objects and images. In fact, only a few surface types generate just one true stigmatic pair. For all other objects (or for objects reflected by other surfaces), the corresponding images have to be approximated, since the reflected light rays do not bundle exactly within a single point.

Like planar mirrors, *convex mirrors* generate virtual images from real objects. This is because light rays always diverge after they are reflected. Rotation-paraboloids (parabolic mirrors), for instance, can generate just one true stigmatic pair (Figure 2.8(b)).

The extended light rays bundle in one virtual point p_i only if p_o is located at infinity. This point is the *focal point* f of the paraboloid. The distance between the focal point and the surface is called the focal distance or *focal length* f. For example, the focal length of a convex mirror is defined as $f = -r/2$, the focal length of a concave mirror is given by $f = r/2$, and the focal length of a planar mirror is $f = 0$, where r is the surface radius.

If p_o is not located at infinity, the extended light rays do not bundle exactly within a single image. Thus, p_i has to be approximated (Figure 2.8(c)). Note, that in this case, images formed by multiple image points appear to be a reduced and deformed version of the reflected object that is represented by the corresponding object points.

In addition to rotation-paraboloids, other mirror surfaces (such as rotation-hyperboloids and prolate ellipsoids) can generate a single true

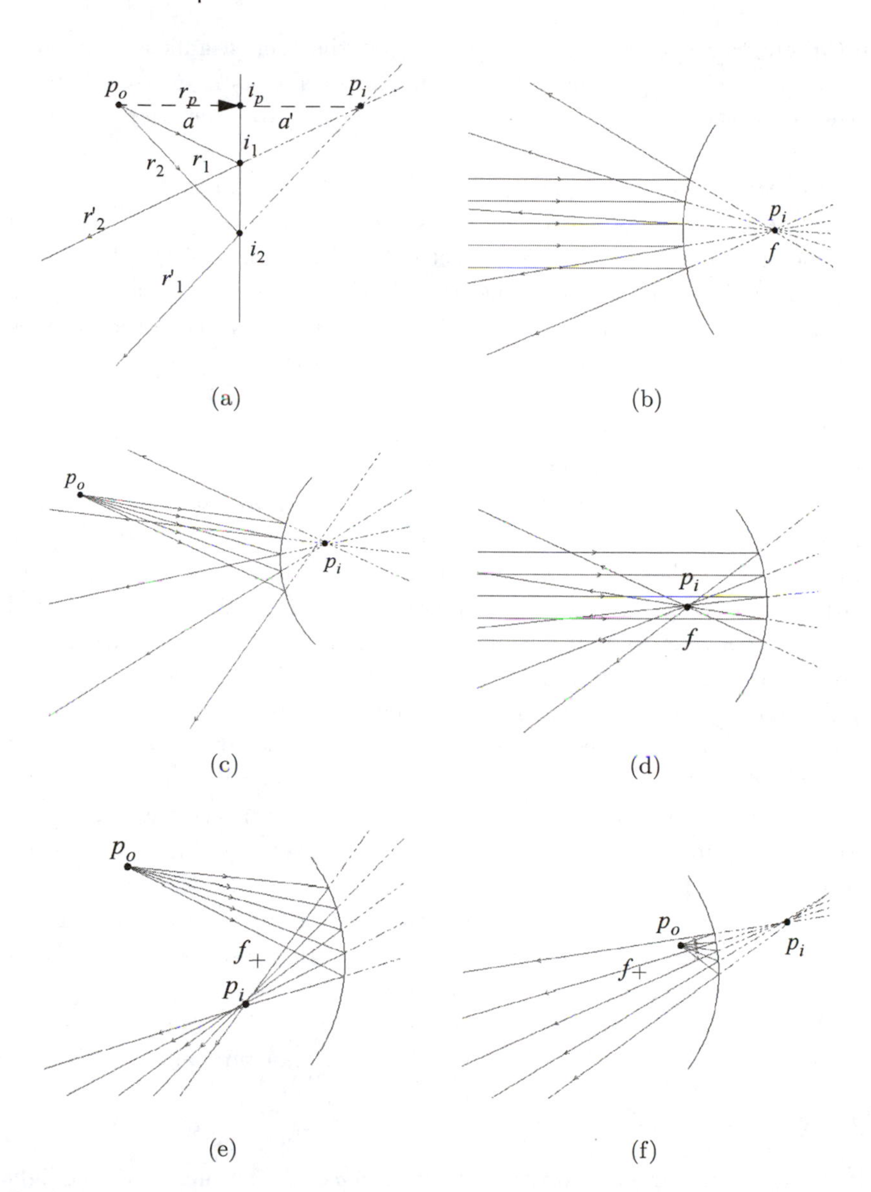

Figure 2.8. (a) Planar mirror; (b) convex parabolic mirror with object at infinity; (c) convex parabolic mirror with an object at a finite distance away from the mirror surface; (d) concave parabolic mirror with object at infinity; (e) concave parabolic mirror with an object at a finite distance behind its focal point; (f) concave parabolic mirror with an object at a finite distance in front of its focal point.

stigmatic pair. In general we can say that the true stigmatic pair, generated by such surfaces, is always their two focal points. Mirror surfaces other than the ones mentioned above do not generate true stigmatic pairs at all.

Concave mirrors can generate both virtual and real images from real objects because the reflected light rays converge or diverge, depending on the location of the object with respect to the focal point. As in the convex case, only the above mentioned surface types can generate just one true stigmatic pair which consists of their two focal points. For other surface types (or for objects that do not match the focal points), images can only be approximated.

Figure 2.8(d) illustrates an example of a concave parabolic mirror, where p_o is located at infinity and p_i is generated at f.

If p_o is not located at infinity, p_i has to be approximated. However, depending on the position of the object with respect to the focal point, the image can be either real or virtual. If, on the one hand, the object p_o is located behind the focal point f (as illustrated in Figure 2.8(e)) the reflected light rays converge and approximately bundle within the real image p_i (also located behind the focal point). Note, that in this case, images formed by multiple image points appear to be an enlarged, flipped, and deformed version of the reflected object that is represented by the corresponding object points.

If, on the other hand, p_o is located between the surface and f (as illustrated in Figure 2.8(f)) the reflected light rays diverge and their extensions approximately bundle within the virtual image p_i. Note, that in this case, images formed by multiple image points appear to be an enlarged and deformed version of the reflected object that is represented by the corresponding object points—yet, it is not flipped.

Note that if $p_o = f$, then p_i is located at infinity (i.e., the reflected light rays are parallel). In this case, p_i is neither real nor virtual.

2.2.4 Refractive Optics

In the case of refractive optical systems (*lenses*), the medium that light rays travel through is inhomogeneous. This means that, with respect to the simplified optical path equation, light rays pass through two different media with different densities and refraction indices ($\eta_1 \neq \eta_2$), where η_1 denotes the refraction index of the medium that surrounds the lens, and η_2 is the refraction index of the lens material. Since the rays are redirected when they change into another medium, their entrance and exit points

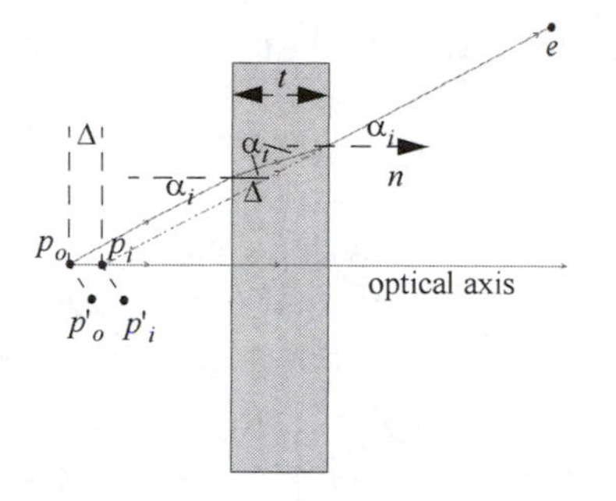

Figure 2.9. Planar lens.

differ ($i_x \neq j_x$). It can be idealized again that a lens is surrounded by air and that the medium air is approximately equivalent to the medium of a vacuum (i.e., $\eta_1 = 1$).

Planar lenses. For the following, we consider a homogeneous medium that is bounded by two plane-parallel panels known as a *planar lens*. Similar to planar mirrors, we can say that for planar lenses, p_o is real while p_i is virtual (Figure 2.9).

Light rays are refracted twice—once at their entrance points and again at their exit points. This is referred to as *in-out refraction*. In the case of planar lenses, the resulting out-refracted light rays have the same direction as the corresponding original rays, but they are shifted by the amount Δ in the direction parallel to their original counterparts. Since the original rays diverge from p_o, the refracted rays also diverge from the lens.

In Chapter 6, we show how to compute Δ for general situations and how to integrate it into fixed function and programmable rendering pipelines to support optical combination with mirror beam combiners. True stigmatic pairs are not generated with planar lenses since the optical path length is not constant for all rays.

An approximation, however, can be made if the lens is centered. An optical system is centered (so-called *centered* or *on-axis optical system*) if the axis of symmetry of all surfaces and the optical axis coincide, where the *optical axis* is given by the center light ray of a light bundle with its direction pointing towards the propagation direction of the light. In this case, we can intersect the extended out-refracted light ray with the optical axis and receive the virtual image p_i. The approximation is a result of the fact that we assume that the offset Δ is constant for all rays that diffuse from p_o. This means that all extended out-refracted light rays intersect at the same point p_i on the optical axis. In the on-axis case, Δ is given by

$$\Delta = t(1 - 1/\eta_2) \tag{2.3}$$

(with $\alpha_i = 0$. Note that this equation, which is commonly referred to in the optics literature, can only be used for on-axis (i.e., centered) optical systems. It is assumed, that a *detector* (e.g., a human eye) has to be located on the optical axis.

For centered optical systems, a further approximation is a result of the assumption that adjacent points appear to transform similarly (from the detector's point of view). Thus, the offset Δ for p_o is the same as for p'_o. The *sine-condition of Abbe* describes this assumption for adjacent point-pairs that are located on the same plane, perpendicular to the optical axis. The *condition of Herschel* expresses the preservation of stigmatism between adjacent point-pairs located on the optical axis. These approximations represent the basis for all centered image-forming optical systems (mirrors and lenses) that do not provide true stigmatism for all points (i.e., for the majority of all optical systems). They describe the approximate preservation of stigmatism within the three-dimensional free-space for centered optical systems. Nevertheless, they introduce aberrations.

The approach described in Chapter 6, on the other hand, is more general and can also be applied for off-axis (i.e., non-centered) situations. Thus, the detector does not have to be located on the optical axis. This is illustrated in Figure 2.9, whereby the detector is indicated with e.

Rather than the on-axis equation, the off-axis equation (see Chapter 6) will be used by the subsequent rendering techniques since we want to assume that the addressed optical systems that are utilized for spatial augmented reality configurations are general and not necessarily centered (i.e., the detector—or the eyes of a human observer, in our case—are not required to be located on the optical axis). Note that the optics used for head-mounted displays is centered, and the correction of aberrations can be precomputed [162, 200].

Since, for a given p_o and a given e, the geometric line of sight $p_o - e$ is known, but the geometric line of sight $p_i - e$, which is required to compute the correct Δ is unknown, we approximate that by $p_o - e = p_i - e$. Since the angular difference between both lines of sight with respect to the plane's normal is minimal, the arising error can be disregarded.

Planar interfacing surfaces. We can also derive the off-axis *out-refraction* behavior of light rays that move from a denser medium into air at the inter-

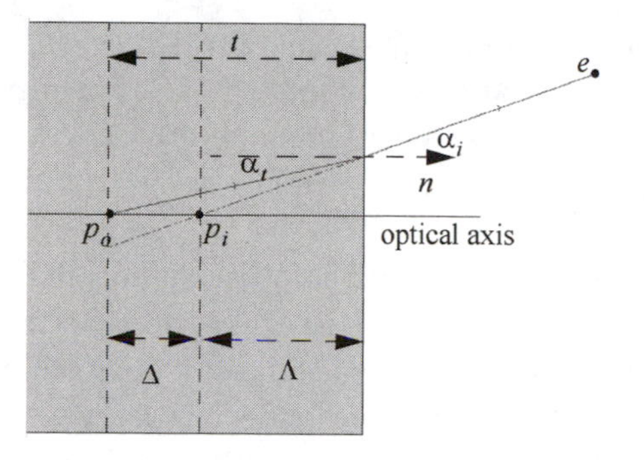

Figure 2.10. Out-refraction at planar interfacing surfaces.

section of a planar interfacing surface. In this case, the rays are refracted only once (Figure 2.10).

For the subsequent derivation, we want to assume a planar lens that extends from the interfacing surface to p_o. With respect to Figure 2.10, we can say that $t = \Lambda + \Delta$. The distance Λ is given by $\Lambda = t(\tan\alpha_t / \tan\alpha_i)$, and consequently $t = \Lambda(\tan\alpha_i / \tan\alpha_t)$.

If p_o is known, then t is the shortest distance between p_o and the plane of the interfacing surface and is given by $t = |f(p_o)| = |np_o + d|$, where $f(x, y, z) = ax + by + cz + d = 0$ and $n = (a, b, c)$ defines the plane ($f(x, y, z)$ is normalized). We can now solve for Δ:

$$\Delta = \Lambda\left(\frac{\tan\alpha_i}{\tan\alpha_t} - 1\right), \tag{2.4}$$

where Λ is constrained by the following boundaries:

$$\lim(\alpha_i \to \frac{\pi}{2}) \Rightarrow \Lambda = 0$$

and

$$\lim(\alpha_i \to 0) \Rightarrow \Lambda = t\left(\frac{\sin\alpha_t}{\sin\alpha_i}\right) = t\left(\frac{1}{\eta_2}\right) = \text{const.}$$

With Equation (2.4), we can derive the commonly referred to refraction ratio for centered planar interfacing surfaces (i.e., for the case that $\alpha_i = 0$): $\Lambda/t = 1/\eta_2$.

This is equivalent to $\eta_2/t = \eta_1/\Lambda$ for a variable refraction index. (Note that we assumed $\eta_1 = 1$ for air.)

This closes the loop and proves that the on-axis refraction computations for centered systems (which are normally discussed in the optics literature) are a special case of our more general off-axis refraction method.

Note that the refraction transformation defined in Chapter 6 can be used for out-refraction and to transform p_o to p_i. For *in-refractions* (in our example, if the object is located within air and the detector is located within the denser medium) the refraction index, as well as the transformation direction, have to be reversed.

In general, we can say that if the object is located within the denser medium and the detector is located within the sparser medium (i.e., the light rays are out-refracted), then the object's image appears closer to the interfacing surface. If, in turn, the object is located within the sparser medium and the detector is located within the denser medium (i.e., the light rays are in-refracted), then the object's image appears further away from the intersecting surface.

However, ray tracers usually determine the specular refraction ray r' of an original ray r as follows:

$$r' = \frac{\eta_1}{\eta_2} r - \left(\cos \alpha_t + \frac{\eta_1}{\eta_2}(nr) \right) n. \tag{2.5}$$

As in the case of planar mirrors, for planar lenses or planar interfacing surfaces, n is constant for all surface points i. However, Equations (2.4) and (2.5) are also valid for non-planar surfaces with individual normal vectors at each surface point. In this case, the intersection i of r with the surface and the normal n at i have to be inserted. As for reflection, the curvilinear behavior of refraction (for planar and for non-planar surfaces) can be well expressed with ray-tracing techniques.

Non-planar lenses. In practice, curved lenses are usually bound by two spherical surfaces. As in the case of spherical mirrors, spherical lenses do not have exact focal points—just areas of rejuvenation which outline approximate locations of the focal points. Consequently, they cannot generate true stigmatic pairs. In contrast to lenses, mirrors are often shaped parabolically to provide exact focal points, and therefore provide one true stigmatic pair. For manufacturing reasons, however, almost all lenses that are used in optical systems are spherical. Stigmatism can only be approximated for such optical systems.

The focal length of a *spherical lens* is defined by (Figure 2.11(a))

$$\frac{1}{f} = (\eta_2 - 1)\left(\frac{1}{r_1} - \frac{1}{r_2} \right) + \frac{(\eta_2 - 1)^2}{\eta_2} \frac{t}{r_1 r_2},$$

where r_1 and r_2 are the two surface radii and t is the central thickness of the lens.

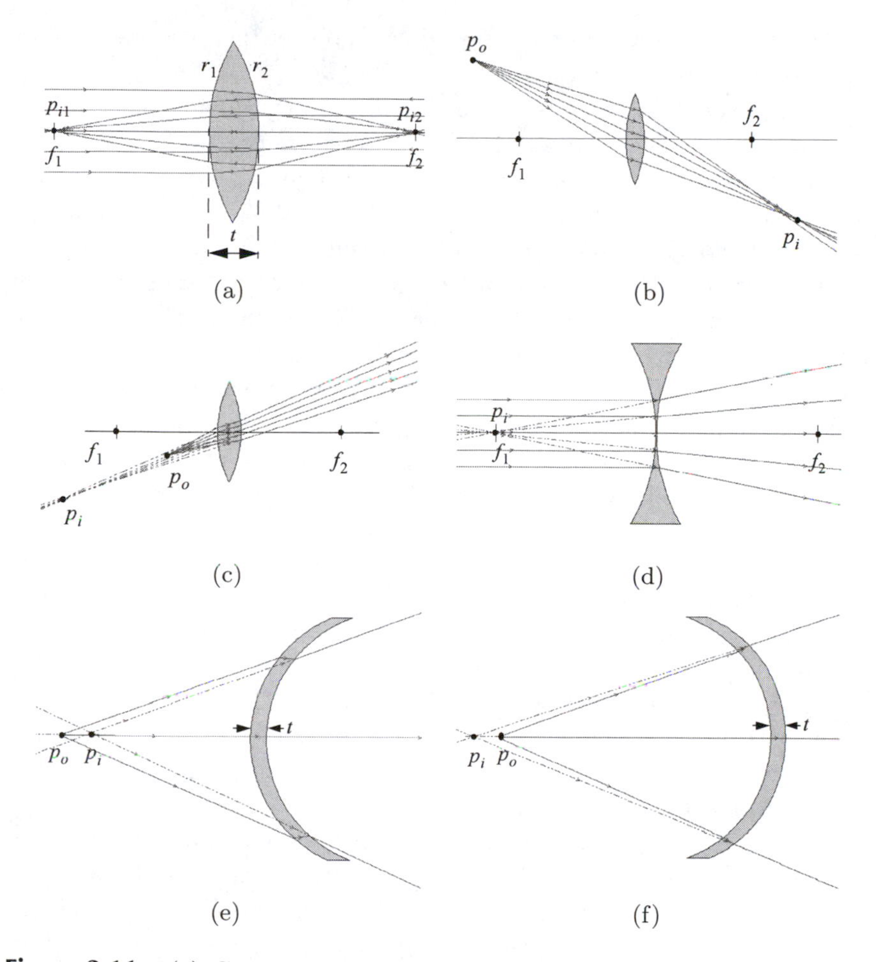

Figure 2.11. (a) Convergent spherical lens with two objects at infinity; (b) convergent spherical lens with an object a finite distance behind a focal point; (c) convergent spherical lens with an object a finite distance in front of a focal point; (d) divergent spherical lens with object at infinity; (e) convex parallel spherical lens with an object a finite distance away; (f) concave parallel spherical lens with an object a finite distance away.

Convergent lenses are mostly bounded by two convex spherical surfaces (Figure 2.11(a)). Since light rays that are emitted from behind a focal point (i.e., further away from the lens) converge after exiting such a lens, real objects that are located behind a focal point form real images which are located behind the opposite focal point (Figure 2.11(b)). Note that in this case images formed by multiple image points appear to be a reduced, flipped, and deformed version of the refracted object that is represented by the corresponding object points.

If, however, the object is located between a focal point and the lens' surface, the light rays diverge after exiting a convergent lens. Thus, their extensions bundle within a virtual image point in front of or behind the same focal point (Figure 2.11(c)). Note that in this case images formed by multiple image points appear to be an enlarged and deformed version of the refracted object that is represented by the corresponding object points—yet, it is not flipped. This behavior is very similar to the behavior of concave mirrors—however, it is reversed.

Divergent lenses are mostly bounded by two concave spherical surfaces (Figure 2.11(d)). In the case of divergent lenses, the exiting light rays always diverge. Thus, real objects always form virtual images—no matter where the object is located. This behavior can be compared with the behavior of convex mirrors; it is also reversed.

Curved lenses that are bounded by two parallel (concave or convex) spherical surfaces can also be considered. In this case, the thickness of the lens t is the same at all surface points. These lenses can only produce virtual images, since exiting light rays always diverge. This is illustrated in Figure 2.11(e) for a convex lens, and in Figure 2.11(f) for a concave lens.

Note that the object-image translation direction of a convex lens is the reverse of the direction of a concave lens. However, in contrast to plane lenses but in correspondence with all other curved lenses, the out-refracted rays do not have the same direction as the corresponding original rays and are not simply shifted parallel to their original counterparts. Thus, Equation (2.3) in combination with approximations, such as the sine-condition of Abbe or the condition of Herschel, do not apply in this case. Rather Equations (2.4) or (2.5) can be used twice—for the in-refraction and for the out-refraction of a ray.

2.3 Visual Depth Perception

In all optical systems, the light rays that are emitted by objects or images are finally perceived by light-sensitive components—the so-called detectors. An example of a detector is a photographic film (or disc or chip) used by cameras to preserve the received light information on a medium. The most common detector for optical systems, however, is the human eye which forwards the detected light information to the brain. The interplay of the two eyes that receive different two-dimensional images of the same environment (seen from slightly different perspectives) enables the brain to reconstruct the depth information. This phenomenon is called *stereoscopic vision*. Stereoscopic vision can be fooled with *stereoscopic displays* which present two artificially generated images of a virtual environment to the eyes of a human observer. As in the real world, these images are interpreted by the brain and fused into a three-dimensional picture.

2.3.1 The Human Eye

The human eye consists of spherical interfacing surfaces and represents a complex optical system itself (Figure 2.12). Its approximate diameter is 25 mm, and it is filled with two different fluids—both having a refraction index of ~1.336. The *iris* is a muscle that regulates the amount of the incoming light by expanding or shrinking. The *cornea* and the elastic biconvex *lens* (refraction index ~1.4) below the iris bundle the transmitted light in such a way that different light rays which are diffused by the same point light source are projected to the same point on the *retina*. Note that the projection is flipped in both directions—horizontally and vertically.

The retina consists of many small conic and cylindrical light-detecting cells called *photoreceptors* (approximate size 1.5 ηm–5 ηm) that are categorized into *rods* and *cones*. The resolution of the eye depends on the density of these cells—which varies along the retina. If the distance between two point projections is too small, only one cell is stimulated and the brain cannot differentiate between the two points. To recognize two individual points, the two stimulated cells have to be separated by at least one additional cell. If the angle between two light rays that are emitted from two different point light sources and enter the eye is below 1.5 arc minutes, the

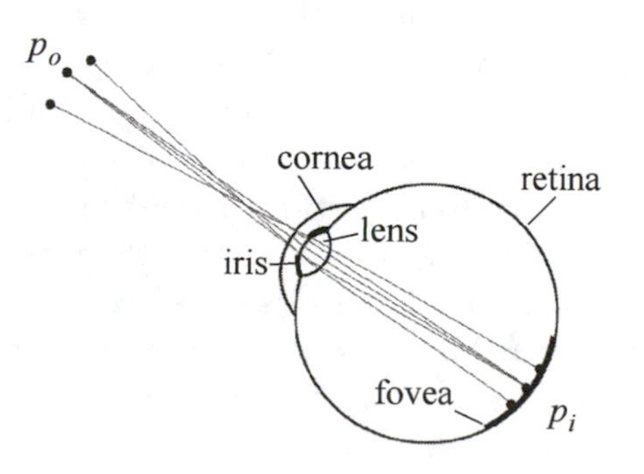

Figure 2.12. The human eye as an optical system.

points cannot be differentiated. The limited resolution of the human eye is the reason why light rays that are emitted from a single object point and pass through an optical system which does not support true stigmatism, still appear to intersect at a single image point. Thus, small aberrations of non-stigmatic and non-absolute optical systems are not detected. Consequently, the observer perceives a single—possibly deformed—image of the object. The area with the highest resolution of detector cells is called the *fovea*.

In addition, the lens adjusts the focus for different distances (i.e., focal lengths) by deforming its shape. The deformation of the lens is called *accommodation*. The human eye can accommodate for focal lengths between ∞ and 100 mm. Objects whose emanating light rays cannot be bundled by the lens on a single point of the retina appear unsharp. This happens, for instance, if the object is located closer to the eye than 100 mm.

2.3.2 Stereoscopic Vision

Two different views of the same object space are required to support depth perception. The perceptual transformation of differences between the two images seen by the eyes is called *stereopsis*. Due to the horizontal eye separation (*interocular distance* = ~6.3 cm), the images that are perceived by the two eyes are slightly shifted horizontally, and are rotated around the vertical axis. Light rays that are emitted from an object project onto different locations on the respective retina. The relative two-dimensional displacement between two projections of a single object onto two different focal planes (i.e., the left and the right eye's retina) is called *retinal disparity*. The stimulation of the detector cells at the corresponding locations is used by the brain to fuse the two images and to approximate the relative distance (i.e., the *depth*) of the object. Note that if the disparity between two projections becomes too large, the perceived images of the object space cannot be fused by the brain and the object space appears twice. This effect is called *diplopia* (double vision).

To facilitate the accurate projection of the light rays onto the proper detector cells, the eyes have to rotate around the vertical axis until they face the focal point. This mechanism is called *vergence*. They can either rotate inwards to focus at close objects (*convergence*) or outwards to focus distant objects (*divergence*). If their alignment is parallel, an object at infinite distance is focused.

The total amount of the environment that can be seen by both eyes is called *monocular field of vision* and extends over a 180° horizontal field of

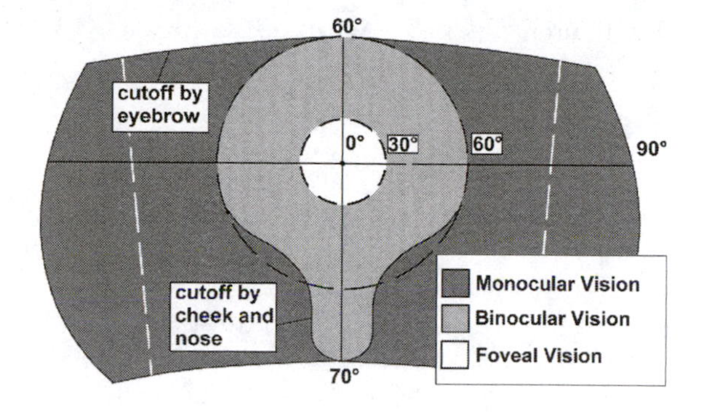

Figure 2.13. The human visual fields.

view and a 130° vertical field of view (Figure 2.13). The portion of this visual field shared by both eyes is known as *binocular field of vision* and extends over a 120° horizontal field of view and a 130° vertical field of view. An even smaller portion within the binocular field of vision is called *foveal field of vision*. It is the area in which both eyes see in focus. It extends over a 60° horizontal and vertical field of view. Note that for stereoscopic vision, only the binocular field of vision is of interest.

2.3.3 Spatial Stereoscopic Presentation

Stereoscopic vision can be fooled with stereoscopic displays (Figure 2.14). For the subsequent explanation, we want to assume spatial stereoscopic displays which allow pixels to be drawn on their rasterized display surface.

Given a fictive object p_o, we can determine the fictive light rays that would be emitted from p_o and intersect the eyes. For this, the eyes' positions need to be known; they are approximated by representing the eyes with single points that are located at the eyeball's center. These rays are projected backwards onto the display surface and result in the positions of the pixels that are finally drawn on the display. The two related (left and right) projections of an object are called the *stereo-pair*. Since the real light rays that are now emitted from the pixels intersect at p_o before they reach the eyes, the brain perceives the fictive object at its corresponding depth—floating in space. Such fictive objects (within the computer graphics community also called *virtual objects*[1]) can be displayed so that they

[1]Not to be confused with the previously used virtual object terminology of the optics community.

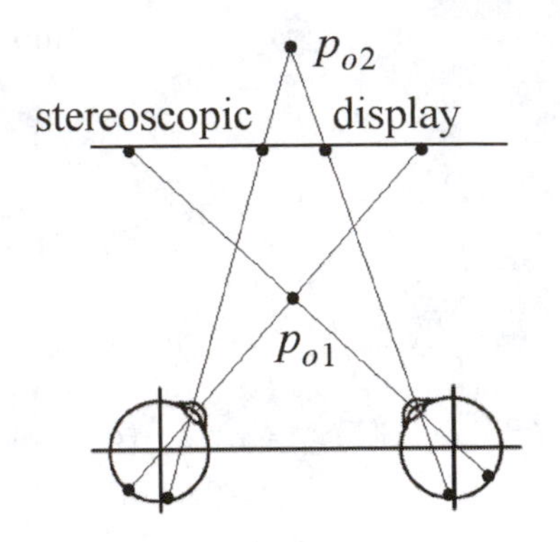

Figure 2.14. Stereoscopic vision with spatial stereoscopic display.

appear in front of (e.g., p_{o1}), on, or behind the display (e.g., p_{o2}). The apparent two-dimensional relative motion of a distant object with respect to a close one as the viewpoint moves is called *parallax*. With respect to planar stereoscopic displays, parallax is usually defined as the distance between one object's pixel projection for the left eye and the same object's pixel projection for the right eye. In case the projections are on the same side as the corresponding eye, the virtual object appears behind the display surface (such as p_{o2}). This situation is called *positive parallax*. Note that the maximum positive parallax occurs when the virtual object is located at infinity. At this point the parallax equals the interocular distance.

If the virtual object appears in front of the display surface (such as p_{o1}), then the projection for the left eye is on the right and the projection for the right eye is on the left. This is known as *negative parallax*. If the virtual object appears half way between the center of the eyes and the display, the negative parallax equals the interocular distance. As the object moves closer to the eyes, the negative parallax increases to infinity and diplopia occurs at some point. In correspondence to this, the case where the object is located exactly on the display surface is called *zero parallax*.

An essential component of stereoscopic displays is the functionality to separate the left and right images for the respective eye when they are displayed. This means, that the left eye should only see the image that has been generated for the left eye, and the right eye should only see the image that has been generated for the right eye. Several mechanical, optical, and physiological techniques exist to provide a proper *stereo separation*.

Note that for stereo pairs that are projected onto a two-dimensional spatial display to form a three-dimensional virtual object, the retinal disparity and the vergence are correct, but the accommodation is inconsistent because the eyes focus at a flat image rather than at the virtual object. With respect to stereoscopic displays, disparity and vergence are considered as the dominate depth cues for stereoscopic viewing.

2.3.4 Classification of Stereoscopic Displays

This section will provide a broad classification of current stereoscopic display approaches (Figure 2.15). Note that we do not claim to present a complete list of existing systems and their variations, but rather focus on the technology that is (or might become) relevant for the concepts described in this book.

Stereoscopic displays can be divided into *autostereoscopic displays* and *goggle-bound displays*. While goggle-bound displays require the aid of additional glasses to support a proper separation of the stereo images, autostereoscopic displays do not. We will discuss goggle-bound displays in the next section and then describe autostereoscopic displays in more detail.

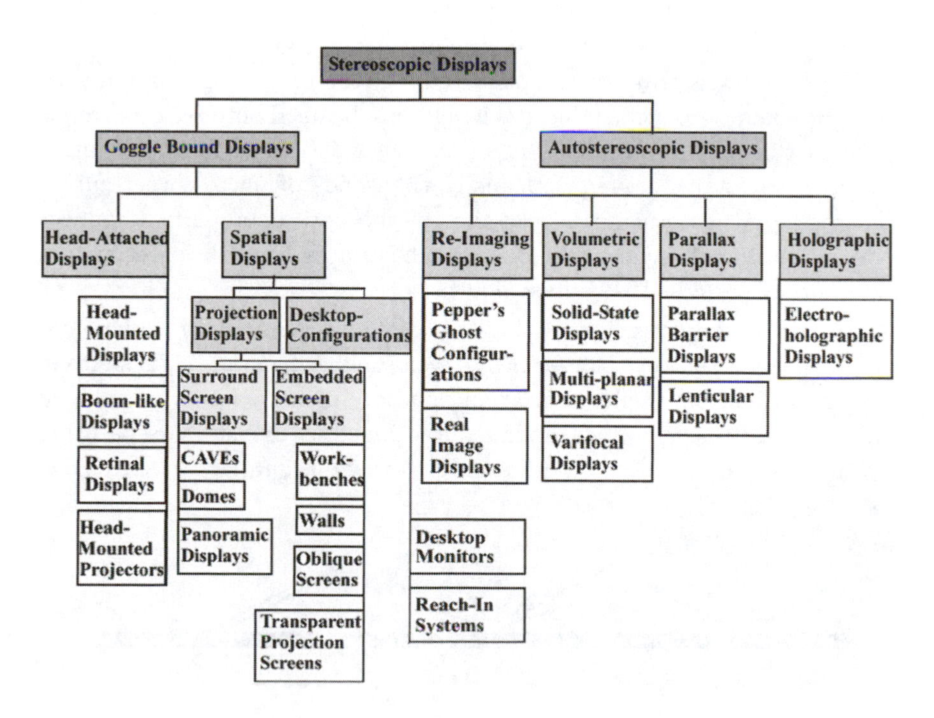

Figure 2.15. Classification of stereoscopic displays.

Goggle-bound displays. Goggle-bound displays require the user to wear additional goggle-like devices in front of the eyes to support a proper separation of the stereo images. They can be divided into *head-attached displays* and *spatial displays*.

Head-attached displays provide individual display elements for each eye, and consequently can present both stereo images simultaneously. Examples for such elements are miniature CRT or *Liquid Crystal Display* (LCD) screens that are used in most head-mounted displays [182, 183] and *BOOM-like displays*. *Retinal displays* [83, 145] utilize low-power lasers to scan modulated light directly onto the retina of the human eye instead of providing screens in front of the eyes. This produces a much brighter and higher resolution image with a potentially wider field of view than a screen-based display. *Head-mounted projective displays* [130, 73, 70] or *projective head-mounted displays* [78] are projection-based alternatives that employ head-mounted miniature projectors instead of miniature displays. Such devices tend to combine the advantages of large projection displays with those of head-mounted displays. Head-attached displays (especially head-mounted displays) are currently the display devices that are mainly used for augmented reality applications.

Head-mounted projective displays redirect the projection frustum with a mirror beam combiner so that the images are beamed onto *retro-reflective surfaces* that are located in front of the viewer (Figure 2.16). A retro-reflective surface is covered with many thousands of micro-corner cubes. Since each micro-corner cube has the unique optical property to reflect light back along its incident direction, such surfaces reflect brighter images than normal surfaces that diffuse light.

Spatial displays use screens that are spatially aligned within the environment. Nevertheless, the user has to wear field-sequential (LCD shutter-glasses) or light-filtering (polarization or color/intensity filters) goggles to support a correct separation of the stereo images. The stereo separation technique for spatial displays is generally known as *shuttering*, since each

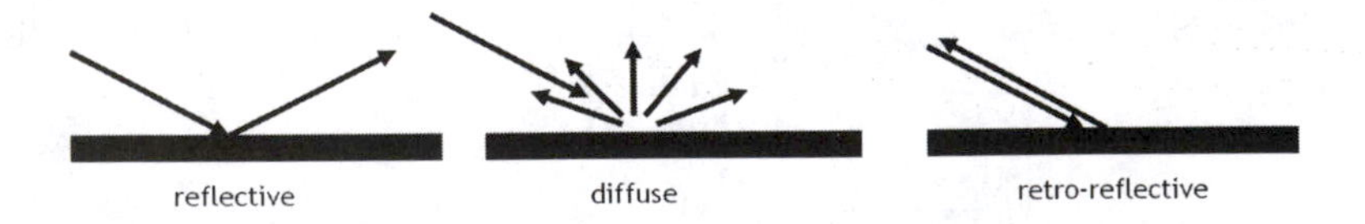

Figure 2.16. Difference between reflection, diffusion, and retro-reflection of light.

of the two stereo images which are presented on the same screen(s) has to be made visible to only one eye (i.e., the other image has to be blocked respectively—by shutting the eye). Depending on the shuttering technology, the stereo images are either presented time sequentially (i.e., with field-sequential goggles) or simultaneously (i.e., with light-filtering goggles).

Several light-filtering techniques exist that are also known as *passive shuttering*. *Anaglyphs*, for instance, encode the stereo images in two different colors (e.g., red/green, red/blue, or red/cyan). Color filters that are integrated into special glasses ensure that only one image reaches the eye while the other one is blocked. A side effect of this technique is that although the brain fuses the stereo images into a three-dimensional picture, this picture is perceived as monochrome. For very simple scenes with well-defined edges, however, it is possible to display a third image which carries the scene color to make a *full color anaglyph*. In this case, the anaglyph effect is only apparent at the edges of objects. *ChromaDepth* is another light-filtering technique that uses color to encode the depth of a scene. This technique is based on the fact that the photoreceptors on the retina focus on light at slightly different angles. ChromaDepth filters alter the angle of the incident light depending on its wavelength and enhance this effect. Such filters are clear and do not change the color of the displayed scene. The disadvantage of this technique is that the presented scene cannot be shown in arbitrary colors. In particular, three-dimensional animations would cause a continuous alternation of colors to encode the depths—which is certainly impractical.

The *Pulfrich effect* is yet another light filtering technique which is based on filtering intensity rather than color. The time from stimulation of the photoreceptors on the retina to sending the processed signal to the brain depends on the lighting conditions. It takes longer to send these signals under low than under bright lighting conditions. Pulfrich glasses shade one eye to realize stereo separation based on this effect. If a movie with horizontal camera motion is watched with Pulfrich glasses, the same video frame will process faster by the non-shaded eye than by the shaded one. If the camera motion is done well, the non-shaded eye will send the image of a later video frame, while the shaded eye simultaneously sends the image of an earlier frame. The horizontal disparity between the two frames is perceived as depth. The Pulfrich effect can only be used for certain video effects that contain a horizontal camera motion, but not for arbitrary graphical scenes.

Finally, *polarization glasses* are the most common passive shuttering technique for stereoscopic projection screens. They are based on the polarization of light described in Section 2.1. The screen polarizes the stereo images in two different directions (e.g., by attaching polarization filters in different orientations in front of two video beamers that display the stereo images simultaneously), while identically-oriented polarization filters integrated in the glasses ensure the correct separation. The advantage of this method is that it does not constrain the color or intensity representation of the displayed scene. However, for projection displays, a special non-organic screen material is required that does not destroy the polarization of the projected light. Regular diffuse projection surfaces cannot be used.

Spatial displays can be further divided into *desktop configurations* and *projection displays*. Using *desktop monitors* as a possible stereoscopic display is the traditional desktop-VR approach (also referred to as *fish tank VR* [199]). Since desktop monitors (i.e., only CRT screens, but not LCD screens) provide the refresh rate of 120Hz that is required for a time-sequential shuttering, LCD shutter glasses are mostly used for stereo separation. This technique is known as *active shuttering*. Note that older applications also use color-filtering glasses (e.g., anaglyphs) to separate monochrome stereo images. Fish tank VR setups are classified as non-immersive since, in contrast to large screens, the degree of immersion is low. *Reach-in systems* [81, 166, 142, 204] represent another type of desktop configuration that consists of an upside-down CRT screen which is reflected by a small horizontal mirror. Nowadays, these systems present stereoscopic three-dimensional graphics to a single user who is able to reach into the presented visual space by directly interacting below the mirror while looking into the mirror. Thus, occlusion of the displayed graphics by the user's hands or input devices is avoided. Such systems are used to overlay the visual space over the interaction space, whereby the interaction space can contain haptic information rendered by a force-feedback device such as a PHANTOM [104]. Due to the small working volume of these devices, their applications are limited to near-field operations.

Projection displays currently use CRT, LCD, LCOS, or digital light projectors (DLP) to beam the stereo images onto single or multiple, planar or curved display surfaces. Two types of projections exist: With *front-projection*, the projectors are located on the same side of the display surface as the observer. Thus, the observer might interfere with the projection frustum and cast a shadow onto the display surface. With *rear-projection* (or *back-projection*), the projectors are located on the opposite side of the display surface to avoid this interference problem. Both active and passive shuttering are used in combination with projection displays. For passive shuttering, at least two projectors are necessary to beam both filtered stereo images simultaneously onto the display surface. Projection displays that beam both stereo images sequentially onto the display surface require field-sequential shutter glasses to separate the stereo images. For active shuttering, only one projector is necessary since the images are projected time sequentially. However, as with desktop monitors, these projectors have to support the required refresh rate of 120Hz to provide a flicker-free update rate of 60Hz per eye. Note that projection screens can either be *opaque* or *transparent*—depending on their application. Depending on the number and the shape of the spatially aligned display surfaces, we can divide projection displays into *surround screen displays* and *embedded screen displays*. Surround screen displays surround the observers with multiple planar (e.g., CAVEs [35], CABINs [63]) or single curved display surfaces (e.g., domes or panoramic displays) to provide an immersive VR experience. Thus, the observers are completely encapsulated from the real environment. Usually, multiple projectors are used to cover the extensive range of the projection surface(s).

In contrast to surround screen displays, embedded screen displays integrate single, or a small number of display surfaces, into the real environment. Thus, the users are not immersed into an exclusively virtual environment, but can interact with a semi-immersive virtual environment that is embedded within the surrounding real environment. Horizontal, *workbench-like* [84, 85] or vertical *wall-like* display screens are currently the most common embedded screen displays. *Oblique* screen displays represent a generalization of embedded screen displays, whereby special display surfaces are not integrated explicitly into the real environment. Rather, the real environment itself (i.e., the walls of a room, furniture, etc.) provides implicit display surfaces.

Although head-attached (especially head-mounted) displays have a long tradition within the VR community, stereoscopic projection displays are currently the dominant output technology for virtual reality applications.

Autostereoscopic displays. Autostereoscopic displays present three-dimensional images to the observer without the need of additional glasses. Four classes of autostereoscopic displays can be found: *re-imaging displays*, *volumetric displays*, *parallax displays*, and *holographic displays*.

Two types of images exist in nature—real and virtual. A real image is one in which light rays actually come from the image. In a virtual image, they appear to come from the reflected image—but do not. In the case of planar or convex mirrors, the virtual image of an object is behind the mirror surface, but light rays do not emanate from there. In contrast, concave mirrors can form reflections in front of the mirror surface where emerging light rays cross—so called "real images." Re-imaging displays project existing real objects to a new position or depth. They capture and re-radiate the light from the real object to a new location in space. An important characteristic of re-imaging displays is that they do not generate three-dimensional images by themselves. Some re-imaging systems use lenses and/or mirrors to generate copies of existing objects. Especially half-silvered mirror setups—called *Pepper's ghost configurations* [196]—are used by theme parks to generate a copy of a real three-dimensional environment and overlay it over another real environment.

Pepper's ghost configurations are a common theatre illusion from around the turn of the century named after John Henry Pepper—a professor of chemistry at the London Polytechnic Institute. At its simplest, a Pepper's ghost configuration consists of a large plate of glass that is mounted in front of a stage (usually with a 45° angle towards the audience). Looking through the glass plate, the audience is able to simultaneously see the stage area and, due to the self-reflection property of the glass, a mirrored image of an off-stage area below the glass plate. Different Pepper's ghost configurations are still used by entertainment and theme parks (such as the Haunted Mansion at Disney World) to present their special effects to the audience. Some of those systems reflect large projection screens that display prerecorded two-dimensional videos or still images instead of real off-stage areas. The setup at London's Shakespeare Rose Theatre, for instance, uses a large 45° half-silvered mirror to reflect a rear-projection system that is aligned parallel to the floor.

Other re-imaging displays use more complex optics and additional display devices. For instance, some re-imaging displays generate a copy of a two-dimensional CRT screen which then appears to float in front of the optics. These types of mirror displays—so-called *real image displays*—generate real images (Figure 2.17).

Several real image displays are commercially available and are frequently used as eye-catchers for product presentation by the advertising industry or to facilitate special on-stage effects by the entertainment industry. On the one hand, they can present real objects that are placed inside the system so that the reflection of the object forms a three-dimensional real image floating in front of the mirror. On the other hand, a display screen (such as a CRT or LCD screen, etc.) can be reflected instead

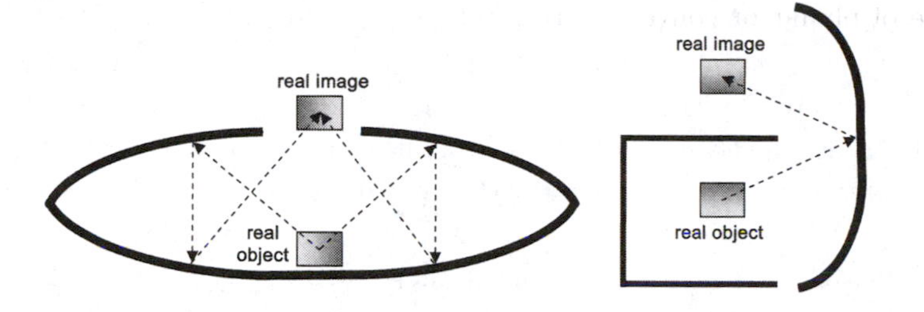

Figure 2.17. Examples of common real-image display configurations using one or two concave mirrors.

[173, 106, 107]—resulting in a free-floating two-dimensional image in front of the mirror optics that is displayed on the screen (some refer to these systems as *pseudo 3D displays* since the free-floating two-dimensional image has an enhanced three-dimensional quality). Usually, prerecorded video images are displayed with such real image displays.

Volumetric displays directly illuminate spatial points within a display volume. In contrast to re-imaging displays, volumetric displays can generate synthetic images of voxelized data or three-dimensional primitives. These types of displays generate images by filling or sweeping out a volumetric image space. *Solid-state devices* are variations of volumetric displays which display voxel data within a translucent substrate by generating light points with an external source (for example with lasers of different wavelengths located outside the substrate that are scanned through the image space) [39].

Multi-planar volumetric displays build volumetric images from a time-multiplexed series of two-dimensional images. These images are displayed with a swiftly moving or spinning display element. This display element can be, for example, a rotating proprietary screen onto which the images are projected [45, 44] (e.g., using an external projector or lasers). Other systems directly move or spin light-generating elements (e.g., light diodes) or use multiple fixed, but switchable, screen planes [180]. In either case, the human visual system interprets these time-multiplexed image slices as a three-dimensional whole.

Varifocal mirror displays [190, 52, 106, 107] are yet another group of volumetric displays. They use flexible mirrors to sweep an image of a CRT screen through different depth planes of the image volume. In some systems the mirror optics is set in vibration by a rear-assembled loudspeaker [52]. Other approaches utilize a vacuum source to manually deform the mirror optics on demand to change its focal length [106, 107]. Vibrating devices, for instance, are synchronized with the refresh-rate of a display system that is reflected by the mirror. Thus, the spatial appearance of a reflected pixel can be exactly controlled, yielding images of pixels that are displayed approximately at their correct depth (i.e., they provide an autostereoscopic viewing and, consequently, no stereo-separation is required). Due to the flexibility of varifocal mirror displays, their mirrors can dynamically deform to a concave, planar, or convex shape (generating real or virtual images). However, these systems are not suitable for optical see-through tasks, since the space behind the mirrors is occupied by the deformation hardware (i.e., loudspeakers or vacuum pumps). In addition, concavely shaped varifocal mirror displays face the same problems as real image displays. Therefore, only full mirrors are used in combination with such systems.

Since view-dependent shading and culling (required to simulate occlusion) of the presented graphics is not supported, volumetric displays are mainly used to present volumetric, wire-frame, or icon-based contents.

Parallax displays are display screens (e.g., CRT or LCD displays) that are overlaid with an array of light-directing elements. Depending on the observer's location, the emitted light that is presented by the display is directed so that it appears to originate from different parts of the display while changing the viewpoint. If the light is directed to both eyes individually, the observer's visual system interprets the different light information to be emitted by the same spatial point.

One example of a parallax display is a *parallax barrier display* (Figure 2.18(a)) that uses a controllable array of light-blocking elements (e.g., a light blocking film or liquid crystal barriers [136]) in front of a CRT screen. Depending on the observer's viewpoint, these light-blocking elements are used to direct the displayed stereo-images to the corresponding eyes.

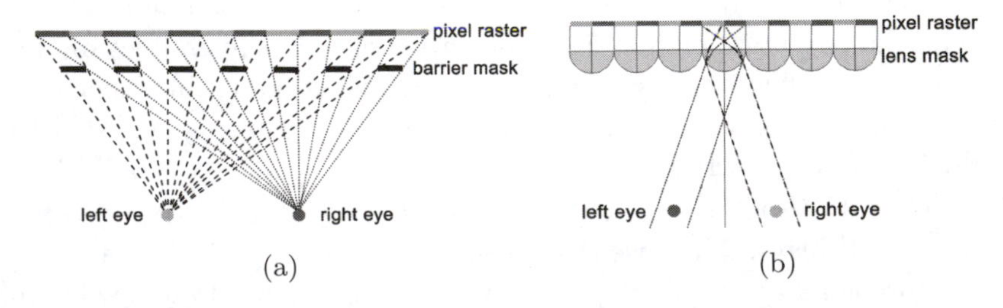

Figure 2.18. Basic parallax display concepts: (a) parallax barrier display; (b) lenticular sheet display.

Another example is a *lenticular sheet display* (Figure 2.18(b)) that uses an array of optical elements (e.g., small cylindrical or spherical lenses) to direct the light for a limited number of defined viewing zones.

Parallax displays can be developed and mass-produced in a wide range of sizes and can be used to display photo-realistic images. Many different types are commercially available and used in applications ranging from desktop screens to cell phone displays to large projection screens [205].

A *hologram* is a photometric emulsion that records interference patterns of coherent light (see Section 2.2). The recording itself stores the amplitude, wavelength, and phase information of light waves. In contrast to simple photographs, which can record only amplitude and wavelength information, holograms can reconstruct complete optical wavefronts. This results in the captured scenery having a three-dimensional appearance that can be observed from different perspectives.

To record an *optical hologram*, a laser beam is split into two identical beams (Figure 2.19). One beam, the *reference wave*, illuminates the holographic emulsion directly while the other beam illuminates the object to

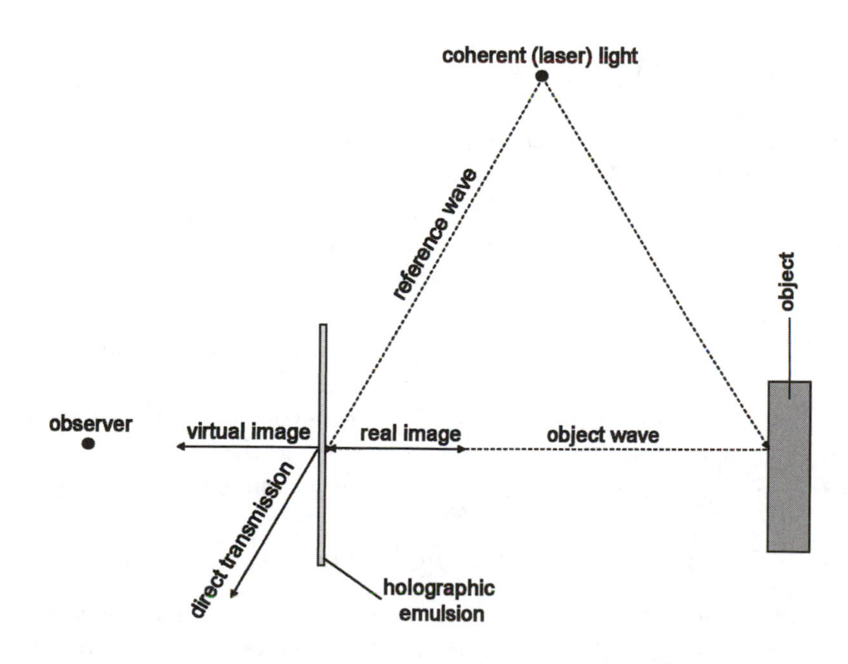

Figure 2.19. Optical holographic recording and reconstruction (example of a transmission hologram). (*Image reprinted from [18] © IEEE.*)

be recorded. The light that is reflected off the object, the *object wave*, together with the reference wave creates an interference pattern (or *interference fringe*) on the emulsion. If the emulsion is illuminated with a copy of the reference wave, it interacts with the recorded interference fringes and reconstructs the object wave, which is visible to the observer.

Computer-generated holograms (CGH) or *electroholography* [96] facilitates the computer-based generation and display of holograms in real time. Holographic fringes can be computed by either rendering multiple perspective images, then combining them into a *stereogram*, or simulating the optical interference and calculating the interference pattern. Once computed, the system dynamically visualizes the fringes with a holographic display. Since creating a hologram requires processing, transmitting, and storing a massive amount of data, today's computer technology still sets CGH limits. To overcome some of these performance issues, researchers have developed advanced reduction and compression methods that create truly interactive CGH. Holographic displays share most of the properties of volumetric displays, but they are still far from producing high-quality three-dimensional images using affordable hardware.

2.4 Rendering Pipelines

The functionality of a graphics engine (implemented in hardware or software) to map a scene description into pixel data is known as the *rendering pipeline*. Since the scene description, such as two-dimensional/three-dimensional vertices, textures coordinates, normals, etc. are passed sequentially through different modules of a render engine, a *function pipeline* is a vivid concept for illustrating the different processing steps.

To describe the full functionality of today's rendering pipelines is beyond the scope of this book. The following sections rather summarize the most important techniques used for the rendering concepts that are described in the following chapters. We presume that the reader has a certain level of knowledge in computer graphics techniques.

2.4.1 Fixed Function Rendering Pipelines

Once configured, traditional rendering pipelines apply the same operations to all data elements that pass through the pipeline. Furthermore, such pipelines require the application of a fixed sequence of basic operations, such as transformations, lighting, texturing, rasterization, etc. After the scene has been rendered, the pipeline can be reconfigured with new parameters. Thus, such pipelines are called *fixed function pipelines*.

The applicability of fixed function pipelines is clearly limited. Scene vertices, for instance, can only be transformed in the same way—allowing only rigid body transformations. Geometry warping, on the other hand, requires per-vertex transformations—a functionality that has to be implemented in software if required. Lighting and shading is another example that is applied in exactly the same way to each scene vertex. During rasterization, approximate values are interpolated for intermediate pixels. Fixed function pipelines do not allow per-pixel operations that would lead to more realistic shading effects.

One of the most important parts of a rendering pipeline is the *transformation pipeline* that applies geometric transformations to the incoming scene vertices and maps them to fragments (or pixels).

Transformation pipeline. Figure 2.20 illustrates OpenGL's transformation pipeline as it is presented by Neider et al. [117]. It is a component of OpenGL's rendering pipeline which is also discussed in [117].

Scene vertices (represented as homogeneous coordinates) are multiplied by the transformation matrix which is a composition of the *model-view matrix* and the *projection matrix*. These transformations are represented as 4×4 homogeneous matrices.

The *model-view transformation* can be conceptually split into the *scene transformation* and the *view transformation*.

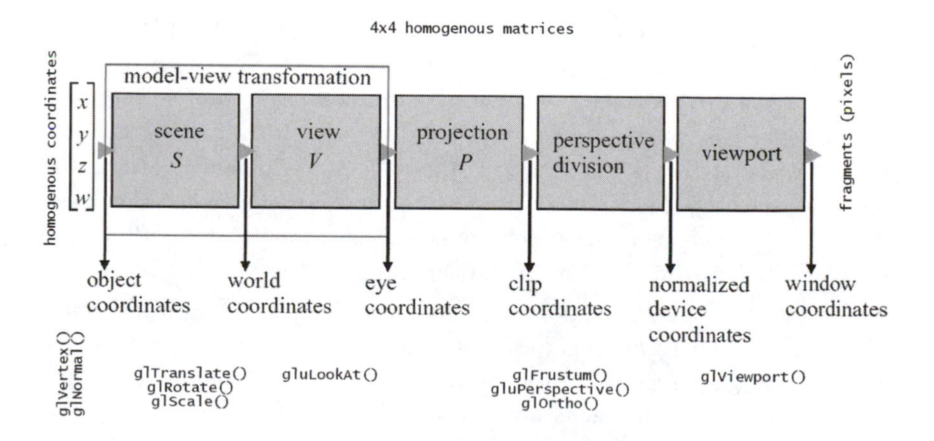

Figure 2.20. OpenGL's transformation pipeline.

The *scene transformation matrix* S can be composed of a sequence of other 4×4 transformation matrices which are all multiplied and result in the model-view matrix. OpenGL offers a number of scene transformation commands which are all translated into corresponding 4×4 matrices and multiplied with the current model-view matrix. Examples of scene transformation commands are `glTranslate()`, `glScale()`, and `glRotate()`. The corresponding implementation examples can be found in Appendix B. The scene transformation maps the vertices from the object coordinate system into the world coordinate system.

The *view transformation matrix* V is either generated by an appropriate composition of `glTranslate()`, `glScale()`, and `glRotate()` commands, or an additional utility library is used that provides an explicit view transformation command which "wraps" the basic commands. These commands are also translated into corresponding $4{\times}4$ matrices and multiplied with the model-view matrix. An example is the `gluLookAt()` command, provided by the OpenGL Utility Library (GLU)—an extension of OpenGL. The view transformation maps the vertices from the *world coordinate system* into the *eye* (or *camera*) *coordinate system.* Note that the inverse transpose modeview transformation matrix is automatically applied to the surface normal vectors.

The *projection transformation matrix* P is generated by calling an appropriate projection command. OpenGL offers two commands: `glFrustum()` (or `gluPerspective()` from the GLU) for a perspective projection and `glOrtho()` for an orthographic projection. They generate the corresponding transformation matrix and multiply it with the projection matrix. The corresponding implementation examples can be found in Appendix B. The projection transformation first maps the vertices from the eye coordinate system into the *clip coordinate system*, and then into the *normalized device coordinate system* after applying the *perspective division* (i.e., $[x/w, y/w, z/w]$). While the clip coordinates are in the range $[-w, w]$, the device coordinates are normalized to the Euclidean space $[-1, 1]$. The normalized component is used for depth handling and is not affected by the viewport transformation.

The *viewport transformation* maps the $[x_{nd}, y_{nd}]$ components of the normalized device coordinates into the *window coordinate system.* OpenGL provides the `glViewport()` command to support this mapping. A homogeneous matrix is not generated for this operation. Rather, the scaling transformation is applied explicitly to the $[x_{nd}, y_{nd}]$ components of the normalized device coordinates. The window coordinates are normalized to the range $a \leq x_w \leq (a + width)$, $b \leq y_w \leq (b + height)$, where $[a, b]$ is the lower left corner and $[width, height]$ are the dimensions of the window in pixels. The window coordinates are computed from the normalized device coordinates as follows:

$$x_w = (x_{nd} + 1)\frac{width}{2} + 1 \quad \text{and} \quad y_w = (y_{nd} + 1)\frac{height}{2} + 1.$$

The following code fragment outlines the correct sequence of operations for configuring an OpenGL transformation pipeline:

```
...

// define window viewport
glViewport(...);

// switch to projection matrix
glMatrixMode(GL_PROJECTION);
glLoadIdentity();

// projection transformation
glFrustum(...); // or gluPerspective(...); or glOrtho(...);

// switch to modelview matrix
glMatrixMode(GL_MODELVIEW);
glLoadIdentity();

// viewpoint transformation
gluLookAt(...);

// scene transformation
glTranslate(...);
glRotate(...);
glScale(...);

//render scene

...
```

Projections. Two basic types of projections exist: on-axis and off-axis projections. As described in Section 2.2, an optical system is called on-axis if it is centered—i.e., if the axis of symmetry of all surfaces and the optical axis coincide. For display surfaces or image planes this is the case if the center of the camera frustum projects onto the center of the image plane (or a display screen). An off-axis system is an optical system where this is

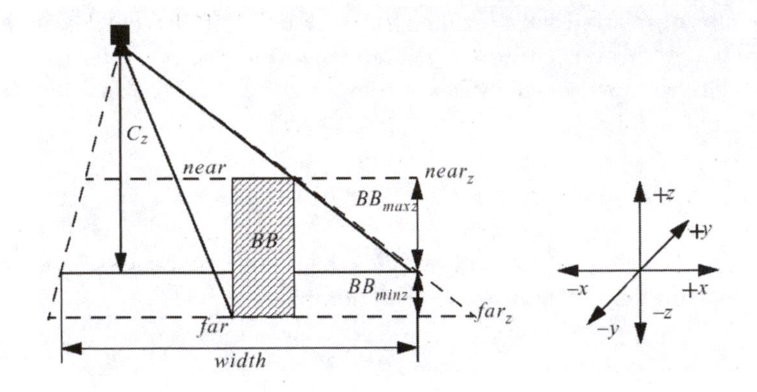

Figure 2.21. Off-axis projection example.

not the case—i.e., (with respect to display surfaces of image planes) where the center of projection projects onto a non-centered position on the image plane (or a display screen).

Figure 2.21 illustrates an *off-axis projection* example. To determine the corresponding viewing frustum, the scene's bounding box (BB) is required to compute the near ($near$) and far (far) clipping planes.

Given the position of the camera C (this is the center of projection), the near and far clipping planes are computed as follows:

$$near_z = C_z - BB_{\max z} - 1 \quad \text{and} \quad far_z = C_z - BB_{\min z} - 1.$$

Note that an offset of ± 1 is required to prevent clipping of the front or backside faces which are aligned with the bounding box's top or bottom.

Assuming an *on-axis situation* (i.e., the camera is located at $C = [0, 0, z]$), the size of the near clipping plane can be determined with a simple triangulation:

$$near_{width2} = \frac{near_{width}}{2} = \frac{width}{2} ratio \quad \text{and}$$
$$near_{height2} = \frac{near_{height}}{2} = \frac{height}{2} ratio,$$

where $width$ and $height$ are the dimensions of the projection plane or drawing window (in the x, y directions), and ratio is given by

$$ratio = \frac{near_z}{C_z}.$$

Since the projection is off-axis, a projectional shift has to be computed before the parameters of the viewing frustum can be determined:

$$shift_x = C_x ratio \text{ and } shift_y = C_y ratio.$$

Using OpenGL's `glFrustum()` function and GLUs `gluLookAt()` function [117], the viewing frustum for our example can be set as follows:

```
...

glMatrixMode(GL_PROJECTION);
glLoadIdentity();
glFrustum(-near_width2  - shift_x, near_width2  - shift_x,
          -near_height2 - shift_y, near_height2 - shift_y,
           near_z, -far_z);

glMatrixMode(GL_MODELVIEW);
glLoadIdentity();
gluLookAt(C_x, C_y, C_z, C_x, C_y, 0, 0, 1, 0);

...
```

Having the camera located at $C = [0, 0, z]$ defines an on-axis situation. Although `glFrustum()` is a function that allows the user to define projection transformations through general frustums, `gluPerspective()` is a simplified version that allows the user to define only on-axis situations by only specifying the near and far clipping planes, the vertical field of view, and the aspect ratio of the frustum.

While on-axis projections are commonly used for head-attached displays, such as head-mounted displays, off-axis projections are used in combination with spatial screens, such as spatial projection screens.

Lighting and shading. In fixed function pipelines, lighting and shading computations are done on a per-vertex basis. Intermediate pixel colors of the shaded surfaces (e.g., a triangle) are interpolated during rasterization.

With respect to Figure 2.22, OpenGL's simplified *shading model* consists of the following components (for point light sources):

- The *attenuation factor* $F = 1/(k_c + k_l r + k_s r^2)$ allows selection of the attenuation mode to be used. The constants k_c, k_l, k_s can have the Boolean values $0, 1$ to enable or disable one or multiple attenuation modes: the constant, linear, or square distance attenuation modes respectively. The variable r is the distance from the vertex v to the

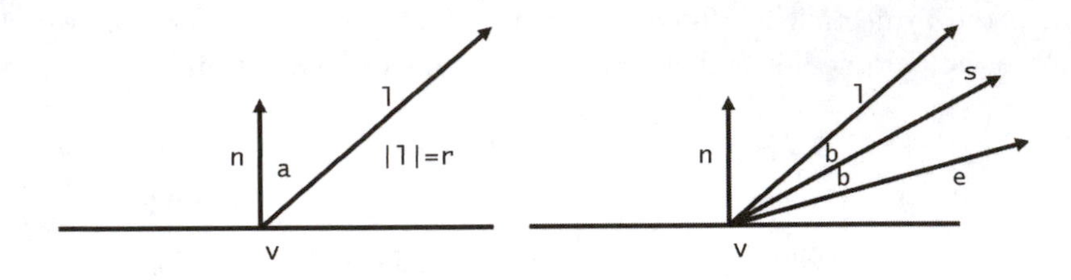

Figure 2.22. OpenGL's simplified lighting model.

light source l, and is automatically computed by the pipeline. While square distance attenuation is physically the most correct mode, usually constant or linear modes are used to avoid a rapid falloff for the intensities within a small distance range from the light source.

- The *emission term* E represents the ability of the vertex to emit light, and is defined within the scene description.

- The *ambient term* $A = A_m A_l$ represents the ambient reflection of light. It consists of the multiplication of the ambient material property A_m and the ambient light property A_l. Both are specified in the scene description.

- The *diffuse term* $D = lnD_m D_l$ defines the diffuse reflection of light at the vertex v. It depends on the angle a between the direction vector to the light source l and the normal vector n at v. Applying Lambert's reflection law $nl = \cos(a)$. The constants D_m and D_l represent the diffuse material and light properties, respectively. Both are set in the scene definition.

- The *specular term* $S = (ns)^i S_m S_l$ computes the specular reflection components. It computes the half-vector $s = (l+e)/|l+e|$ between the vector pointing towards the view point (or camera) e, and the light vector l. The exponent i allows for modification of the shininess of the surface, and S_m and S_l are the specular material and light properties, which are also defined in the scene description.

Putting all these components together, the simplified lighting equation for n point light sources can be summarized as

$$C = E + \sum_{i=0}^{n-1} F_i(A_i + D_i + S_i). \tag{2.6}$$

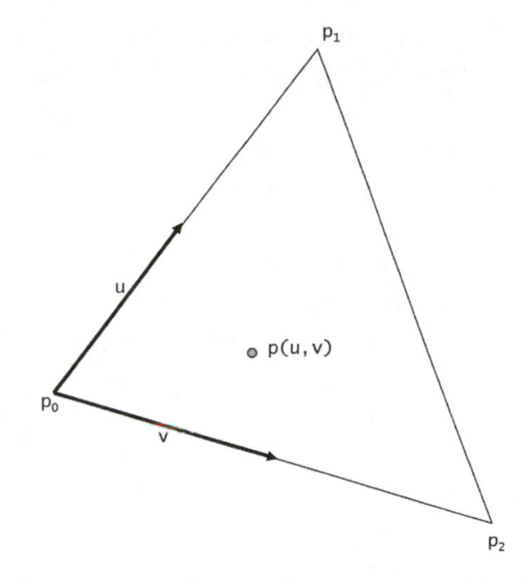

Figure 2.23. Linear interpolation of vertex properties within a triangle.

Equation (2.6) can be used to compute the light intensities at each vertex of a scene triangle. Depending on the selected shading model, the intensities of the intermediate pixels inside the triangles are either interpolated between the corresponding three vertices (*Gouraud shading*) or they are simply copied (*flat shading*). These two simple shading models are supported by fixed function pipelines, such as standard OpenGL. More advanced models, such as *Phong shading* that computes individual shading values depending on interpolated surface normals, require per-pixel operations. This is not supported by simple fixed function pipelines.

Different values are computed for each vertex, such as lighting intensities, texture coordinates, normals, alpha values, etc. They are called *vertex parameters*. In fixed function pipelines, some of them are linearly interpolated during rasterization to receive approximate values for pixels within triangles to avoid the enormous amount of individual computations.

A linear interpolation of a vertex property $p_{u,v}$ at the *barycentric coordinates* $u, v = [0, 1]$ between the three constant corner values p_0, p_1, p_2 is given by

$$p(u, v) = (1 - u)(1 - v)p_0 + (1 - v)up_1 + (1 - u)vp_2.$$

To cover the entire the triangle, its area is usually sampled at discrete u, v positions during rasterization.

Buffers. Several memory buffers are used in conventional graphics systems, like OpenGL, to support different operations. One of the most important buffers is the *frame buffer* which is the memory that holds the pixel information after rasterization. It is a two-dimensional matrix-like memory structure whose size matches the image resolution. The frame buffer stores different values at each individual pixel matrix element, such as the red, green, blue, and alpha components of the pixel. A graphics system usually provides many different frame buffers.

The most common situation is that only two frame buffers are provided to support a smooth display of frames during animations. One buffer, the *back buffer*, is used by the application for direct rendering. The other buffer, the *front buffer*, is used for the direct output to the screen. Once the back buffer has been filled with pixel information, it is copied (or *swapped*) to the front buffer and displayed on the screen. To display the next frame, the application can overwrite the back buffer while the image of the previous frame (stored in the front buffer) is still displayed on the screen. Once the application has finished the rendering, the front buffer content is exchanged by swapping the back buffer content again. This sequence ensures flicker-free animation sequences, since the image assembly of new frames in done in the background. The combination of back and front buffers is called *double buffering*. Applications that allow stereoscopic rendering sometimes require two double buffers for rendering flicker-free stereo pairs. Some graphics cards do support this by providing two double buffers, called *quad buffers*. Quad buffers consist of back-left, back-right, front-left, and front-right buffer sections, and their operations are equivalent to double buffers.

Another relevant buffer is the *depth buffer* (or *z-buffer*). The depth buffer has the same two-dimensional size as the frame buffer. At every pixel position it stores the relative distance from the camera to the object that projects onto the pixel. The depth test ensures that if multiple objects project onto the same pixel, only the one that is closest to the camera appears in the frame buffer. The range of depth values influences the precision of the depth test since the bit-depth of each entry is normally fixed. That means that a larger range has a lower precision than a smaller range.

The *stencil buffer* is a buffer with the same dimensions as the frame buffer. It can be used to stencil out regions in the frame buffer depending on specific rules that can be influenced by the content of the frame of the depth buffer. In a simple case, the stencil buffer stores binary values, where zero could indicate that the application cannot render into the corresponding pixel in the frame buffer. Consequently, the application can first render a dynamic mask into the stencil buffer. During the actual rendering pass, the application renders into the frame buffer, but pixels that are blocked by the stencil buffer cannot be filled; this is called a *stencil test*. In general, each stencil value is not binary, but supports several *stencil levels*. A variety of rules can be used depending on the content of the depth buffer and frame buffer for rendering the stencil mask and making the stencil test. Logical operations can also be applied to define the required stencil rule.

The *texture memory* is a buffer that does not have to have the same size as the frame buffer. It stores image information that can be used for texture mapping a surface with pre-generated color and structure information. Textures can be generated on-line (i.e., dynamically during rendering) or off-line (e.g., by loading them from a file). There are two possibilities to generate them dynamically. For example, an image can be rendered into the back-buffer. The back-buffer is not swapped to the front-buffer, but is copied into the texture memory, and then overwritten by the application. The transfer from back-buffer to texture memory is called *read-back*. This operation usually requires a lot of time that leads to a lower total frame rate—especially if the back-buffer is located on the graphics card and the texture memory is part of the main memory (in this case the entire back-buffer has to be transferred from the graphics card, over the bus to the main memory, and back). The time required for a read-back operation increases with the size of the texture.

An alternative is a direct *render-to-texture* operation which is supported by newer graphics cards to avoid a large delay caused by memory read-backs. In this case, the application can generate a texture by directly rendering into a special auxiliary buffer that can be addressed during texture mapping. Usually, this buffer is located on the graphics card, and a bus transfer is not necessary. Such an off-line buffer is sometimes referred to as a *P-buffer*.

Additional *accumulation buffers* allow the combination (accumulation) of a series of pre-rendered images. A variety of different operations, such as addition and subtraction, is normally supported to achieve the desired effect. It is not possible to render directly into accumulation buffers. Consequently, a time-consuming read-back is required.

All these buffers work hand-in-hand within a rendering pipeline and are important tools for realizing higher-level operations, such as multi-pass rendering techniques.

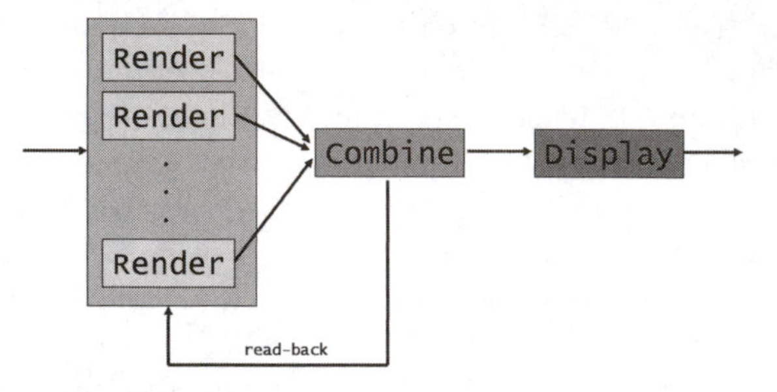

Figure 2.24. Multi-pass rendering concept.

Multi-pass rendering. Simple rendering pipelines process their data in a linear and fixed sequence, called a *rendering pass*. The scene is transformed by the transformation pipeline, is lightened, texture-mapped, and rasterized into the frame buffer under consideration of the depth buffer and the stencil buffer. Finally it is displayed on the screen. To achieve more advanced effects, a single rendering pass might not be sufficient. Some applications create more than one image during multiple rendering passes that are combined into a final image before displaying it (Figure 2.24).

In this case, portions of the rendering pipeline are executed multiple times and the resulting images are stored in auxiliary buffers, such as the P-buffer or the texture memory. They might serve as input for subsequent rendering passes. When all effects have been computed, the different results are assembled in the final image which is then displayed on the screen. A popular example of multi-pass rendering is shadow mapping, in which shadow masks are generated in a first rendering step. These maps are stored in textures that are applied to the scene geometry during a second rendering pass.

Multi-pass rendering is an essential tool for creating many common illumination effects, such as shadows and reflections, but also for supporting special display concepts—as we will see in the following chapters. Below, we give three examples of basic multi-pass rendering techniques: *projective textures*, *shadow mapping*, and *cube mapping*.

Projective Textures. Textures are normally mapped onto scene geometry by assigning each scene vertex to a texture coordinate. Texture coordinates are two-dimensional and refer to individual pixel positions (called *texels*) in a texture. In many cases they are normalized to a range of $[0, 1]$ (this is also the case in OpenGL). Thus, the texture coordinates $0, 0$ and $1, 1$ refer to the lower and the upper right texel in a texture. Since texture coordinates

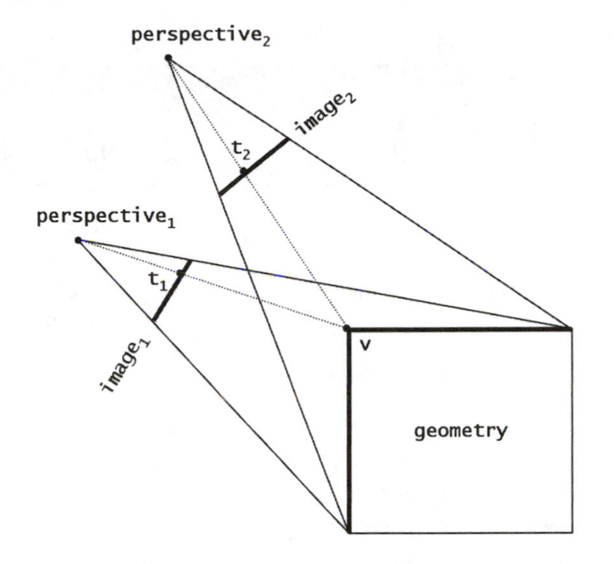

Figure 2.25. Projective texture mapping principle.

are vertex properties, they are linearly interpolated within triangles as previously described.

Besides assigning texture coordinates to scene vertices manually, they can be generated automatically during rendering. *Projective texture mapping* [168] is an important automatic texture generation method that allows for the warping of an image between two different perspectives (Figure 2.25).

Consider Figure 2.25: Given the texture image$_1$, the challenge is to project it out from perspective$_1$ in such a way that each texel (e.g., t_1) maps perspectively correct to a corresponding scene vertex (e.g., v). One can imagine a slide projector that displays a slide onto a curved surface. This can be achieved by computing a texture coordinate for each scene vertex that assigns the correct texel to it (e.g., t_1 to v in our example). This dynamic computation of texture coordinates is exactly what projective texture mapping does. It is very simple to explain: If the scene geometry is rendered from perspective$_1$ through a transformation pipeline as outlined in Figure 2.20—but without viewport transformation—the results are projected device coordinates for each vertex. They are normalized to a range of $[-1, 1]$. We mentioned previously, that texture coordinates are normalized to a range of $[0, 1]$. Consequently, an additional mapping between device and texture coordinates is required. This mapping simply consists

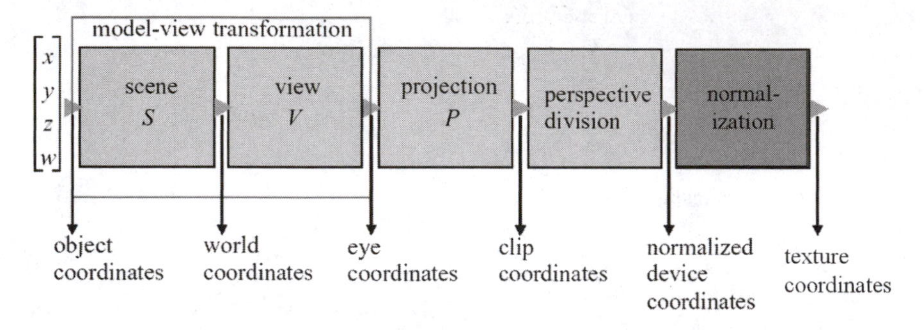

Figure 2.26. Projective texture mapping transformation pipeline.

of a two-dimensional scale transformation by 0.5, and a translation by 0.5 in the x/y-directions. Thus, to compute the correct texture coordinate for each scene vertex, a transformation pipeline has to be configured that consists of the first four steps of a regular transformation pipeline (as outlined in Figure 2.20 but without viewport transformation) plus two additional correction transformations (which we refer to as normalization). This is outlined in Figure 2.26.

With such a transformation pipeline, the scene geometry can simply be rendered. The results are not pixel positions that map to the frame buffer, but texture coordinates that map to a texture.

OpenGL has its own *matrix stack* for automatic texture coordinate generation. Consequently, the regular matrix stacks for model, view, and projection transformations remain unaffected by these operations.

If a projective texture map has been created for perspective$_1$ in our example, the textured scene geometry can be rendered from a different perspective (e.g., perspective$_2$). The result is that each texel projected out from perspective$_1$ appears at the correct projection in perspective$_2$. Thus, t_1 in perspective$_1$ is warped into t_2 in perspective$_2$ in such a way that both texels/pixels are spatially bound to vertex v. The same is true for surface points, texels, and texture coordinates within triangles—they are linearly interpolated.

The code fragment below describes how to set up a projective texture map with OpenGL.

```
...

// generate texture in 1st pass
// projective texture parameters init
```

```
GLfloat eyePlaneS[]={1.0, 0.0, 0.0, 0.0};
GLfloat eyePlaneT[]={0.0, 1.0, 0.0, 0.0};
GLfloat eyePlaneR[]={0.0, 0.0, 1.0, 0.0};
GLfloat eyePlaneQ[]={0.0, 0.0, 0.0, 1.0};
glTexParameteri(GL_TEXTURE_2D, GL_TEXTURE_WRAP_S, GL_CLAMP);
glTexParameteri(GL_TEXTURE_2D, GL_TEXTURE_WRAP_T, GL_CLAMP);
glTexEnvi(GL_TEXTURE_ENV, GL_TEXTURE_ENV_MODE, GL_MODULATE);
glTexGeni(GL_S, GL_TEXTURE_GEN_MODE, GL_EYE_LINEAR);
glTexGenfv(GL_S, GL_EYE_PLANE, eyePlaneS);
glTexGeni(GL_T, GL_TEXTURE_GEN_MODE, GL_EYE_LINEAR);
glTexGenfv(GL_T, GL_EYE_PLANE, eyePlaneT);
glTexGeni(GL_R, GL_TEXTURE_GEN_MODE, GL_EYE_LINEAR);
glTexGenfv(GL_R, GL_EYE_PLANE, eyePlaneR);
glTexGeni(GL_Q, GL_TEXTURE_GEN_MODE, GL_EYE_LINEAR);
glTexGenfv(GL_Q, GL_EYE_PLANE, eyePlaneQ);

// projection matrix = texture matrix (perspective_1)
glMatrixMode(GL_TEXTURE);
glLoadIdentity();

// normalization
glTranslatef(0.5, 0.5, 0.0);
glScalef(0.5, 0.5, 1.0);

// projection (on-axis): CG = center of gravity of scene,
// C = center of projection (camera)
float len, delta;
len = sqrt( (CG[0] - C[0]) * (CG[0] - C[0])
          + (CG[1] - C[1]) * (CG[1] - C[1])
          + (CG[2] - C[2]) * (CG[2] - C[2]) );
delta = radius * (len - radius)/len;
glFrustum(-delta, delta, -delta,
           delta, len - radius, len + radius);

// viewing transformation
gluLookAt(C[0], C[1], C[2], CG[0], CG[1], CG[2], 0, 0, 1);
glMatrixMode(GL_MODELVIEW);

// enable texture mapping and automatic texture
// coordinate generation
glEnable(GL_TEXTURE_2D);
glEnable(GL_TEXTURE_GEN_S);
glEnable(GL_TEXTURE_GEN_T);
glEnable(GL_TEXTURE_GEN_R);
glEnable(GL_TEXTURE_GEN_Q);
```

```
// render scene in 2nd pass (from perspective_2)
// disable texture mapping and automatic texture
// coordinate generation
glDisable(GL_TEXTURE_2D);
glDisable(GL_TEXTURE_GEN_S);
glDisable(GL_TEXTURE_GEN_T);
glDisable(GL_TEXTURE_GEN_R);
glDisable(GL_TEXTURE_GEN_Q);

...
```

Shadow mapping. Shadows provide essential queues that help to interpret a scene more efficiently and enhance realism. Because of this, shadowing techniques are hardware-supported by today's graphics cards and can be generated at interactive frame rates.

Shadow mapping is a two-pass technique that generates hard shadows (i.e., shadows that are generated from a single point-light source) in a three-dimensional scene.

Shadow mapping uses projective texture mapping (Figure 2.27(a)). In a first rendering pass the scene is rendered from the perspective of the light source without swapping the frame back buffer into the front buffer. Only the depth buffer content is read back into the texture memory. This texture is usually referred to as a *shadow map*—although it indicates the illuminated scene area rather than the portion that lies in shadow. In the second pass, the scene is rendered from the perspective of the camera. The

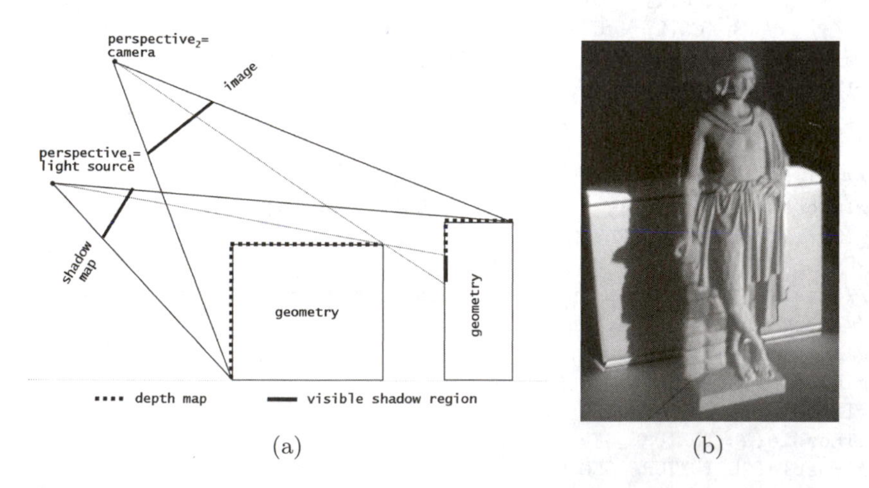

(a) (b)

Figure 2.27. (a) Shadow-mapping principle; (b) example in an AR context.

shadow map is warped into the perspective of the camera via projective texture mapping (as described in the previous section). This allows us to determine two depth values for each visible scene point: The first one can be read out of the warped shadow map. It represents the distance d_1 of scene points to the light source that are not in shadow (i.e., those that are visible to the light source). Reading the correct value out of the shadow map is enabled by the projective texture mapping which results in corresponding texture coordinates for each scene point at the end of the projective texture mapping transformation pipeline (Figure 2.26). The second depth value is computed during the second rendering pass—also during projective texture mapping. The normalization step in the projective texture mapping transformation pipeline (Figure 2.26) transforms only the x, y coordinates from normalized device coordinates into texture coordinates. The z component remains unchanged. Thus it represents the distance d_2 from every (for the camera visible) scene point to the light source—no matter if they are visible to the light source or not. Comparing these two values allows us to determine the shadow areas ($d_2 > d_1$) and the illuminated areas ($d_1 \leq d_2$).

The texture resolution and the precision of the depth buffer are important for good quality shadow maps. Both parameters can be controlled by optimizing the frustum values (i.e., clipping planes for depth buffer resolution and field of view for texture resolution) of the light source and the camera with respect to the scene dimensions. For dynamic scenes, these values have to be adjusted every frame to avoid artifacts that result from a too low depth-buffer or texture resolution with respect to specific scene dimensions. It is important that the near and far clipping planes have the same distance in both frustums—otherwise the depth values are not mapped to the same space, and the depth test will fail.

The following code fragment describes how to set up a shadow map with OpenGL.

```
...

// render scene in 1st pass (perspective_1=light source
// perspective) copy the shadow-map into texture memory
// (read-back)
glBindTexture(GL_TEXTURE_2D, Depth_Map_Id);
glCopyTexImage2D(GL_TEXTURE_2D, 0, GL_DEPTH_COMPONENT32_SGIX,
                 0, 0, ShadowMapSize, ShadowMapSize, 0);
glDepthFunc(GL_LEQUAL);

// projective texture parameters init -> same as above
```

```
...
GLfloat sm_border[]={1.0, 1.0, 1.0, 1.0};
glTexEnvf(GL_TEXTURE_ENV, GL_TEXTURE_ENV_MODE, GL_MODULATE);
glTexParameterf(GL_TEXTURE_2D, GL_TEXTURE_WRAP_S,
                GL_CLAMP_TO_BORDER_ARB);
glTexParameterf(GL_TEXTURE_2D, GL_TEXTURE_WRAP_T,
                GL_CLAMP_TO_BORDER_ARB);
glTexParameterfv(GL_TEXTURE_2D, GL_TEXTURE_BORDER_COLOR,
                 sm_border);
glTexParameteri(GL_TEXTURE_2D, GL_TEXTURE_COMPARE_SGIX,
                GL_TRUE);
glTexParameteri(GL_TEXTURE_2D,
                GL_TEXTURE_COMPARE_OPERATOR_SGIX,
                GL_TEXTURE_LEQUAL_R_SGIX);

// projection matrix = texture matrix (perspective_1=light
// source perspective) -> same as above
...

// enable texture mapping and automatic texture coordinate
// generation
glEnable(GL_TEXTURE_2D);
glEnable(GL_TEXTURE_GEN_S);
glEnable(GL_TEXTURE_GEN_T);
glEnable(GL_TEXTURE_GEN_R);
glEnable(GL_TEXTURE_GEN_Q);

// render scene in 2nd pass (from perspective_2=perspective
// of camera) disable texture mapping and automatic texture
// coordinate generation
glDisable(GL_TEXTURE_GEN_S);
glDisable(GL_TEXTURE_GEN_T);
glDisable(GL_TEXTURE_GEN_R);
glDisable(GL_TEXTURE_GEN_Q);
glDisable(GL_TEXTURE_2D);

...
```

Figure 2.27(b) shows an example of shadow-mapping applied in an optical see-through AR context. How shadows can be thrown from real objects onto virtual ones and vice versa will be described in Chapter 7.

Cube Mapping. Just like projective texture mapping and shadow mapping, cube mapping is yet another multi-pass technique that is based on automatic texture generation. It simulates reflections of an environment scene

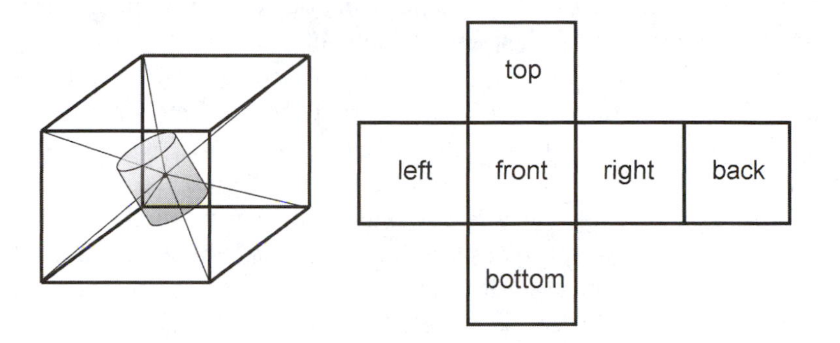

Figure 2.28. Cube mapping principle.

on the surface of a particular object. It has been derived from more general environment mapping techniques [21] and was adapted to the capabilities of graphics hardware to support interactive frame rates.

A cube map is a texture that consists of six sub-images. These images are generated by placing a camera in the center of gravity of an object (called reflecting object) that has to reflect the surrounding environment scene. Six perspective images of the environment scene (without the reflecting object) are then generated in six separate rendering passes. During each pass, the camera points in a different direction. We can imagine a cube that encloses the reflecting object. The cube is aligned with the world coordinate system and its center matches the center of gravity of the reflecting object (Figure 2.28). The parameters of the cameras that generate the six perspective views are chosen in such a way that each perspective camera is spanned by one face of the cube. Thus, their directions all differ by 90 degrees, and their x/y field-of-view is also 90 degrees. The six textures together cover the entire surrounding environment.

During the second rendering pass, the scene (including the reflecting object) is rendered from the regular camera's perspective. When the reflecting object is being rendered, cube mapping is enabled and an approximated reflection of the surrounding environment on the reflecting object's surface is visible. This is achieved by determining the correct cube face and texture coordinates within each cube-face texture. To do this quickly, a reflection vector at the cube map's center is computed for each reflecting object's vertex. This vector depends on the regular camera's position (i.e., the view vector). The reflection vector is then used to look up the correct sub-image and texture coordinates.

Note that this method is only an approximation of reflection, in contrast to the physical reflection principles described in Section 2.2. There are two assumptions made in cube mapping that are physically incorrect, but lead to interactive frame rates. First, the reflection vectors are computed for a single point only at the center of the reflecting object and not for each of its surface points. Second, the reflection vectors are not traced until the corresponding reflected scene object is found, but they are used to compute an intersection with perspective scene images arranged in a cubical manor around the reflecting object. These perspective images are generated for a single point at the center of the reflecting object and are fixed in direction (e.g., aligned with the world coordinate system).

The code fragment which follows describes how to set up a cube map in the correct order with OpenGL.

```
...

GLuint cm_s = 512;     // texture resolution of each face
GLfloat cm_dir[6][3]; // direction vectors
GLfloat cm_up[6][3];  // up vectors
GLfloat cm_c[3];      // viewpoint / center of gravity
GLenum cubeFace[6] = {
                      GL_TEXTURE_CUBE_MAP_POSITIVE_X_EXT,
                      GL_TEXTURE_CUBE_MAP_NEGATIVE_X_EXT,
                      GL_TEXTURE_CUBE_MAP_NEGATIVE_Z_EXT,
                      GL_TEXTURE_CUBE_MAP_POSITIVE_Z_EXT,
                      GL_TEXTURE_CUBE_MAP_POSITIVE_Y_EXT,
                      GL_TEXTURE_CUBE_MAP_NEGATIVE_Y_EXT
};

// define cube map's center cm_c[] -> center of object
// (in which scene has to be reflected)
// set up cube map's view directions in correct order
cm_dir[0][0] = cm_c[0] + 1.0; //right
cm_dir[0][1] = cm_c[1] + 0.0; //right
cm_dir[0][2] = cm_c[2] + 0.0; //right
cm_dir[1][0] = cm_c[0] - 1.0; //left
cm_dir[1][1] = cm_c[1] + 0.0; //left
cm_dir[1][2] = cm_c[2] + 0.0; //left
cm_dir[2][0] = cm_c[0] + 0.0; //bottom
cm_dir[2][1] = cm_c[1] + 0.0; //bottom
cm_dir[2][2] = cm_c[2] - 1.0; //bottom
cm_dir[3][0] = cm_c[0] + 0.0; //top
cm_dir[3][1] = cm_c[1] + 0.0; //top
cm_dir[3][2] = cm_c[2] + 1.0; //top
```

```
cm_dir[4][0] = cm_c[0] + 0.0; //back
cm_dir[4][1] = cm_c[1] + 1.0; //back
cm_dir[4][2] = cm_c[2] + 0.0; //back
cm_dir[5][0] = cm_c[0] + 0.0; //front
cm_dir[5][1] = cm_c[1] - 1.0; //front
cm_dir[5][2] = cm_c[2] + 0.0; //front

// set up cube map's up vector in correct order
cm_up[0][0]=0.0; cm_up[0][1]=-1.0; cm_up[0][2]= 0.0; //+x
cm_up[1][0]=0.0; cm_up[1][1]=-1.0; cm_up[1][2]= 0.0; //-x
cm_up[2][0]=0.0; cm_up[2][1]=-1.0; cm_up[2][2]= 0.0; //+y
cm_up[3][0]=0.0; cm_up[3][1]=-1.0; cm_up[3][2]= 0.0; //-y
cm_up[4][0]=0.0; cm_up[4][1]= 0.0; cm_up[4][2]= 1.0; //+z
cm_up[5][0]=0.0; cm_up[5][1]= 0.0; cm_up[5][2]=-1.0; //-z

// render the 6 perspective views (first 6 render passes)
for(int i=0;i<6;i++){
  glClear (GL_COLOR_BUFFER_BIT | GL_DEPTH_BUFFER_BIT);
  glViewport(0,0,cubemap_size, cubemap_size);
  glMatrixMode(GL_PROJECTION);
  glLoadIdentity();
  gluPerspective(90.0, 1.0,0.1, INF);
  glMatrixMode(GL_MODELVIEW);
  glLoadIdentity();
  gluLookAt(cm_c[0], cm_c[1], cm_c[2],
            cm_dir[i][0], cm_dir[i][1], cm_dir[i][2],
            cm_up[i][0], cm_up[i][1], cm_up[i][2]);

  // render scene to be reflected
  // read-back into corresponding texture map
  glCopyTexImage2D(cubeFace[i], 0, GL_RGB, 0, 0,cm_s,cm_s,0)
}

// cube map texture parameters init
glTexEnvf(GL_TEXTURE_ENV, GL_TEXTURE_ENV_MODE, GL_MODULATE);
glTexParameteri(GL_TEXTURE_CUBE_MAP_EXT, GL_TEXTURE_WRAP_S,
                GL_CLAMP);
glTexParameteri(GL_TEXTURE_CUBE_MAP_EXT, GL_TEXTURE_WRAP_T,
                GL_CLAMP);
glTexParameterf(GL_TEXTURE_CUBE_MAP_EXT, GL_TEXTURE_MAG_FILTER,
                GL_LINEAR);
glTexParameterf(GL_TEXTURE_CUBE_MAP_EXT, GL_TEXTURE_MIN_FILTER,
                GL_NEAREST);
glTexGeni(GL_S, GL_TEXTURE_GEN_MODE, GL_REFLECTION_MAP_NV);
glTexGeni(GL_T, GL_TEXTURE_GEN_MODE, GL_REFLECTION_MAP_NV);
glTexGeni(GL_R, GL_TEXTURE_GEN_MODE, GL_REFLECTION_MAP_NV);
```

```
// enable texture mapping and automatic texture coordinate
// generation
glEnable(GL_TEXTURE_GEN_S);
glEnable(GL_TEXTURE_GEN_T);
glEnable(GL_TEXTURE_GEN_R);
glEnable(GL_TEXTURE_CUBE_MAP_EXT);

// render object in 7th pass (in which scene has to be
// reflected)
// disable texture mapping and automatic texture coordinate
// generation
glDisable(GL_TEXTURE_CUBE_MAP_EXT);
glDisable(GL_TEXTURE_GEN_S);
glDisable(GL_TEXTURE_GEN_T);
glDisable(GL_TEXTURE_GEN_R);

...
```

How cube mapping is applied to enhance realism for optical see-through AR in which real objects are being reflected in virtual ones is described in Chapter 7.

2.4.2 Programmable Rendering Pipelines

Fixed function pipelines transform all incoming data in the same way within each frame. It is not possible to reconfigure them to treat each entity differently. This supports rigid body transformations, but non-linear warping is not possible.

As we will see in Chapter 6, fixed function pipelines are sufficient to pre-distort images which are displayed in combination with optical elements that provide the same transformation characteristic—such as planar mirrors. Other optical elements, such as curved mirrors or lenses, require a curvilinear pre-distortion that cannot be realized with early fixed function pipelines.

A new generation of rendering pipelines is called *programmable rendering pipelines*. Basically they work in the same way as fixed function pipelines. The general difference, however, is that programmable pipelines can be configured with small programs—called *shaders*—instead of with simple homogeneous matrices only. These shaders accept parameters that are passed to them from the main application. This concept allows a much greater flexibility, because arbitrary and element-individual transformations can be realized. Even control structures, such as loops or conditions,

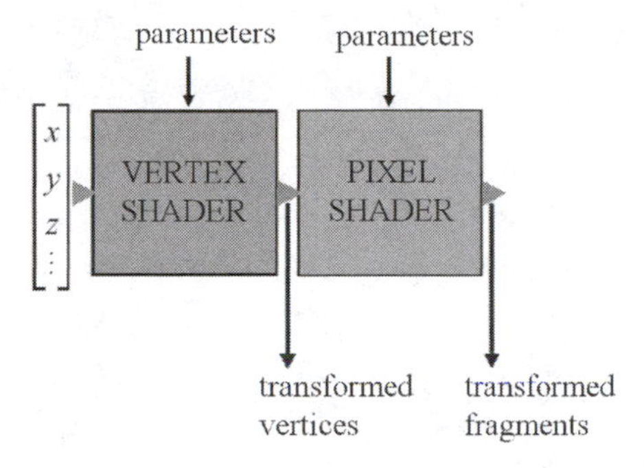

Figure 2.29. Simple programmable rendering pipeline.

can be used in shaders. Such pipelines are important, because they do hardware-support curvilinear transformations, and consequently real-time pre-distortion images displayed in combination with view-dependent optics, such as curved mirrors and lenses.

Currently, two different types of shaders exist: *vertex shaders* and *pixel shaders*. As illustrated in Figure 2.29 they are arranged in a pipeline configuration.

The following code fragment illustrates a simple frame of a vertex shader as implemented in NVIDIA's C for Graphics (Cg) language [47]:

```
struct input { /*vertex parameters*/
  float4 position : POSITION;
  float4 normal : NORMAL;
  ...
};

struct output { /*fragment parameters*/
  float2 texcoord : TEXCOORD;
  float4 position : POSITION;
  float4 color : COLOR;
  ...
};

output main(input VERTEX_IN /*, additional parameters*/)
{
  output PIXEL_OUT;

  /* perform computations with VERTEX_IN and
     additional parameters that result in PIXEL_OUT */
```

```
  return PIXEL_OUT;
}
```

In Chapter 6, we will see examples for vertex shaders that are used for image pre-distortion in combination with curved mirrors and lenses.

The next code fragment gives a simple frame for a pixel shader—also implemented in Cg [47]:

```
struct input { /* fragment parameters */
  float2 texcoord : TEXCOORD;
  float4 position : POSITION;
  float4 color : COLOR;
  ...
};

struct output { /* fragment parameters */
  float4 color : COLOR;
  ...
};

output main( input PIXEL_IN /*, additional parameters */) {
  output PIXEL_OUT;

  /* perform computations with PIXEL_IN and
     additional parameters that result in PIXEL_OUT */

  return PIXEL_OUT;
}
```

In Chapter 7 we discuss an example application of a pixel shader used for real-time color correction that enables a projection on textured surfaces.

2.5 Summary and Discussion

There are two fundamental views on light: light as waves and light as rays. In reality, they are both just abstractions of what light really is, but the two views are of advantage when using them to explain how optical systems work. The subgroup of optics that applies the wave concept is called *wave optics*, while the one that abstracts light as rays is called *geometric optics*.

Our starting point of geometric optics was Snell's laws for reflection and refraction which led to image forming optical systems. Mirrors and lenses—as the two major components of optical systems—and their image-forming behavior for different surface types have been described in detail and appropriate geometric object-image transformations have been presented. We have seen, that only Cartesian surfaces (such as rotation-paraboloids, rotation-hyperboloids, and prolate ellipsoids) can generate stigmatic image pairs. Only planar mirrors, however, provide true stigmatism between all object-image pairs and represent absolute optical systems. The other surface types, which provide stigmatism for a limited number of points (normally only for their focal points) are difficult to produce. This is the reason why most optical instruments approximate stigmatic image formation, and therefore introduce aberrations.

We noted that the human eye itself is a complex optical system. Being the final link of an optical chain, the eyes can detect two perspectively different versions of the formed images and send signals to the brain which fuses them to a three-dimensional picture. Disparity, vergence, and accommodation are the main mechanisms to support stereoscopic viewing. However, due to the limited retinal resolution of the eyes, small aberrations of non-stigmatic optical systems are not detected. Consequently, the eyes perceive a single consistent image of the corresponding object—even if the light rays that are emitted by the object do not intersect exactly at the image point after travelling through the optical system. We have seen how stereoscopic viewing can be fooled with graphical stereoscopic displays by presenting different two-dimensional graphical images to both eyes. Common passive and active stereo-separation techniques, such as shuttering, polarization, anaglyphs, ChromaDepth and the Pulfrich effect have been described, and we explained how they work based on the wave-optical or physiological and psychological observations we made earlier.

We will learn in Chapter 4 that two different augmented reality display concepts exist to superimpose graphics onto the user's view of the real world. *Video see-through*, on the one hand, makes use of video-mixing to integrate graphics into a displayed video-stream of the environment. *Optical see-through*, on the other hand, applies optical combiners (essentially half-silvered mirrors or transparent LCD displays) to overlay rendered graphics optically over the real view on the physical environment. Both techniques have advantages and disadvantages, as we will see later.

In this chapter, we have given a broad classification of today's stereoscopic displays and described several classes of autostereoscopic and goggle-bound techniques. In general, we can say that most autostereoscopic displays do not yet support optical see-through augmented reality applications. This is generally due to the technological constraints of the applied optics. Exceptions are some mirror-based re-imaging displays. Although

Figure 2.30. First prototype of an autostereoscopic, multi-user capable optical see-through display implemented by the Bauhaus University-Weimar.

an indirect "window on the world" view on the real environment supported by video-mixing would be feasible, autostereoscopic displays are still hardly used for augmented reality tasks. We believe that this will change for both optical and video see-through with the availability of cost-effective parallax displays. (We will present examples in upcoming chapters.) Figure 2.30 shows a first example of a multi-user optical see-through display that uses an autostereoscopic parallax screen.

While autostereoscopic displays do not require additional devices to address most visual depth cues, goggle-bound displays strongly depend on head-worn components to support a proper separation of the presented stereoscopic images. Video see-through and optical see-through head-mounted displays are today's dominant AR display devices. However, they entail a number of ergonomic and technological shortcomings, which will also be discussed later. To overcome these shortcomings and to open new application areas, the virtual reality community orientated itself more and more away from head-mounted displays and towards projection-based spatial displays such as surround screen displays and embedded screen displays. Compared to head-mounted displays, projection-based devices provide a high and scalable resolution, a large and extendable field of view, an easier eye accommodation, and a lower incidence of discomfort due to simulator sickness. They lack in mobility and in optical see-through capabilities though, and have limitations in the number of users they can support simultaneously. Head-mounted projector displays might represent a compromise that tends to combine the advantages of HMDs with those of projection displays. However, they suffer from the imbalanced ratio between heavy optics (and projectors) that results in cumbersome and uncomfortable devices and ergonomic devices with a low image quality. Currently, this is a general drawback of all head-attached displays that are dependent on miniature display elements. All these issues might improve in the future, with emerging technology, such as miniature projectors, new optics, *Light Emitting Polymers* (LEPs), or *Organic Light Emitting Diodes* (OLEDs), etc.

Projection-based spatial displays in combination with video-mixing support a more immersive "window on the world" viewing. Video-mixing, however, still banishes the users from the real environment and, in combination with such devices, allows only a remote interaction. Compared to an optical combination, video-mixing also has a number of technological shortcomings. Especially for projection-based display systems, problems that are related to the video-mixing technology, such as a time delayed video-presentation (due to the time required to capture and pre-mix the video streams), a reduced resolution of the real environment (due to the limited resolution of the cameras), and a strong limitation of head movements (due to restricted movements of the cameras) handicap the implementation of interactive and flexible augmented reality applications on this basis.

Augmented reality concepts are being proposed that suggest to detach the display technology from the user and to embed it into the real environment instead. This follows the evolution of virtual reality technology. In analogy to the term spatial virtual reality, this paradigm became known as spatial augmented reality. This concept is the focus of this book.

Just like for mobile AR displays, spatial augmented reality displays consist of two main optical components; *light generators* (such as projectors or screens) and *optical combiners* (such as mirrors, semi-transparent screens, and lenses). In the following chapters, we will describe how graphics can be rendered in real-time to support view-dependent and view-independent, on- and off-axis spatial AR displays. We learn how image distortion caused by non-trivial display geometries or optical reflection and refraction can be graphically neutralized in such a way that the formed images appear orthoscopic, stereoscopically, and perspectively correct and undistorted to an observer.

While absolute optical systems (i.e., planar mirrors) provide an affine optical mapping, affine geometry transformations that are integrated into traditional fixed function pipelines can be used for neutralization. For optical elements that require curvilinear transformations (like curved mirrors and lenses), however, an image warping is more efficient and more flexible than a geometric transformation. Programmable rendering pipelines in combination with multi-pass techniques neutralize such optical effects at interactive frame rates. These techniques will be discussed in more detail in Chapter 6. Finally, projective texture mapping and derived techniques are essential tools for projector-based augmentation and projector-based illumination methods, discussed in Chapter 5 and Chapter 7, respectively.

3

Augmented Reality Displays

Augmented reality displays are image-forming systems that use a set of optical, electronic, and mechanical components to generate images somewhere on the *optical path* in between the observer's eyes and the physical object to be augmented. Depending on the optics being used, the image can be formed on a plane or on a more complex non-planar surface.

Figure 3.1 illustrates the different possibilities of where the image can be formed to support augmented reality applications, where the displays are located with respect to the observer and the real object, and what type of image is produced (i.e., planar or curved).

Head-attached displays, such as retinal displays, head-mounted displays, and head-mounted projectors, have to be worn by the observer. While some displays are *hand-held*, others are *spatially aligned* and completely detached from the users. Retinal displays and several projector-based approaches produce curved images—either on the observer's retina or directly on the physical object. Most of the displays, however, produce images on planes—called *image planes*—that can be either head-attached or spatially aligned. Images behind real objects cannot be formed by a display that is located in front of real objects. In addition, if images are formed by a display located behind a real object, this object will occlude the image portion that is required to support augmentation.

When *stereoscopic rendering* is used to present mixed (real and virtual) worlds, two basic fusion technologies are currently being used: *video-mixing* and *optical combination*.

While video mixing merges live record video streams with computer-generated graphics and displays the result on the screen, optical combination generates an optical image of the real screen (displaying computer graphics) which appears within the real environment or within the viewer's

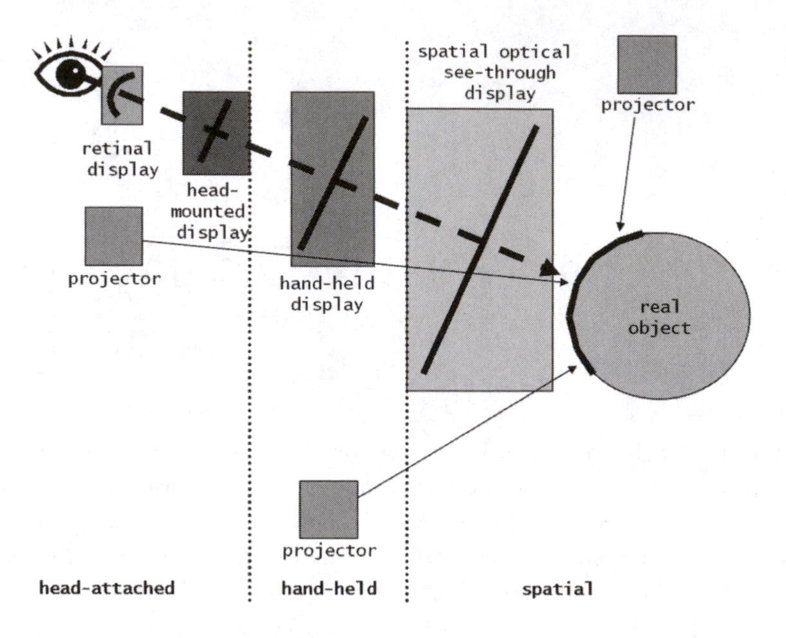

Figure 3.1. Image generation for augmented reality displays.

visual field while observing the real environment. Both technologies have a number of advantages and disadvantages which influence the type of application they can address.

Today, most of the stereoscopic AR displays require the user to wear some sort of goggles to provide stereo separation. Autostereoscopic approaches, however, might play a dominant role in the future of AR.

In this chapter, we discuss several types of augmented reality displays. Note that we present examples in each display category, rather than to provide a complete list of individual devices.

3.1 Head-Attached Displays

Head-attached displays require the user to wear the display system on his or her head. Depending on the image generation technology, three main types exist: retinal displays that use low-power lasers to project images directly onto the retina of the eye, head-mounted displays that use miniature displays in front of the eyes, and head-mounted projectors that make use of miniature projectors or miniature LCD panels with backlighting and project images on the surfaces of the real environment.

3.1.1 Retinal Displays

Retinal displays [83, 145, 89] utilize low-power semiconductor lasers (or, in the future, special light-emitting diodes) to scan modulated light directly onto the retina of the human eye, instead of providing screens in front of the eyes. This produces a much brighter and higher-resolution image with a potentially wider field of view than a screen-based display.

Current retinal displays share many shortcomings with head-mounted displays (see Section 3.1.2). However, some additional disadvantages can be identified for existing versions:

- Only monochrome (red) images are presented since cheap low-power blue and green lasers do not yet exist.
- The sense of ocular accommodation is not supported due to the complete bypass of the ocular motor system by scanning directly onto the retina. Consequently, the focal length is fixed.
- Stereoscopic versions do not yet exist.

The main advantages of retinal displays are the brightness and contrast, and low-power consumption—which make them well suited for mobile out-

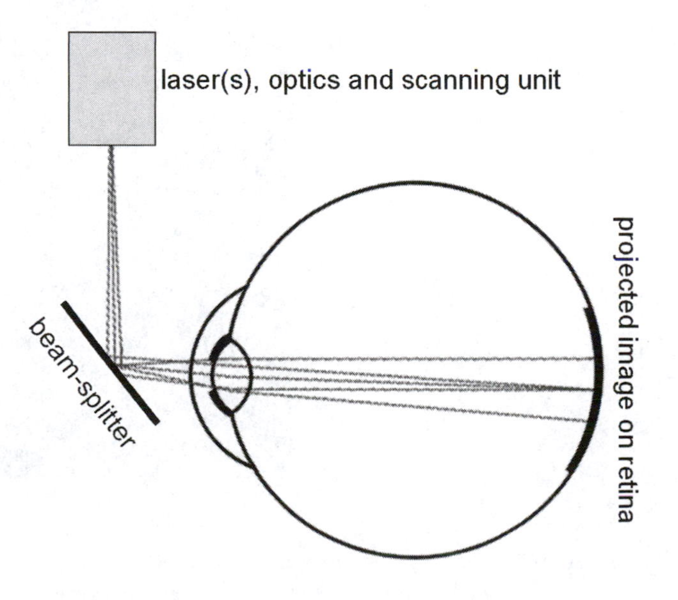

Figure 3.2. Simplified diagram of a retinal display.

door applications. Future generations also hold the potential to provide dynamic re-focus, full-color stereoscopic images, and an extremely high resolution and large field-of-view.

3.1.2 Head-Mounted Displays

Head-mounted displays are currently the display devices which are mainly used for augmented reality applications. Two different head-mounted display technologies exist to superimpose graphics onto the user's view of the real world: *video see-through* head-mounted displays that make use of video-mixing and display the merged images within a closed-view head-mounted display, or *optical see-through* head-mounted displays that make use of optical combiners (essentially half-silvered mirrors or transparent LCD displays). A comparison between these two general technologies can be found in [163].

Several disadvantages can be noted in the use of head-mounted displays as an augmented reality device. Note that most of these shortcomings are inherited form the general limitations of head-attached display technology.

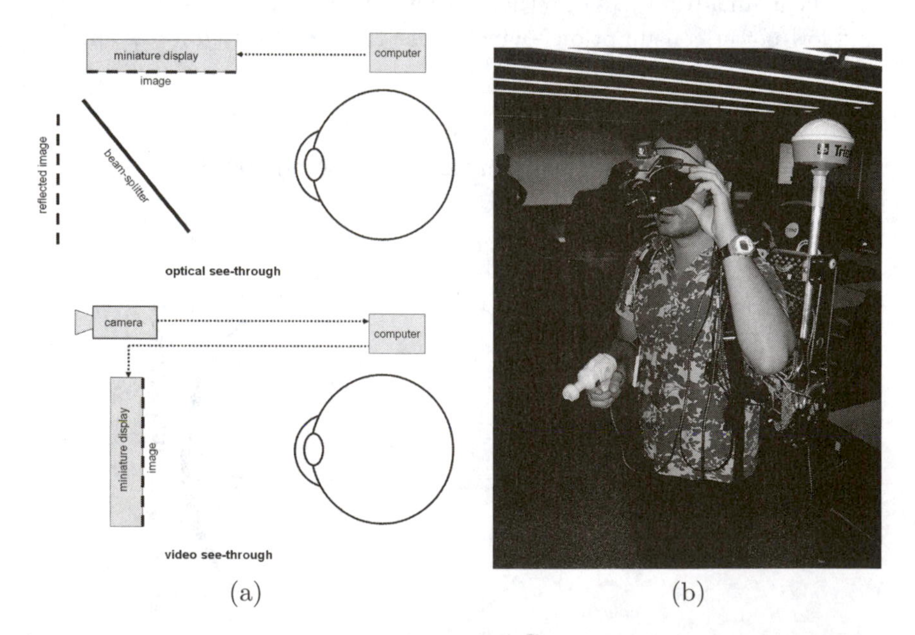

Figure 3.3. (a) Simplified optical see-through; (b) video see-through concepts. Example of video see-through head-mounted display setup with camera, backpack computer, and GPS antenna.

- Lack in resolution that is due to limitations of the applied miniature displays. In the optical see-through case, only the graphical overlays suffer from a relatively low resolution, while the real environment can be perceived in the resolution of the human visual system. For video see-through devices, both the real environment and the graphical overlays are perceived in the resolution of the video source or display.

- Limited field of view that is due to limitations of the applied optics.

- Imbalanced ratio between heavy optics (that results in cumbersome and uncomfortable devices) and ergonomic devices with a low image quality.

- Visual perception issues that are due to the constant image depth. For optical see-through, since objects within the real environment and the image plane that is attached to the viewer's head are sensed at different depths, the eyes are forced to either continuously shift focus between the different depth levels, or perceive one depth level as unsharp. This is known as the *fixed focal length problem*, and it is more critical for see-through than for closed-view head-mounted displays. For video see-through, only one focal plane exists—the image plane.

- Optical see-through devices require difficult (user- and session-dependent) calibration and precise head tracking to ensure a correct graphical overlay. For video see-through, graphics can be integrated on a pixel-precise basis, but image processing for optical tracking increases the end-to-end system delay.

- Increased incidence of discomfort due to simulator sickness because of head-attached image plane (especially during fast head movements) [133].

- Conventional optical see-through devices are incapable of providing consistent occlusion effects between real and virtual objects. This is due to the mirror beam combiners that reflect the light of the miniature displays interfering with the transmitted light of the illuminated real environment. To solve this problem, Kiyokawa et al. [79] use additional LCD panels to selectively block the incoming light with respect to the rendered graphics.

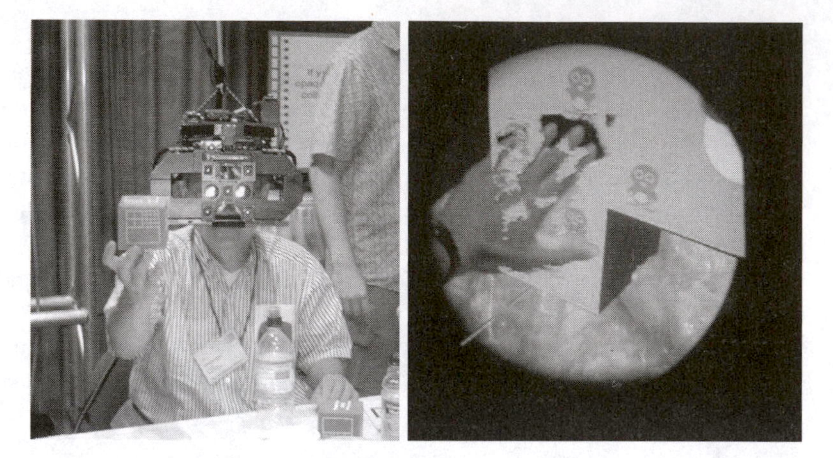

Figure 3.4. Osaka University's ELMO–An optical see-through head-mounted display that provides mutual occlusion by using a see-through LCD panel in front of the HMD optics. (*Image courtesy of Kiyoshi Kiyokawa [79] © 2002.*)

Head-mounted displays are currently the dominant display technology within the AR field. They support mobile applications and multi-user applications if a large number of users need to be supported.

Some variations of head-mounted displays exist that are more attached to the real environment than to the user. Optical see-through boom-like displays (e.g., Osaka University's ELMO [79]) or video see-through, application-adopted devices (e.g., the head-mounted operating microscope [48]) represent only two examples.

3.1.3 Head-Mounted Projectors

Head-mounted projective displays (HMPDs) [130, 73, 70] redirect the frustum of miniature projectors with mirror beam combiners so that the images are beamed onto retro-reflective surfaces (see Chapter 2) that are located in front of the viewer. A retro-reflective surface is covered with many thousands of micro corner cubes. Since each micro corner cube has the unique optical property to reflect light back along its incident direction, such surfaces reflect brighter images than normal surfaces that diffuse light. Note that this is similar in spirit to the holographic films used for transparent projection screens. However, these films are back-projected while retro-reflective surfaces are front-projected.

Other displays that utilize head-attached projectors are *projective head-mounted displays* (PHMDs) [78]. They beam the generated images onto

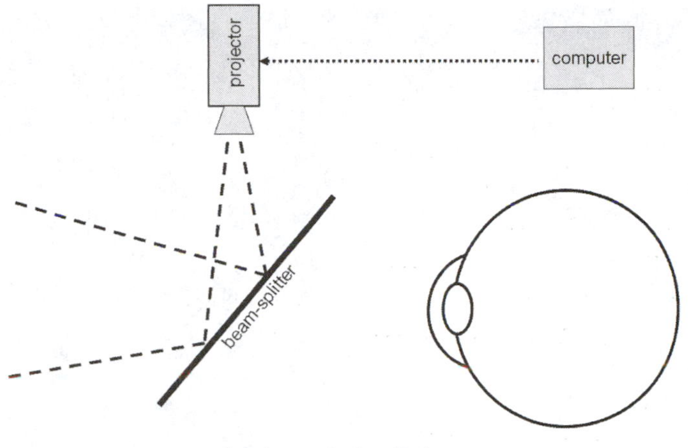

(a)

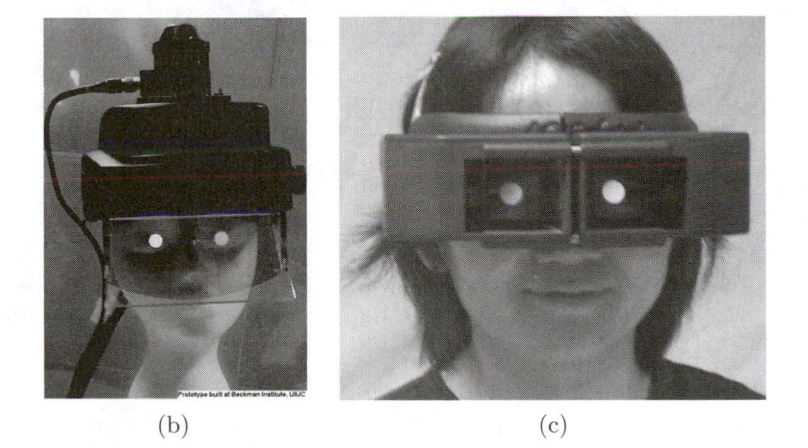

(b) (c)

Figure 3.5. (a) Simplified head-mounted projector concept; (b) and (c) examples of head-mounted projector prototypes (*Image (b) courtesy of Hong Hua and Jannick Rolland [70] © 2001.*)

regular ceilings, rather than onto special surfaces that face the viewer. Two half-silvered mirrors are then used to integrate the projected stereo image into the viewer's visual field. Their functioning is generally different than head-mounted projectors. It can be compared to optical see-through head-mounted displays. However, in this case the images are not displayed on miniature screens, but projected to the ceiling before being reflected by the beam splitter.

Figure 3.6. Example of how HMPDs are used to make things transparent—Optical Camouflage. (*Image courtesy of Inami et al. [73].*)

Head-mounted projective displays decrease the effect of inconsistency between accommodation and convergence that is related to HMDs. Both head-mounted projective displays and projective head-mounted displays also address other problems that are related to HMDs. They provide a larger field of view without the application of additional lenses that introduce distorting arbitrations. They also prevent incorrect parallax distortions caused by inter-pupil distance (IPD) mismatch that occurs if HMDs are worn incorrectly (e.g., if they slip slightly from their designed position). However, current prototypes also have the following shortcomings:

- The integrated miniature projectors/LCDs offer limited resolution and brightness.

- Head-mounted projective displays might require special display surfaces (i.e., retro-reflective surfaces) to provide bright images, stereo separation, and multi-user viewing.

- For projective head-mounted displays, the brightness of the images depends strongly on the environmental light conditions.

- Projective head-mounted displays can only be used indoors, since they require the presence of a ceiling.

Such displays technically tend to combine the advantages of projection displays with the advantages of traditional head-attached displays. However, the need to cover the real environment with a special retro-reflective film material make HMPDs more applicable for virtual reality applications, rather than for augmenting a real environment.

3.2 Hand-Held Displays

Conventional examples of *hand-held displays*, such as Tablet PCs, personal digital assistants [51, 54, 53, 131, 194], or—more recently—cell phones [112] generate images within arm's reach. All of these examples combine processor, memory, display, and interaction technology into one single device, and aim at supporting a wireless and unconstrained mobile handling. Video see-through is the preferred concept for such approaches. Integrated video cameras capture live video streams of the environment that are overlaid by graphical augmentations before displaying them.

However, optical see-through hand-held devices also exist. Stetton et al [174] for instance, have introduced a device for overlaying real-time tomographic data. It consists of an ultrasound transducer that scans ultrasound slices of objects in front of it. The slices are displayed time-sequentially on a small flat-panel monitor and are then reflected by a planar half-silvered mirror in such a way that the virtual image is exactly aligned with the scanned slice area. Stereoscopic rendering is not required in this case, since the visualized data is two-dimensional and appears at its correct three-dimensional location.

Hand-held mirror beam combiners can be used in combination with large, semi-immersive or immersive screens to support augmented reality tasks with rear-projection systems [8]. Tracked mirror beam combiners act as optical combiners that merge the reflected graphics, which are displayed on the projection plane with the transmitted image of the real environment.

Yet another interesting display concept is described in Raskar [158]. It proposes the application of a hand-held and spatially aware video projector to augment the real environment with context sensitive content. A combination of a hand-held video projector and a camera was also used by Foxlin and Naimark of InterSense to demonstrate the capabilities of their optical tracking system. This concept represents an interesting application of AR to the fields of architecture and maintenance.

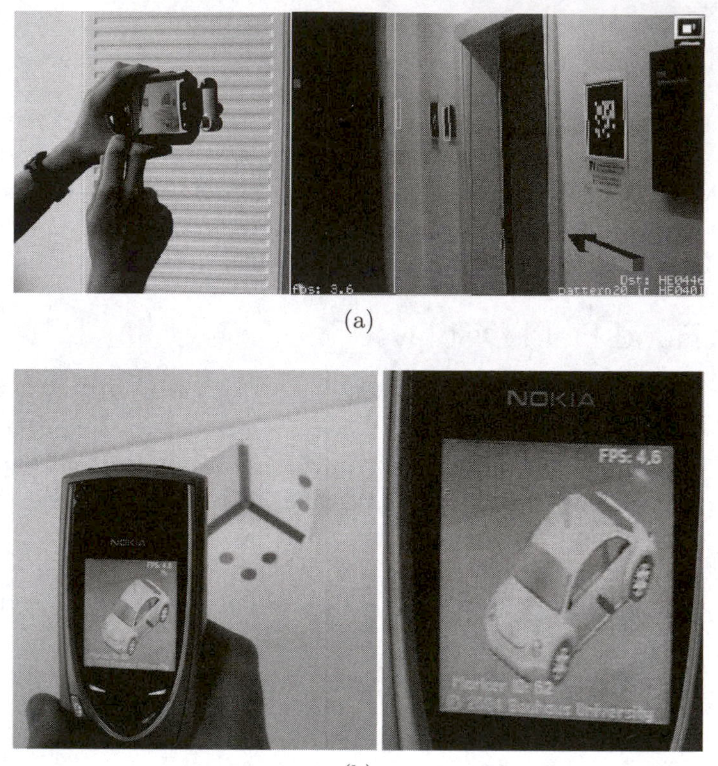

(a)

(b)

Figure 3.7. (a) AR application running on a PocketPC; (b) a first prototype on a conventional consumer cell phone. (*Images: (a) courtesy of Graz University of Technology [194]; (b) reprinted from [112] © IEEE.*)

There are disadvantages to each individual approach:

- The image analysis and rendering components is processor and memory intensive. This is critical for low-end devices such as PDAs and cell phones and might result in a too high end-to-end system delay and low frame rates. Such devices often lack a floating point unit which makes precise image processing and fast rendering even more difficult.

- The limited screen size of most hand-held devices restricts the covered field-of-view. However, moving the image to navigate through an information space that is essentially larger than the display device supports a visual perception phenomenon that is known as *Parks*

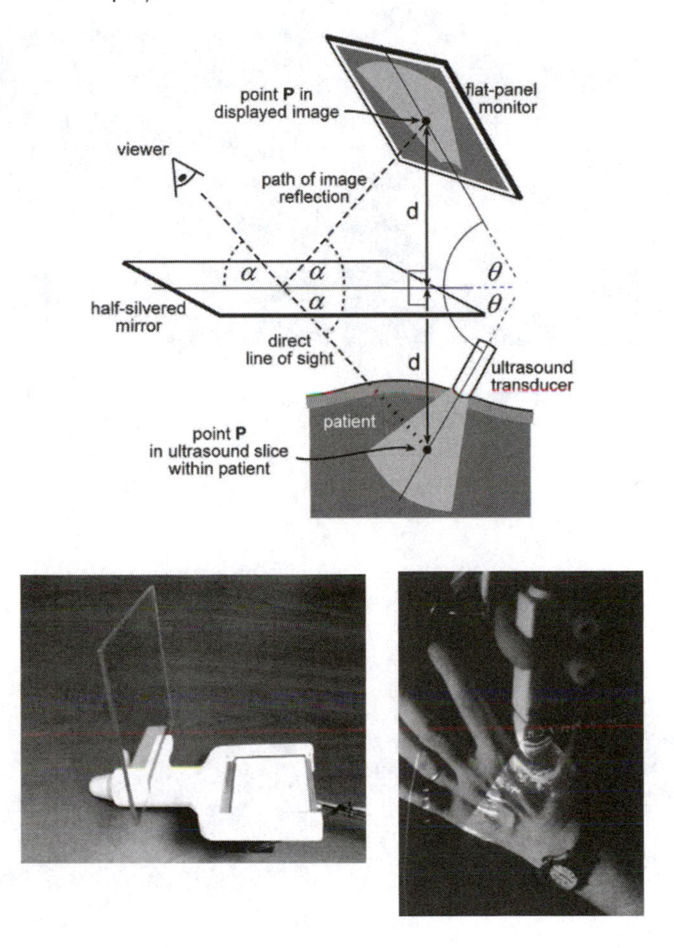

Figure 3.8. Example of a hand-held mirror display—The Sonic Flashlight. (*Image courtesy of George Stetten [174].*)

effect [129]. That is, moving a scene on a fixed display is not the same as moving a display over a stationary scene because of the persistence of the image on the viewer's retina. Thus, if the display can be moved, the effective size of the virtual display can be larger than its physical size, and a larger image of the scene can be left on the retina. This effect was also pointed out by the early work of Fitzmaurice [49] on mobile VR devices.

- The optics and image sensor chips of integrated cameras in consumer hand-held devices are targeted to other applications and,

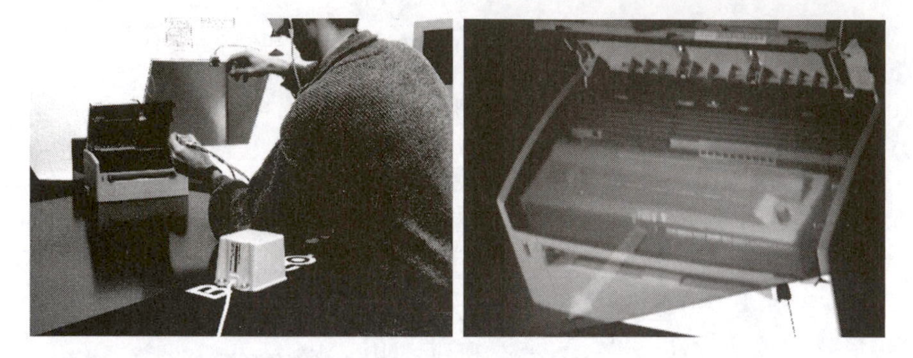

Figure 3.9. The Transflective Pad—a hand-held mirror beam combiner in combination with a large rear-projection screen. (*Image reprinted from [8] © Blackwell.*)

(a)

(b)

Figure 3.10. (a) AR Flashlight—augmenting the world with a tracked hand-held projector; (b) context aware iLamp—augmenting of an identified surface (*Images: (a) courtesy of InterSense Inc. © 2002; (b) reprinted from [158] © ACM.*)

consequently, provide a limited quality for image processing tasks (e.g., usually high barrel distortion). For example, they do not allow for modifying focus. Fixed focus cameras can only be effective in a certain depth range. This also applies to the image output of hand-held projectors if they are used to augment surfaces with a certain depth variance, since projected images can only be focused on a single plane.

- Compared to head-attached devices, hand-held devices do not provide a completely hands-free working environment.

Hand-held devices, however, represent a real alternative to head-attached devices for mobile applications. Consumer devices, such as PDAs and cell phones, have a large potential to bring AR to a mass market.

More than 500 million mobile phones have been sold worldwide in 2004. It has been estimated that by the end of the year 2005 over fifty percent of all cell phones will be equipped with digital cameras. Today, a large variation of communication protocols allows the transfer of data between individual units, or access to larger networks, such as the Internet. Leading graphics board vendors are about to release new chips that will enable hardware-accelerated 3D graphics on mobile phones including geometry processing and per-pixel rendering pipelines. *Variable-focus liquid lenses* will enable dynamic and automatic focus adjustment for mobile phone cameras, supporting better image processing. Some exotic devices even support auto-stereoscopic viewing (e.g., Sharp), GPS navigation, or scanning of RFID tags. Due to the rapid technological advances of cell phones, the distinction between PDAs and mobile phones might be history soon. Obviously, compact hand-held devices, such as PDAs—but especially mobile phones—are becoming platforms that have the potential to bring augmented reality to a mass market. This will influence application areas, such as entertainment, edutainment, service, and many others.

3.3 Spatial Displays

In contrast to body-attached displays (head-attached or hand-held), spatial displays detach most of the technology from the user and integrate it into the environment. Three different approaches exist which mainly differ in the way they augment the environment—using either video see-through, optical see-through, or direct augmentation.

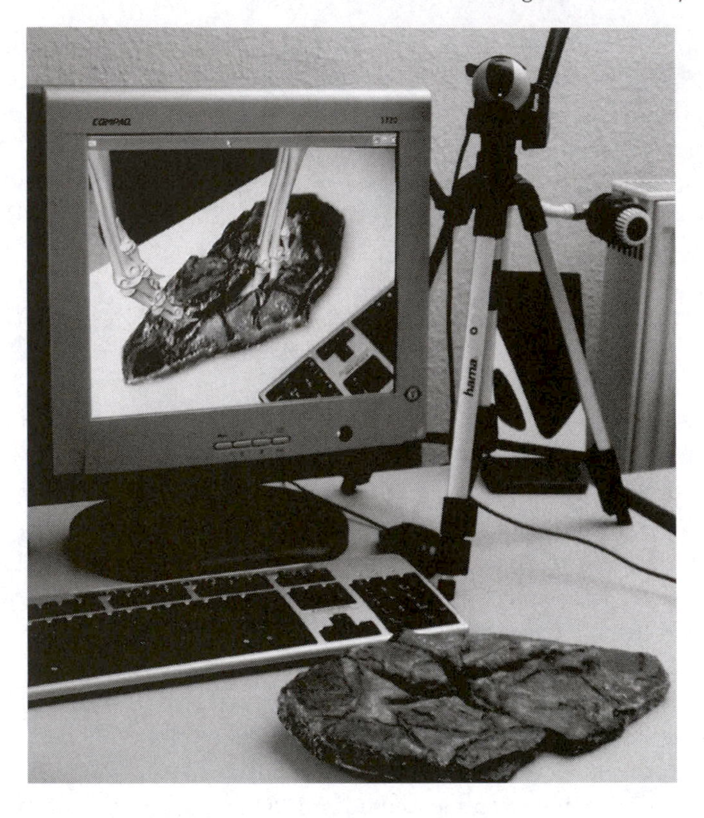

Figure 3.11. Example for a screen-based video see-through display. The locomotion of a dinosaur is simulated over a physical footprint. (*Image reprinted from [13] © IEEE.*)

3.3.1 Screen-Based Video See-Through Displays

Screen-based augmented reality is sometimes referred to as *window on the world* [46]. Such systems make use of video mixing and display the merged images on a regular monitor.

Like fish tank virtual reality systems which also use monitors, window on the world setups provide a low degree of immersion. Within an augmented reality context, the degree of immersion into an augmented real environment is frequently expressed by the amount of the observer's visual field (i.e., the field of view) that can be superimposed with graphics. In the case of screen-based augmented reality, the field of view is limited and restricted to the monitor size, its spatial alignment relative to the observer, and its distance to the observer.

For screen-based augmented reality, the following disadvantages exist:

- Small field of view that is due to relatively small monitor sizes. However, the screen-size is scalable if projection displays are used.

- Limited resolution of the merged images (especially dissatisfying is the limited resolution of the real environment's video image). This is a general disadvantage of video see-through [163].

- Mostly remote viewing, rather than supporting a see-through metaphor.

Screen-based augmentation is a common technique if mobile applications or optical see-through does not have to be supported. It represents probably the most cost efficient AR approach, since only off-the-shelf hardware components and standard PC equipment is required.

3.3.2 Spatial Optical See-Through Displays

In contrast to head-attached or hand-held optical see-through displays, spatial optical see-through displays generate images that are aligned within the physical environment. Spatial optical combiners, such as planar [9, 10] or curved [9, 16] mirror beam combiners, transparent screens [124, 19], or optical holograms [18] are essential components of such displays. This class of augmented reality displays is described in much more detail in Chapter 6.

Spatial optical see-through configurations have the following shortcomings.

- They do not support mobile applications because of the spatially aligned optics and display technology.

- In most cases, the applied optics prevents a direct manipulative interaction with virtual and real objects that are located behind the optics. Exceptions are reach-in configurations—either realized with see-through LCD panels [167] or mirror beam combiners [56].

- The number of observers that can be supported simultaneously is restricted by the applied optics. Multi-user displays such as the Virtual Showcase [9] and its variations support four and more users.

- A mutual occlusion between real and virtual environment is not supported for the same reasons as for optical see-through head-mounted displays. Projector-based illumination techniques which solve this problem [12] are discussed in Chapter 7.

- Due to the limited size of screens and optical combiners, virtual objects outside the display area are unnaturally cropped. This effect is known as *window violation* and is also present for fish tank and semi-immersive virtual reality displays.

The general advantages of spatial optical see-through displays are easier eye accommodation and vergence, higher and scalable resolution, larger and scalable field of view, improved ergonomic factors, easier and more stable calibration, and better controllable environment (e.g., tracking, illumination, etc.).This can lead to more realistic augmented environments. Chapter 6 will discuss these issues in more detail.

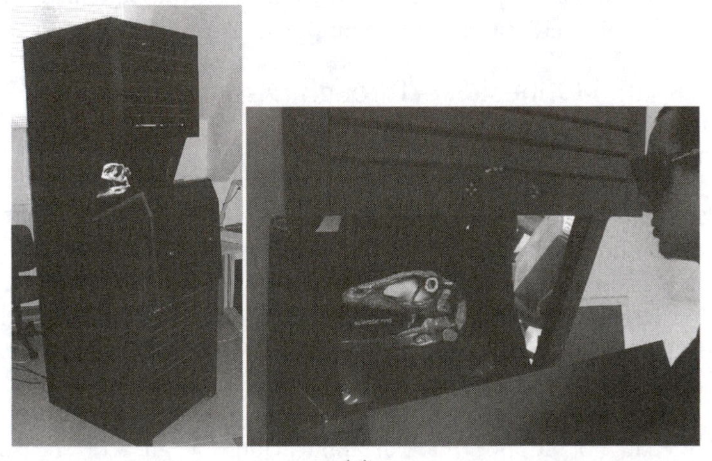

(a)

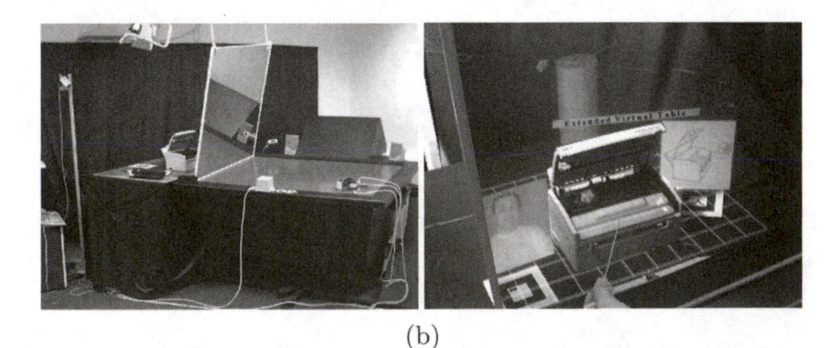

(b)

Figure 3.12. (a) A monitor-based single-user Virtual Showcase variation; (b) the Extended Virtual Table using a large beam splitter and projection screen. (*Images: (a) reprinted from [9] © IEEE; (b) reprinted from [10] © MIT Press.*)

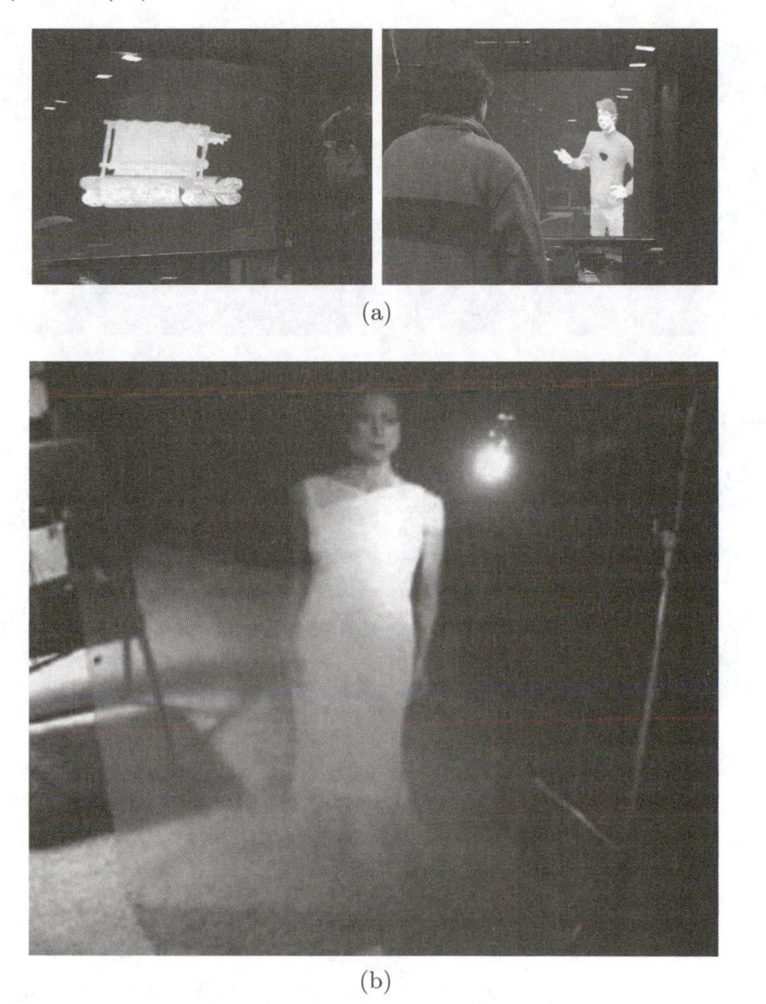

(a)

(b)

Figure 3.13. (a) Ogi's invisible interface; (b) transparent projection screen. (*Images: (a) courtesy of Ogi et al. [124]; (b) courtesy of Laser Magic Productions [86].*)

3.3.3 Projection-Based Spatial Displays

Projector-based spatial displays use front-projection to seamlessly project images directly on a physical objects' surfaces instead of displaying them on an image plane (or surface) somewhere within the viewer's visual field. Single static [147, 192, 13] or steerable [138], and multiple [152, 155, 157, 19, 20] projectors are used to increase the potential display area and enhance

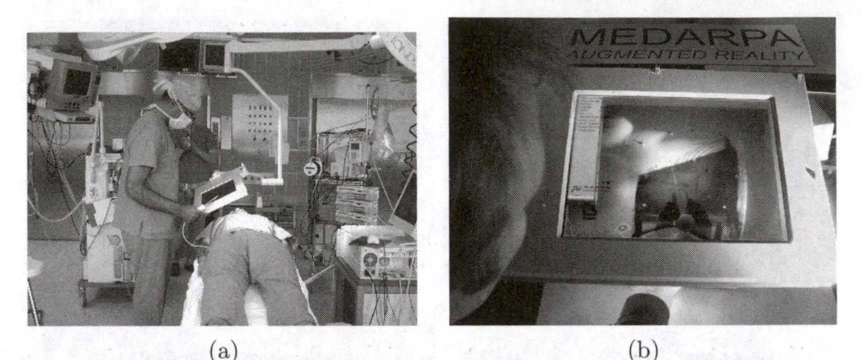

(a) (b)

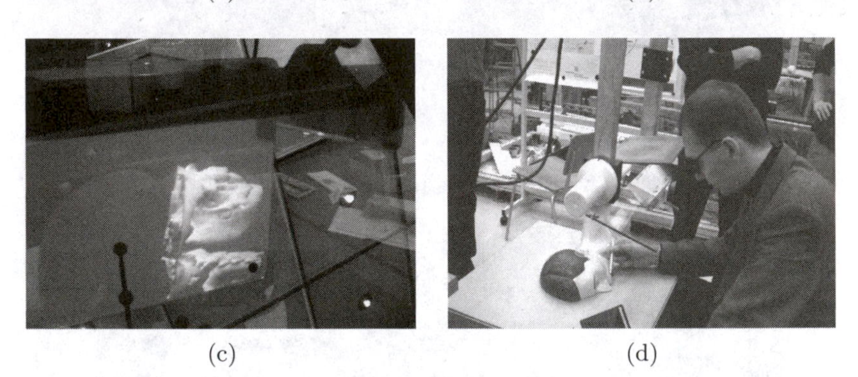

(c) (d)

Figure 3.14. (a) and (b) Example of a screen-based AR display using a see-through LCD panel—The AR window; (c) and (d) example of a mirror-based AR display using optical see-through beam splitters (*Images: (a) and (b) courtesy of ZGDV e.V. [167] © 2005; (c) and (d) courtesy of Manfred Bogen [56].*)

the image quality. Chapters 5 and 7 will discuss these concepts in more detail.

A stereoscopic projection and, consequently, the technology to separate stereo images is not necessarily required if only the surface properties (e.g., color, illumination, or texture) of the real objects are changed by overlaying images. In this case, a correct depth perception is still provided by the physical depth of the objects' surfaces.

However, if three-dimensional graphics are to be displayed in front of or behind the object's surfaces, a view-dependent, stereoscopic projection is required as for other oblique screen displays [150].

Projector-based spatial displays introduce several new problems:

- Shadow-casting of the physical objects and of interacting users that is due to the utilized front-projection. (Multi-projector configurations can solve this problem.)

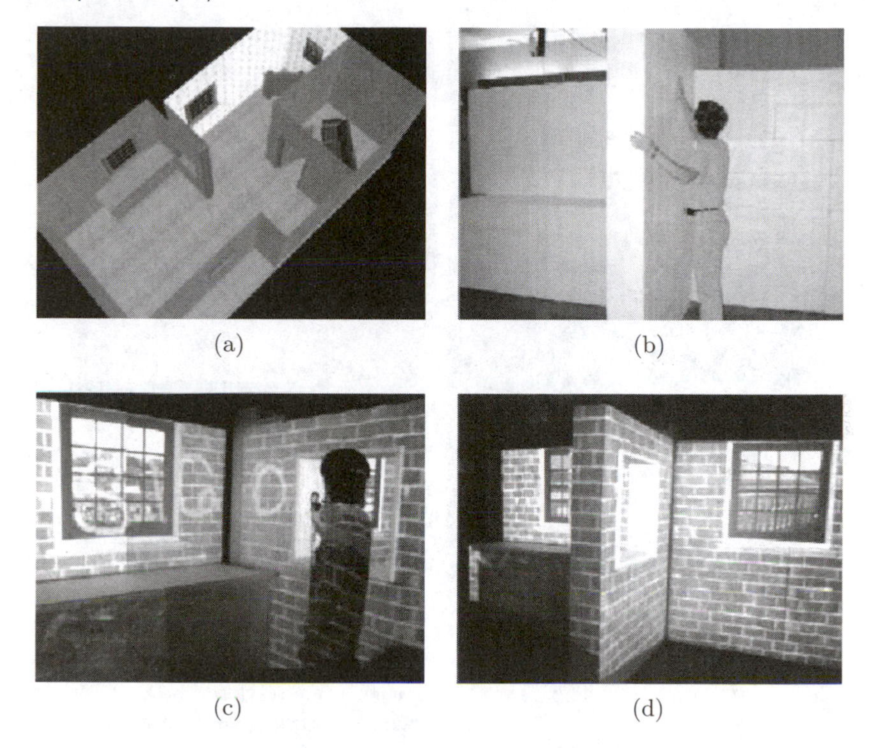

(a) (b) (c) (d)

Figure 3.15. Spatially augmenting a large environment: (a) virtual model; (b) physical display environment constructed using Styrofoam blocks; (c) and (d) augmented display. Note the view-dependent nature of the display, the perspectively correct view through the hole in the wall, and the windows. (*Images courtesy of Kok-Lim Low [93].*)

- Restrictions of the display area that is constrained to the size, shape, and color of the physical objects' surfaces (for example, no graphics can be displayed beside the objects' surfaces if no projection surface is present). Multi-projector configurations can solve this problem.

- Restriction to a single user in case virtual objects are displayed with non-zero parallax. (Multi-user projector configurations can solve this problem.)

- Conventional projectors only focusing on a single focal plane located at a constant distance. Projecting images onto non-planar surfaces causes blur. Exceptions are laser projectors which do not suffer from this effect.

Figure 3.16. ShaderLamps with Taj Mahal: The wooden white model is illuminated; the scanned geometry of the Taj Mahal is augmented to add texture and material properties; The geometry is then registered to the real Taj Mahal and displayed from the projector's viewpoint. (*Images reprinted from [155] © Springer-Verlag; see Plate II.*)

- Increased complexity of consistent geometric alignment and color calibration as the number of applied projectors increases.

On the other hand, projector-based spatial displays overcome some of the shortcomings that are related to head-attached displays: an improved ergonomics, a theoretically unlimited field of view, a scalable resolution, and an easier eye accommodation (because the virtual objects are typically rendered near their real world location).

As we will see in Chapters 5 and 7, real-time geometric, photometric and radiometric image compensation enables the augmentation of geometrically complex, colored and textured everyday surfaces that do not need to be optimal for a projection.

3.4 Summary and Discussion

In this chapter, we gave an overview of different display technologies suitable for augmented reality applications. We categorize these approaches

Figure 3.17. The Everywhere Display concepts (*Images courtesy of Claudio Pinhanez [138], IBM Corporation © 2001–2002.*)

into three main classes: head-attached, hand-held, and spatial displays. Different technologies have their individual advantages and disadvantages. While head-attached and hand-held displays support mobile applications, spatial displays do not. However, spatial displays can currently provide higher quality and more realism than can be achieved with current mobile devices. This is due to two reasons. First, a limited environment with a fixed display can be controlled better than a large environment with a moving display. This mainly applies to tracking and illumination, which are difficult to realize for arbitrary mobile applications. Second, most of the technology is not body-attached. Consequently, the size and weight of spatial displays does not conflict with ergonomic factors. The power consumption of spatial displays is also not a critical factor, as is the case for mobile devices. All of this leads to spatial displays that provide a higher resolution, a larger field of view, brighter and higher contrast images, a better scalability, an improved focus situation, and a consistent augmentation (e.g., consistent lighting and occlusion, better image registration, etc.) than possible with head-attached or hand-held displays. For applications that do not require mobility, spatial displays are a true alternative.

For historic reasons, head-mounted displays are mainly used for augmented reality applications today. They serve as an all-purpose platform for all sorts of scenarios. Many of these scenarios are not addressed efficiently because the technology has not been adapted to the application. This might be one reason for the fact that the breakthrough of augmented reality has not happened yet. We can compare this to the virtual reality field—which is more or less related to augmented reality. In the beginnings of virtual reality, head-mounted displays were also the dominant display technology. They are still used today, but they co-exist as part of a large pallet of different display types, including non-immersive fish tank VR screens as well as semi-immersive and immersive projection displays in all variations. Today, the VR community and end-users have the possibility to choose the display technology that best suits their application demands. This is probably one reason for the success of VR in recent years.

A similar development process can also shape the future of augmented reality. The display examples that are presented in this chapter are only a few examples of different possibilities. It is clear that some of them will be established inside and outside the field, and others will not. Cell phones, for instance, have an extremely high potential for being the breakthrough for mobile augmented reality. An AR application developed for such a platform can directly address a mass market, since no specialized and expensive hardware is required. Spatial displays, on the other hand, have the potential of merging real and virtual with very high quality. Thus, a visible separation between physical and synthetic environments might disappear completely one day. The advantage of this evolution is obvious. The right technology can be chosen to address the demands of an application with the highest possible efficiency. Thus, mobile AR and spatial AR go hand-in-hand.

4

Geometric Projection Concepts

In this chapter, we review the fundamentals of image formation and projection using a projector. Traditionally, a projector is used to create flat and usually rectangular images. In this sense, the three-dimensional computer graphics rendering algorithms used for a CRT or a flat LCD panel can be directly used for a projector without modification. However, a projector and display surface can be used in a variety of geometric configurations. In this chapter, we introduce a general framework that allows image synthesis under these geometric variations. The framework leads to a rendering framework and a better understanding of the calibration goals. We describe the issues involved and how the proposed algorithms are used in various types of display environments.

4.1 Geometric Model

Consider the conceptual framework for a camera in computer vision. A camera model defines the relationship between three-dimensional points and two-dimensional pixels. These are extremely useful geometric abstractions based on simple approximations, e.g., based on a pin-hole camera model that ignores optical or quantization issues. We introduce a similar conceptual framework for projector-based environments to express the geometric relationship between the display components.

Let us consider the problem of rendering images of three-dimensional virtual objects using a projector. There are various possibilities. The user could be moving or be static. The projector can be in an arbitrary position illuminating the display surface in a front-projection setup or rear-projection setup. The display surface can be planar or non-planar. The displayed virtual object can be in front of, behind, or on the display surface. The proposed framework can be used under any one of the display configurations.

4.1.1 Geometric Relationship

Consider the components in Figure 4.1. What is their interrelation regardless of the specific display configuration?

> The geometric framework defines the geometric relationship among the (tracked) user T, projector P, and the display surface D so that, for any arbitrary three-dimensional point V on the virtual object, we can express the mapping between V and the projector pixel m_p in the rendered image of the object.

Further, the transformation from the three-dimensional virtual object space to the two-dimensional projector image space can be described via an intermediate point on the display surface. Let us express the projector, P by its center of projection, P_c, and retinal image plane, P_R.

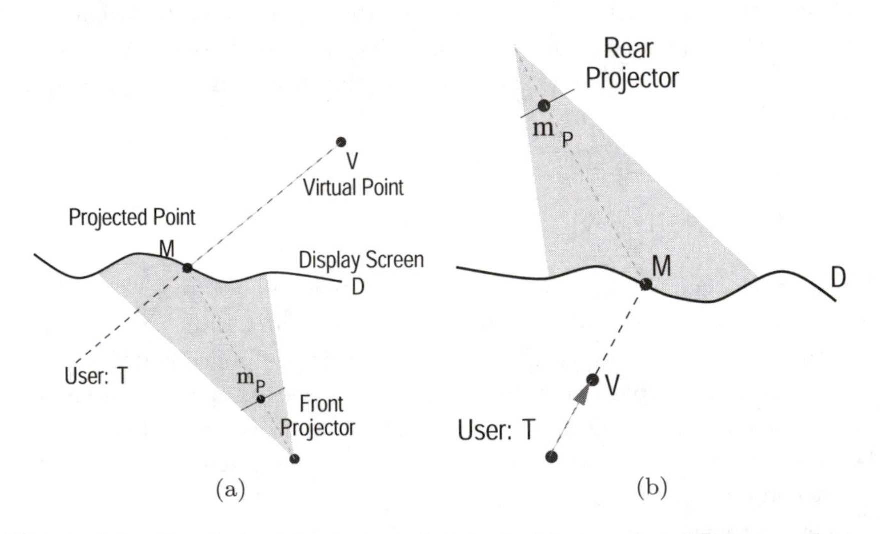

Figure 4.1. Rendering the image of virtual object under different projector display configurations: (a) a front projection display with a point on the virtual object behind the screen; (b) a rear-projection system with the virtual point in front of the screen. The display surface in both cases is non-planar.

> The projection m_p of a virtual point V is the intersection of the ray P_cM with projector image plane P_R, where M is the intersection of the ray TV with the display surface D.

The process can be described with more details in two logical steps. For a tracked user at T, we need to present the image of V to the user along the ray TV.

1. First compute where V would be displayed on the display surface D by intersecting ray TV with D, shown in Figure 4.1 as M.

2. Then find the pixel, m_p, that illuminates M using the analytic projection model of the projector.

Thus, V and m_p are related by the projection of M which in turn is defined by the ray TV and the surface D. The projection is undefined if the ray TV does not intersect D or if ray P_cM does not intersect image plane $P_{\mathcal{R}}$.

Similar to the geometric relationship in computer vision and image-based rendering, this simple abstraction can be exploited to describe forward, backward and mutual relationships between three-dimensional points and two-dimensional pixels. In the simplest case, in perspective projection or scan-conversion during image synthesis, which are both forward mapping stages, this relationship can be used to find the pixel coordinates $\{m_p\}$ of virtual points $\{V\}$. On the other hand, using backward mapping between the pixels $\{m_p\}$ and points on the display surface $\{M\}$, one can compute the user view of the displayed image. Backward mapping is also useful for

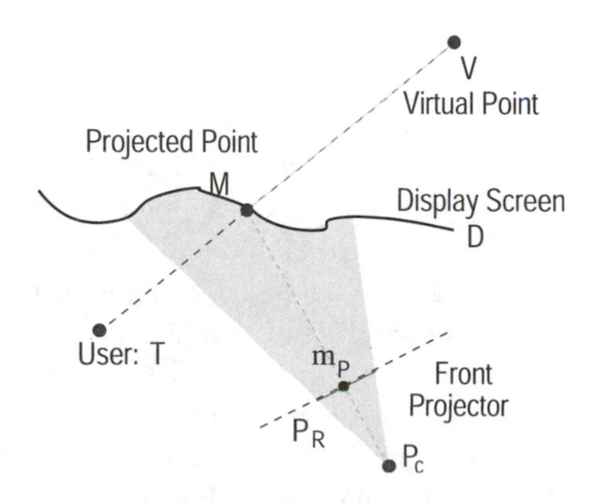

Figure 4.2. Relationship between geometric components.

texture mapping and, as described in Chapter 2, it can be used in projective textures. Mutual relationships, such as "in front of", between points in the virtual scene can be expressed for visibility computation in the user's view. Mutual relationship between views can also be used to determine corresponding pixels that render the same virtual feature point in two or more overlapping projectors. As described later, this mapping is critical for geometric registration in seamless multi-projector displays.

The proposed framework is general enough to describe the image synthesis process of most projector-based applications. Any specific projector-based application is a special case of this general description.

4.1.2 Geometric Components

Irrespective of which three-dimensional virtual scene is being displayed, the rendering process for a given projector involves three geometric components that define the display configuration:

1. Projection model for the projector;
2. Shape representation of the *display portal* surface;
3. User location.

The display portal for a given projector is the segment of the display surface illuminated by that projector. The user views the displayed virtual objects through this portal. Although the following discussion is valid for diverse representations for these three components, we will make some practical assumptions that simplify the discussion and the algorithms that follow.

Projector model. The projector image generation can be approximated using a pin-hole projection model, similar to the pin-hole camera model. In practice, this is a good approximation if the aperture is small and the radial distortion in the optical system is minimal. The advantage of assuming a pin-hole model is that the image generation can be mimicked using the projection matrix of the traditional graphics pipeline. The projection parameters for a pin-hole device can be defined by the traditional 3×4 three-dimensional perspective projection matrix [25, 119, 103, 91, 44]. Let us denote a two-dimensional pixel by $m = [u, v]^T$ and a three-dimensional point by $M = [X, Y, Z]^T$. The corresponding points in homogeneous coordinates are represented by $\tilde{m} = [u, v, 1]^T$ and $\tilde{M} = [X, Y, Z, 1]^T$. The relationship between the projector pixel m and the corresponding 3D point M illuminated on the display surface can be described as follows:

$$w\,\tilde{m} = F\,[R \;\; t]\,\tilde{M},$$

where w is an arbitrary scale factor, R and t represent the external parameters (i.e., transformation between the world coordinates system and the projector coordinate system), and F represents the projector *intrinsic* matrix. F can be expressed as

$$\begin{bmatrix} \alpha & \gamma & u_0 \\ 0 & \beta & v_0 \\ 0 & 0 & 1 \end{bmatrix},$$

where $(u0, v0)$ are the coordinates of the principal point, α and β are the scale factors in image u and v axes of the projector frame buffer and γ is the parameter describing the skew of the two image axes. In many cases, there is no need to explicitly represent the intrinsic and external parameters. We can use the 3×4 concatenated projection matrix $\tilde{P} = F\,[R\;t]$. It depends on 11 parameters (twelve minus a scale factor) and completely specifies the idealized pin-hole projection for a projector:

$$w\tilde{m} = \tilde{P}\tilde{M}. \tag{4.1}$$

In this sense, the pin-hole projector is the dual of a pin-hole camera. The projection matrix for both devices is defined by the same set of parameters. In the traditional graphics pipeline, the 3×4 matrix $\tilde{P}$ is upgraded to a 4×4 matrix P. The additional row allows computation of depth values for visibility queries.

Display portal. The display surface, D, can be represented using a piecewise planar approximation, $\hat{D}$. A common method is to express the surface using a polygonal mesh. The mesh is in turn defined by a set of vertices (with local surface orientation) and connectivity between those vertices. There is a simple quality versus cost trade-off: a mesh with more vertices usually allows a better approximation but requires more storage and additional computation. The advantage of using a polygonal mesh representation is again its compatibility with the geometric primitives used in graphics hardware.

User location. The user location is represented simply by a three-dimensional position in the world coordinate system. For *stereo* or multiple first-person rendering, the user location may be different for each eye or each view.

Finally, we are ready to express the relationship between a virtual point V and its projection m_p in the final image, in terms of the projection matrix $\tilde{P}$, display surface $\hat{D}$, and the (tracked) user location T. If $TM = k\ TV$ and $k > 0$

$$\begin{aligned} \tilde{m_p} &\cong \tilde{\mathcal{P}}\tilde{M} \\ &\cong \tilde{\mathcal{P}}[TV \wedge \hat{D}] \end{aligned}$$

The binary operator $\wedge$ denotes the intersection of a ray and a surface and $\cong$ indicates equality up to scale. The condition $k > 0$, ensures the display portal and virtual object are on the same side of T. When $1 \geq k > 0$, the virtual object is behind the screen with respect to the user, and when $k > 1$, the virtual object is in front of the screen.

4.2 Rendering Framework

Given the geometric relationship as described in Section 3.1 and the assumptions stated previously, we can define the projection of a single virtual point. What approach should we use to compute a complete image of a given virtual object? We describe a new rendering framework that addresses the basic problems in image generation: transformation and projection to compute final pixel coordinates, visibility, and color calculations.

A simple but naive approach would be to compute the required intersection and projection for every point on the virtual object. However, a more sophisticated approach can take advantage of known computer graphics techniques such as scan conversion, ray-casting, image-based rendering with pixel reprojection, or a look-up using a stored light field.

4.2.1 Rendering Components

Our goal is to render a perspectively correct image of a virtual three-dimensional scene for a moving user on an *irregular surface* using a casually aligned projector. An irregular surface is typically a curved or non-planar surface which may also be discontinuous (Figure 4.3). A casually aligned projector is a projector in a roughly desirable but not necessarily predetermined configuration.

In addition to the three geometric components, projector parameters, display portal, and user location, the rendering framework involves the following components:

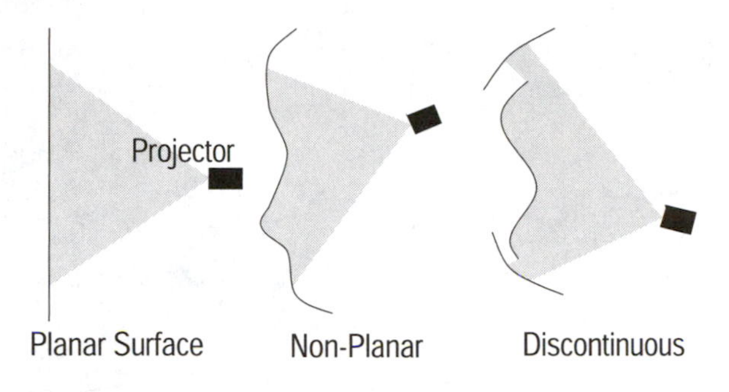

Figure 4.3. Types of irregular display surfaces.

- Virtual object, the input model for the rendering. It is a collection of geometric or image primitives sufficient to create novel views of that object.
- Desired view, the perspective view presented to the user. It is an intermediate view.
- Projected image, which is the final rendering result displayed by the projector.

4.2.2 Rendering Strategy

We propose a rendering strategy which is described with the following two logical steps and then presented in more detail in the rest of this section.

(1) Compute desired view through the display portal and map the view back on to the portal's idealized display surface;

(2) Compute projector's view of that augmented portal.

This surprisingly straightforward strategy simplifies existing rendering techniques and enables new algorithms. Note that, depending on the choice of forward or backward mapping, the sequence of the two logical steps may be reversed. Figure 4.4 shows the rendering process corresponding to the situation in Figure 4.1(a). The following discussion, however, is valid for any projector-display configuration.
The two logical steps are now described in more detail.

(a) First, in the frustum of the desired view, we represent the ray TV using two parameters on an arbitrary projection plane. This is equivalent to computing the image of V from the user location T on that

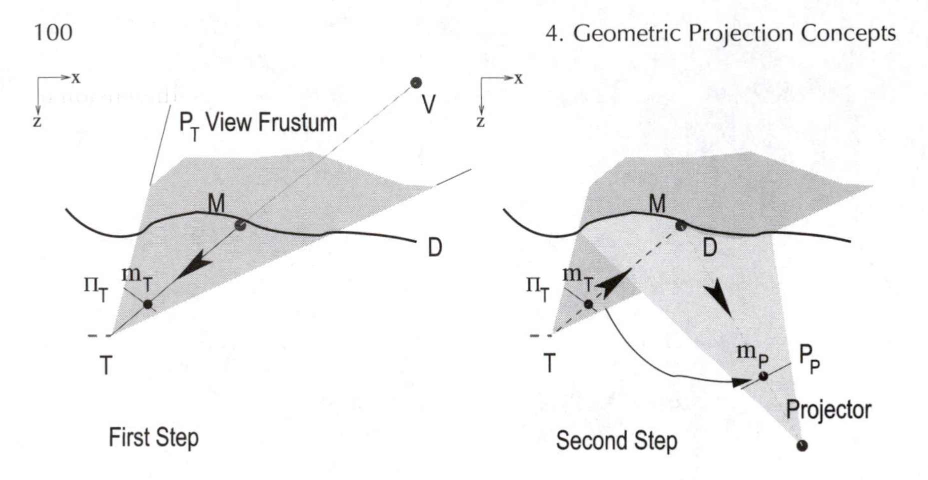

Figure 4.4. The Rendering Framework. The two rendering steps corresponding to the display configuration shown in Figure 4.1(a).

projection plane. In Figure 4.4 this projection is denoted by the projection matrix P_T and the projected pixel is denoted by m_T on the image plane Π_T. Pixel m_T is shaded according to the nearest object point along the ray TV. To transfer pixel m_T to a projector pixel, we again need to use the intermediate point on the display portal. The pixel m_T is transferred to the nearest point on the display portal along the ray Tm_T. Finally, the display portal is augmented with the image transferred from Π_T via the center of projection T.

(b) In the second step, we find the image of the augmented display portal. Referring to Equation (4.1), let us say that the projector's internal and external parameters are mimicked by the projection matrix P_P. If we render the augmented display portal with this projection matrix, the point M is transferred to pixel m_P.

The two projection matrices are sufficient to describe the pixel projection, visibility, color computations, and view frustum culling operations in traditional computer graphics. We consider three examples to demonstrate how the rendering strategy can be used effectively to construct conceptually simple rendering algorithms.

4.2.3 Example 1: Non-Planar Surface

In the general case, (i.e., when the irregular display surface is represented using a piecewise planar representation), we can use a two-pass rendering method. The first pass exploits the forward mapping of three-dimensional virtual points to pixels in the desired view. The second pass involves backward mapping to find projector pixel colors using a variant of conventional texture mapping.

In the first pass, the desired image for the user is computed and stored as a texture map. In the second pass, the texture is effectively projected from the user's viewpoint onto the polygonal model of the display surface. The display surface model, with the desired image texture mapped onto it, is then rendered from the projector's viewpoint. In OpenGL API, this is achieved in real time using projective textures [168]. When this rendered image is projected, the user will see a correct perspective view of the virtual object.

The additional rendering cost of the second pass is independent of the complexity of the virtual model [148]. However, due to limited resolution of the texture used during the second pass, the resultant warped image may exhibit resampling artifacts. From a practical implementation standpoint, the artifacts are minimized if the image plane of the view frustum chosen in the first pass is parallel to the best fit plane of the display surface.

This two-pass technique is sufficient to describe the rendering process for various projector-display configurations. However, many simplifications are possible under more restricted situations such as when the user is expected to be at a fixed location (e.g., a *sweet-spot,*) when the display surface is planar, or when the display portal matches the virtual object.

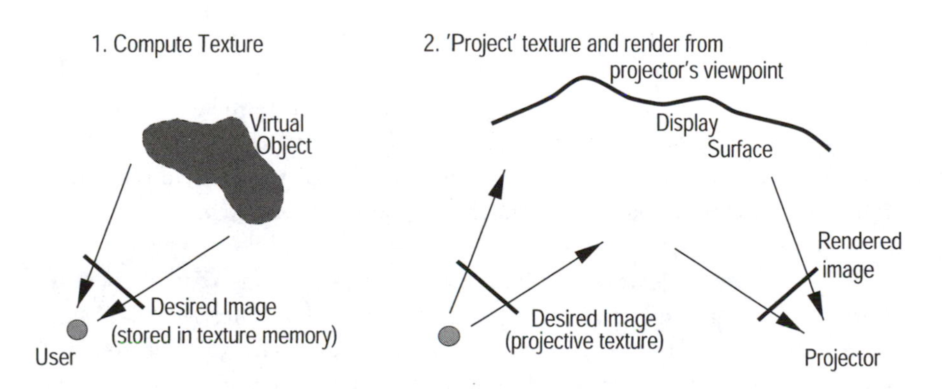

Figure 4.5. Two-pass rendering for a non-planar display surface involves forward and backward mapping.

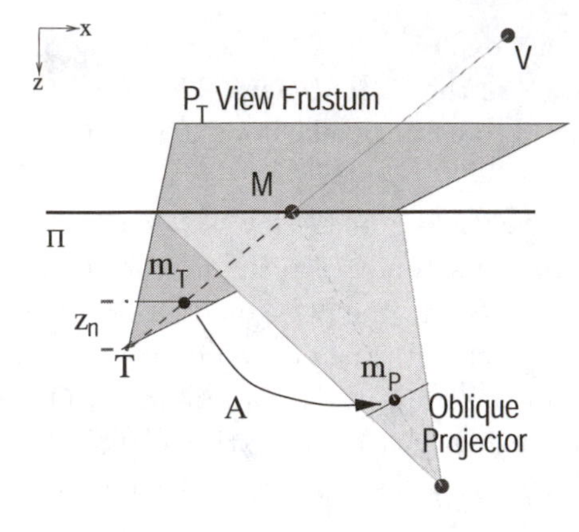

Figure 4.6. Rendering for a planar surface. The pixels m_T and m_P are related by a homography.

4.2.4 Example 2: Planar Surface

Consider rendering images for a head-tracked user in CAVE [35]. In the paper, the authors describe a relatively complex mechanism to render the images. How can we use our simple two-step rendering strategy? In CAVE, each projector illuminates a planar rectangular display portal. Hence, we can use a simplification of the rendering steps. We propose a strategy using forward mapping for both rendering steps. In the first step, the view through the display portal mapped back on the display portal can be rendered directly using a simple off-axis projection matrix P_T. The view frustum for P_T is specified by the center of projection, T, and the rectangular display portal. Thus, the image plane Π_T is parallel to the display portal. (In OpenGL API, we can use the `glFrustum()` function call to set up the matrix.) For the second step, we need to consider the relationship between the display portal and projector image. In CAVE, the projector optical axis is orthogonal to the display portal. In addition, the projected image rectangle exactly matches the display portal rectangle. Hence, the projector projection matrix P_P is the identity, and we do not need to find the internal parameters of the projector.

What if the projector is oblique with respect to the display portal? How can we create a perspectively correct image for a head-tracked user when a projector creates a keystoned quadrilateral image (Figure 4.6)? As described later, we exploit the well known homography between the projector image and the display portal. Hence, the transformation P_P in the second step can be expressed in terms of the homography.

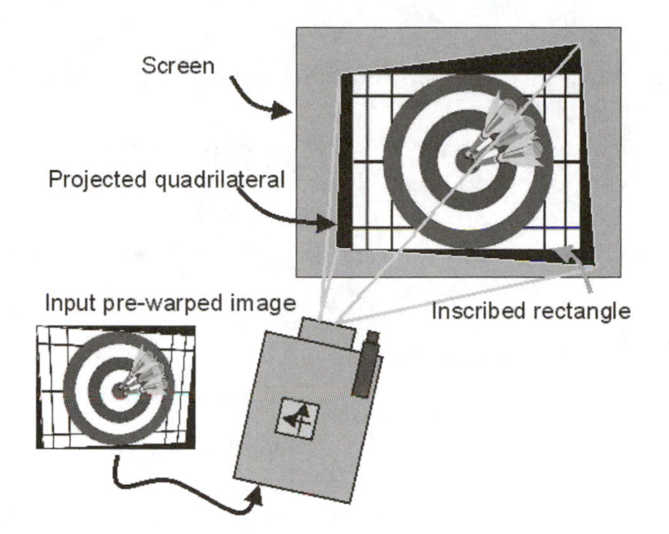

Figure 4.7. Given three angles of relative orientation, how should the input image be pre-warped to display an axis-aligned rectangular image? (*Image reprinted from [154] © IEEE.*)

4.2.5 Example 3: A Self Correcting Projector

Suppose we want to create a projector that illuminates a "correct" image on a vertical planar surface even when it is oblique. The user is not head-tracked, the input is a rectangular image and a "correct" displayed image in this case is a rectangular image that is aligned with the world horizon and vertical. In Figure 4.7, the goal is to display the desired image inside the inscribed rectangle by pre-warping the input image. The required keystone correction parameters are dependent on the three parameters of rotation, R_{w-p}, between the world coordinate system and the projector coordinate system. The three are essentially angles between the projector optical axis and the normal of the vertical plane [154]. One can detect the three angles with gravity-based tilt sensors and a camera. The question is, given the three angles, how can we pre-warp the input rectangular image so that it appears correct? At first this may appear as a difficult problem, and one may be tempted to first compute the planar perspective distortion based on the three angles. However, using the two-step rendering strategy, we can create a conceptually simple solution.

The first step is trivial. Since the user is not head-tracked and the display portal is expected to match the rectangular input image, P_T is the identity. The input image is simply texture mapped on a normalized rectangle i.e., within coordinates $(-1 : 1, -1 : 1, -1)$. The second step is more interesting. Given the projector internal matrix, F, and the rotation, R_{w-p}, we can simply use $P_P = F[R_{w-p} \ \ 0]$.

The second step essentially imitates the display situation. What is the projector's view of a world axis-aligned rectangle on a vertical wall? If we render such a view and, with the same projector, project the resultant image back on the vertical wall, we will get back the exact rectangle.

4.3 Calibration Goals

To improve the flexibility of the display setup, we need to reduce the rigid constraints on the components. Once the geometric parameters of the display environment are estimated, the system can tune itself to generate images that look perspectively correct. Thus, we are substituting the hardware-oriented tasks, e.g., electromechanical adjustment of rigid infrastructure, with intelligent but computationally intensive software algorithms. From a geometric viewpoint, there are three main calibration goals in using the proposed rendering framework: parameter estimation, geometric registration, and intensity blending. The calibration procedures for individual applications are governed by these goals.

4.3.1 Parameter Estimation

The accurate estimation of geometric quantities—user location, shape of the display portal, and intrinsic and extrinsic parameters of the projector—is essential for generating images without distortion or artifacts. The user location is reported by optical or magnetic tracking sensors [141, 201]. The shape of the display surface is computed using a depth extraction system with a camera in the loop or using a mechanical or laser scanning device. It is also possible to use the projector in two roles: as a source of active structured light for depth extraction and as a device for displaying perspectively correct images. The process of estimating projector parameters is similar to camera calibration, given the duality.

In addition, relationships among coordinate systems of multiple devices are required. They include rigid transformation between tracker and projector coordinate systems and pixel correspondence between overlapping projectors.

While the techniques for estimating such parameters are well known, it is very difficult to compute them to a high degree of precision. It is safe to assume that the estimated parameters will have non-zero errors. The goal then is to devise robust techniques to display images that perceptually approach the desired view for the user.

4.3.2 Geometric Registration

There are two types of registration that are crucial for displaying correct imagery in augmented reality: (1) precise alignment between the projected image and the features on the display surface, and (2) geometric agreement between image features in overlapping images projected by multiple projectors.

For each projector, we need to ensure that the projected image is registered with respect to the intended display surface features. In other words, after estimation of the display surface and projection matrix P_P of the projector, if a three-dimensional point M on the display surface maps to pixel $\tilde{m_p} \cong P_P\tilde{M}$, then for correct alignment, pixel m_p should physically illuminate the three-dimensional point M. For planar or panoramic display applications, the display surface is typically chosen to be locally smooth so that small errors are not perceptible. In more complex setups, such as the one described in Chapter 8 (Figure 8.2), the image features in the rendering of the virtual model of the Taj Mahal need to be registered with corresponding shape features on the physical model of the Taj Mahal.

When multiple projectors are used, any small error for individual projector images will lead to disagreement between projected image features in an overlap region. This, in turn, will lead to visible gaps and discontinuities in projected imagery. To achieve seamless images with accurate geometric alignment, we need to ensure that an image of a virtual three-dimensional point V from two or more projectors project at the same point on the display surface. If pixel m_{p^i} for each projector $i = 1, 2..n, n \geq 2$, illuminates the same point M on the display surface, then they must all correspond to a common pixel m_T in the desired image.

However, we do not compute explicit projector-to-projector pixel correspondence to achieve the goals. The geometric registration is achieved implicitly using the proposed rendering framework. This greatly reduces the cumbersome task of the electromechanical alignment process and eliminates two-dimensional image space warps. By imitating the internal parameters of the projector in the rendering parameters, we can essentially represent the complete display process.

4.3.3 Intensity Normalization

Regions of the display surface that are illuminated by multiple projectors appear brighter, making overlap regions very noticeable to the user. In traditional multi-projector setups, overlapping projector images are feathered near the edges. This is also known as *soft-edge blending*.

In the ideal case, the weights can be easily determined once we have registered all of the projectors. In practice, there are two types of deviations that can affect the blended images.

1. *Geometric misregistration*: Due to small errors in estimated parameters of the geometric components, the projected images do not match exactly at their respective edges. This type of static misregistration is due to (a) calibration errors and (b) deviation from the assumed idealized analytic model of the display components. In addition, over time, electro-mechanical vibrations disturb the positions of projectors.

2. *Intensity variation*: The color response of two or more projectors is likely to be different due to non-linearities. Further, over time, the response may change.

Hence, there is a need to achieve a smooth transition in the overlap. The intensities in the resulting superimposition then would have reduced sensitivity to the static calibration errors and dynamic variations. The strategy is to assign weights $\in [0, 1]$ to attenuate pixel intensities in the overlap region.

Some choices are shown in Figures 4.8–4.11. In the top two rows of Figure 4.8(a), black indicates a 0 and grey indicates a 1. The other shades of grey indicate weights between 0 and 1. The figures show three choices for intensity weight distribution.

(a) Contribution from a single projector at every point in the overlap. The overlap is segmented into distinct sub-regions and each sub-region is illuminated by a single projector. All the intensity weights are either 0 or 1.

(b) Equal contribution from all projectors in the overlap. In a region with n overlapping projectors, the weight for all the pixels in that region is $1/n$.

(c) Linear ramps in the overlap, where the weights are 0 near the boundary and increase linearly away from the boundary.

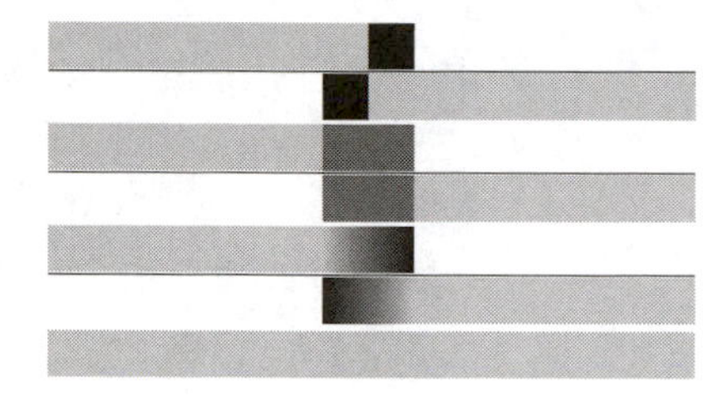

Figure 4.8. Ideal case, accurate registration. Top two rows, choice (a): contribution from a single projector for each point. Rows 3 and 4, choice (b): equal contribution from each projector in the overlap. Rows 5 and 6, choice (c): linear ramps in the overlap. Bottom row: Per pixel addition of contributing weights, all methods result in same solution.

Figure 4.9. Choice (a): Contribution from a single projector at every point in overlap. Top three rows show result of misregistration in case of projectors moving away from each other. Top row, projector moves to left. Second row, projector moves to right. Third row, resulting addition shows a dark gap. Bottom three rows show extra overlap results in a spiked region that is twice as bright.

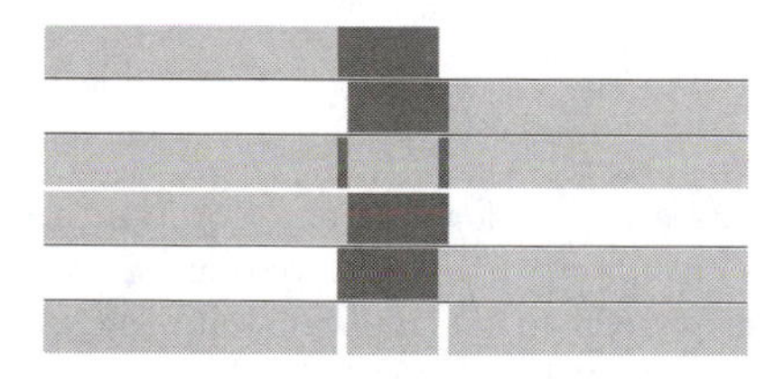

Figure 4.10. Choice (b): Equal contribution from both projectors at every point in overlap. Top three rows show projector moving away from each other, resulting in a visible valley of lowered intensity. Bottom three rows show extra overlap results in a narrow brighter regions.

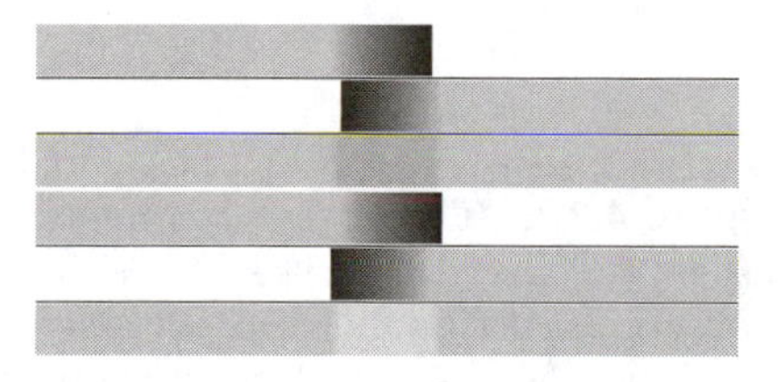

Figure 4.11. Choice (c): Linear ramps in overlap. Top three rows show projector moving away from each other, resulting in an overlap region with overall slightly lower intensity. Bottom three rows show extra overlap results in slightly brighter overlap regions.

In the ideal case of accurate geometric registration, all three cases behave the same. In the presence of small misregistration (either a gap or an additional overlap), as can be easily observed, feathering is preferred over other solutions that create visible gaps, valleys, or any other sharp variation. The specific intensity feathering algorithms are more complex and described later.

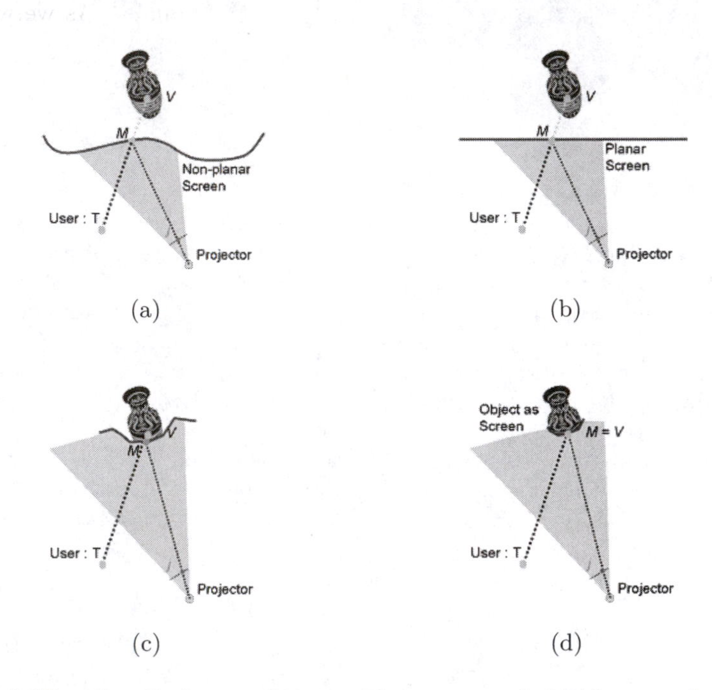

Figure 4.12. The display continuum: The same rendering framework can be used for different displays surfaces, from (a) arbitrary and unrelated curved and non-planar surface and (b) traditional planar screen to (c) surfaces that are close to virtual objects and (d) identical to the virtual object.

4.4 Display Environments and Applications

In Figure 4.12, we show various types of display surfaces for a single projector application. Whether the display surface is non-planar, planar, or we are projecting on a closed object, we can use the same rendering framework. The framework can be also be used in applications where a set of projectors are used together. As described earlier, the setup could have a static or moving user and front or rear projection screen. In the next chapters, we will learn about the specific techniques for spatially immersive displays.

4.5 Summary and Discussion

In this chapter, we reviewed the fundamental geometric concepts in using a projector for displaying images. Our goal is to understand how a projector can be used beyond creating a flat and rectangular image. As we will see in the following chapters, a projector can be treated as a dual of a camera to create a myriad of applications on planar or non-planar surfaces, as well as for creating images on closed objects in combination with mirrors and holograms. For the sake of geometric understanding, we have treated the projector as a pin-hole projection device, similar to a pin-hole camera. Then, the geometric relationship between the two-dimensional pixels in the projector frame buffer and the three-dimensional points in the world illuminated by those pixels is described using perspective projection. However, in practice, a projector has a non-zero aperture and involves optical elements. This introduces issues of depth of field (focus), radial distortion, and chromatic aberration. In this book, we do not go into the details of these issues; however, while describing projector-based applications, we provide practical guidelines to minimize the impact of these issues.

The proposed geometric framework leads to a simpler rendering technique and a better understanding of the calibration goals. In the next chapter, we describe the applications of the framework to form a concrete basis for explanation and analysis.

5

Creating Images with Spatial Projection Displays

In this chapter, we present the calibration and rendering schemes for various types of display surfaces. We first describe the strategy for planar displays, arbitrary non-planar displays, and quadric curved displays. Then, we focus on the specific projector-based augmentation problem where images are projected not on display screens, but directly onto real-world objects.

5.1 Planar Displays

In Section 4.2.4 we described a single-pass technique to display onto planar surfaces. That technique is an improvement over the more general two-pass rendering method for non-planar surfaces. Here, we present a complete set of techniques to demonstrate the application of the single-pass method. First, we show how the technique can be integrated with depth-buffer-

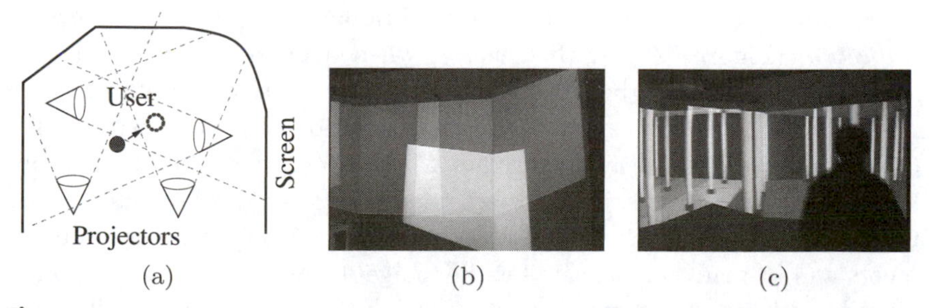

Figure 5.1. (a) The display techniques allow for a wide range of configurations of roughly aligned projectors and displays; (b) example of arbitrary projector overlaps before calibration; (c) viewer in the final display environment. (*Images reprinted from [151]* © *IEEE.*)

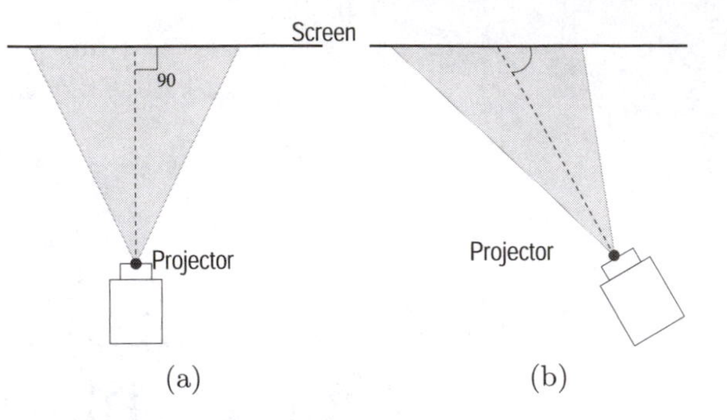

Figure 5.2. (a) Traditional projectors are orthogonal and create rectangular images; (b) oblique projectors create keystoned images.

based visibility computation. Then, we extend the concept of homography, due to planar homographies, to multi-projector tiled displays, and finally we describe some practical methods to calibrate and render such displays.

Projectors are typically mounted so that their optical axis is perpendicular to the planar display surface. Such configurations are also used in immersive environments to render perspectively correct imagery for a head-tracked moving user. They include CAVE [35], PowerWall [143] ($m \times n$ array of projectors), or ImmersiveDesk (back-lit and tilted desktop workbenches) [36, 146, 1]. By design, typical display systems try to maintain the image plane parallel to the plane of the display surface. However, this leads to the need for constant electro-mechanical alignment and calibration of the projectors, screens, and the supporting structure.

When the projector optical axis is not perpendicular to the display screen, the resultant image is keystoned and appears distorted (Figure 5.2(b)). As mentioned in Chapter 3, we will call this type of projection an *oblique projection* and the traditional projection an *orthogonal projection* (Figure 5.2(a)). In general, when a projector is roughly positioned, it is likely to be oblique. Even if the projector is orthogonally positioned in large display systems, after a period of time it can become oblique due to mechanical or thermal variations. We can address the problem of rendering perspectively correct images with oblique projectors using the rendering framework. Our goal is to avoid frequent mechanical adjustments and instead, compensate for the image distortion using the graphics pipeline. Although techniques to prewarp the images to avoid visible distortion exist, the technique described here achieves the results without additional cost of rendering or affecting the visual quality. The planar case of the rendering framework uses the homography between the points on the display screen and the projector pixels, induced due to the planarity of the screen. By using the homography during rendering, we show that oblique projectors can be used to easily create immersive displays. The main advantage is that we can use the traditional graphics pipeline with a modified projection matrix and an approximation of the depth buffer.

5.1.1 Previous Approach

In most cases, the problem of oblique projection is avoided simply by design. The orthogonal projectors create well-defined rectangular images. In some projectors the projected cone appears to be off-axis, but the display screen is still perpendicular to the optical axis. In workbenches [36], mirrors are used, but the resultant image is designed to appear rectangular (corresponding to rectangular frame buffers). Great efforts are taken to ensure that the projector image plane is parallel to the display screen and that the corner pixels are matched to predefined locations on the screen. Similar to the situation with current panoramic displays mentioned in the previous section, this leads to expensive maintenance and frequent adjustments of the projectors, screens, and the supporting structure. The cost of the system itself becomes very high because the supporting structure needed is usually very large and rigid.

When the projectors are oblique, a popular technique is to use a two-pass rendering method to prewarp the projected image. In the first pass, one computes the image of the 3D virtual scene from the viewpoint of the head-tracked user. The result is stored in texture memory and, in the second pass, the image is warped using texture mapping. This warped image, when displayed by the projector, appears perspectively correct to the user. Such two-pass rendering techniques are used for planar surfaces [181] as well as for irregular display surfaces [38, 147, 149].

The second rendering pass increases the computation cost and, in the case of immersive displays, it will also increase the rendering latency. In addition, texture mapping of the limited resolution image of the result of the first pass leads to resampling artifacts such as aliasing and jaggies. Due to the loss in the visual quality, such two-pass techniques are likely to work well only when high resolution projectors are used.

5.1.2 Single Projector

When the 3D surface points illuminated by all projector pixels lie in a plane, we can use an algorithm that avoids the additional texture mapping stage.

The following method achieves the result using a single rendering pass. A complete application of this algorithm, including the issue of visibility determination, is discussed later.

Let us re-visit the problem of rendering images of a virtual point V for a user at T. In this case, the display surface is represented by a plane Π, however, the optical axis of the projector may be oblique with respect to the display plane. Since the projector pixel m_P illuminates the point M on the screen, the corresponding rendering process should map point V to pixel m_P. This can be achieved in two steps of the rendering framework. First, compute the image of V from the user location T, which we denote by m_T using the view frustum P_T. Then, find the mapping between m_T and m_P (Figure 4.6).

A simple observation is that the two images of a point on the plane (such as M) are related by a homography induced by the plane of the screen. Hence, the two images of any virtual point V, m_T computed using P_T and m_P of the projector, are also related by the same homography. The homography is well known to be a 3×3 matrix defined up to scale, which we denote by A [44]. This observation allows us to create a new projection matrix for rendering. The projection matrix is a product of an off-axis projection matrix, $\tilde{P_T}$ (from the user's viewpoint) and the 3×3 homography matrix, A. There are various ways to create the projection matrix $\tilde{P_T}$ and the corresponding homography matrix A. We describe a method that updates P_T as the user moves, but the homography matrix remains constant.

Orthogonal projection. The first step in creating image m_T for V for a given user location is the same as the process for displaying images assuming orthogonal projectors (Figure 5.3). The method is currently used in many immersive display systems. Here we describe a more general technique. This technique also deals with a possibly non-rectangular projected image due to keystoning.

Let us consider the definition of $\tilde{P_T}$ for the user at T. Without loss of generality, let us assume that the display plane Π is defined by $z = 0$. First, create an (world coordinate) axis-aligned rectangle S on Π bounding the keystoned quadrilateral illuminated by the projector (Figure 5.4). The projection matrix is specified by the view frustum pyramid created by T and the four corners of S. Thus, T is the center of projection and S on Π is the retinal plane. As the user moves, S remains the same, but T, and hence the projection matrix $\tilde{P_T}$, are updated.

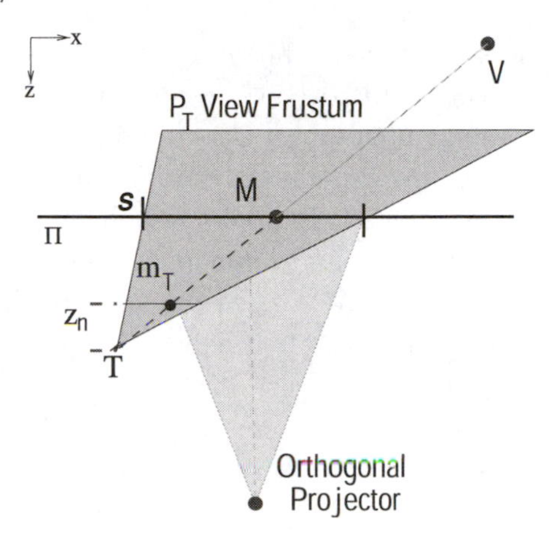

Figure 5.3. For orthogonal projectors, a simple off-axis projection can be used.

If the projector is orthogonal, S can be the same as the rectangular area illuminated by the projector. However, even if the area is not rectangular (because, say, the shape of the frame buffer chosen for projection is not rectangular), the projection matrix $\tilde{P_T}$ can be used to render correct images.

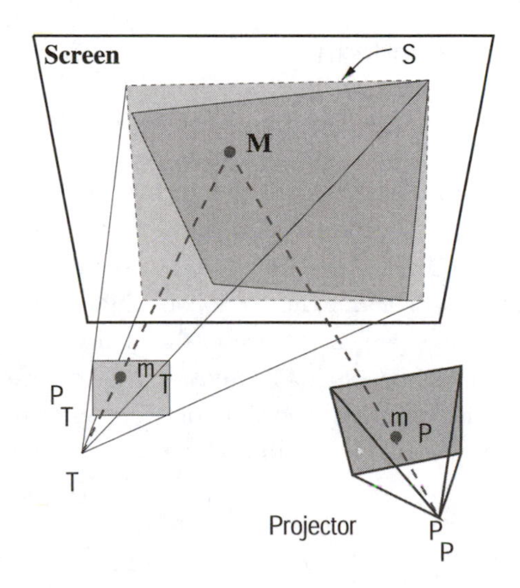

Figure 5.4. Defining P_T using axis-aligned bounding rectangle S.

Homography. In the second step, the image calculated assuming an orthogonal projector is corrected to compensate for the oblique projection using homography. The homography—between an image created with $\tilde{P_T}$ and the image to be rendered in the projector's frame buffer—is induced due to the plane of the screen. The homography matrix, $A_{3\times 3}$, is well known to be defined up to scale and maps pixel coordinates from one image to a second image [44, 33]. We need to compute the eight unknown parameters of $A_{3\times 3}$ that relate the two images of V: m_T due to $\tilde{P_T}$, and its image in the projector, m_P.

$$
\begin{aligned}
m_P &\cong A_{3\times 3} m_T, \\
\begin{bmatrix} m_{Px} \\ m_{Py} \\ 1 \end{bmatrix} &\cong \begin{bmatrix} a_{11} & a_{12} & a_{13} \\ a_{21} & a_{22} & a_{23} \\ a_{31} & a_{32} & 1 \end{bmatrix} \begin{bmatrix} m_{Tx} \\ m_{Ty} \\ 1 \end{bmatrix}.
\end{aligned}
$$

where the symbol $\cong$ denotes equality up to a scale factor. Note that the choice of view frustum for $\tilde{P_T}$ makes this homography independent of the user location and hence remains constant. If the three-dimensional positions of points on Π illuminated by four or more pixels of the projector are known, the eight parameters of the homography matrix, $A = [a_{11}, a_{12}, a_{13}; a_{21}, a_{22}, a_{23}; a_{31}, a_{32}, 1]$, can be easily calculated. Since the homography matrix remains constant, it can be computed off-line.

5.1.3 Single-Pass Rendering

We would ideally like to compute the pixel coordinates m_P directly from the 3D virtual point V. This can be achieved by creating a single 3×4 projection matrix, $A\tilde{P_T}$. Let us consider how this approach can be used in the traditional graphics pipeline. We need to create a 4×4 version of $\tilde{P_T}$, denoted by P_T and of A, denoted by $A_{4\times 4}$. As a traditional projection matrix, P_T transforms 3D homogeneous coordinates into 3D normalized homogeneous coordinates (4 element vectors). Typically, one can obtain the pixel coordinates after the perspective division. We create the new matrix, $A_{4\times 4}$ to transform these normalized 3D coordinates (without perspective division) to the projector pixel coordinates but try to keep the depth values intact:

$$
A_{4\times 4} = \begin{bmatrix} a_{11} & a_{12} & 0 & a_{13} \\ a_{21} & a_{22} & 0 & a_{23} \\ 0 & 0 & 1 & 0 \\ a_{31} & a_{32} & 0 & 1 \end{bmatrix}.
$$

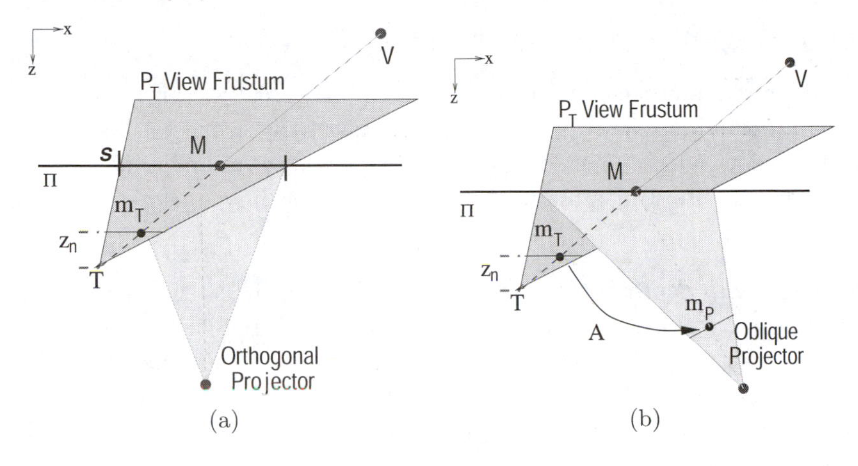

Figure 5.5. (a) First step using simple off-axis projection matrix P_T; (b) in the second step, warp using homography A. The modified projection matrix is AP_T.

The complete projection matrix is $P'_T = (A_{4\times4}P_T)$. (It is easy to verify that $[m_{Px}, m_{Py}, m_{Pz}, 1]^T \cong A_{4\times4}P_T[V, 1]^T$.) This achieves the desired rendering and warping using a single projection matrix without additional cost. It is more efficient than using the general rendering framework which requires two rendering passes. Since the image in the frame buffer is generated in a single pass, there are no resampling artifacts. Finally, when the image is projected on any surface coplanar with Π, the displayed virtual object appears perspectively correct. This approach is sufficient to render correct images of virtual three-dimensional points.

However, if the visibility computation is based on the traditional depth buffer or z-buffer, one important modification is necessary. In this modified algorithm there is a third logical step. In the first step, we compute the image of the virtual three-dimensional scene assuming the projector is orthogonal (Figure 5.5(a)). Then, we warp this image to compensate for the oblique projection using the homography between the display plane and the projector's image plane (Figure 5.5(b)). The depth values are also affected due to the warping, and hence, in a third step, they are transformed so that they can be used for visibility computations.

However, all three steps can be implemented using a single 4×4 projection matrix as in traditional rendering using graphics hardware. In this section, we focus on an application that involves a front-projection display system, but the techniques are also applicable to rear-projection systems such as CAVE or Immersive Workbenches.

5.1.4 Visibility Using Depth Buffer

Although the approach described in Section 5.1.2 creates correct images of virtual three-dimensional points, it is important to note that the traditional depth buffer cannot be effectively used for visibility and clipping. Let us consider this problem after we have rendered the image as described in Section 5.1.3.

Problems with depth buffer. As we saw earlier, a virtual point V is rendered using a projection matrix $P'_T = A_{4\times4}P_T$ (Figure 5.5(b)). The depth values of virtual points between the near and far plane due to P_T are mapped to $[-1, 1]$. Assume $[m_{Tx}, m_{Ty}, m_{Tz}, m_{Tw}]^T = P_T[V, 1]^T$ and the corresponding depth value is $m_{Tz}/m_{Tw} \in [-1, 1]$. After homography, the new depth value, is actually

$$m_{Pz} = \frac{m_{Tz}}{(a_{31}m_{Tx} + a_{32}m_{Ty} + m_{Tw})},$$

which (1) may not be in $[-1, 1]$, resulting in undesirable clipping with near and far plane of the view frustum, and (2) is a function of pixel coordinates.

Let us consider the problem of interpolation during scan conversion in traditional rendering. The depth values (as well as color, normal, and texture coordinates) at a pixel need to be computed using a hyperbolic interpolation. Homogeneous coordinates along with a 4×4 perspective projection matrix (such as P_T) achieve hyperbolic interpolation of depth values using a simpler linear interpolation (of the reciprocal of the values). In our case, when using a projection matrix $P'_T = A_{4\times4}P_T$, we are performing two successive hyperbolic interpolations; first for the perspective projection and second for the homography. At first, it may seem that a single 4×4 matrix would not be able to achieve the required interpolation. Fortunately, while the first matrix P_T introduces a hyperbolic relationship in the depth (or 'z') direction, the homography matrix $A_{4\times4}$ leads to a hyperbolic increment along the x and y coordinate axis of the viewing coordinate system. Thus both depth as well as screen space transformation can be achieved with a single 4×4 matrix.

Thus, although the homography does not introduce any inaccuracies in interpolation of depth values in the frame buffer, the clipping with near and far planes creates serious problems; some parts of virtual objects may not be rendered.

Approximation of depth buffer. We can achieve the rendering and warping in a single pass, however, using an approximation of the depth buffer.

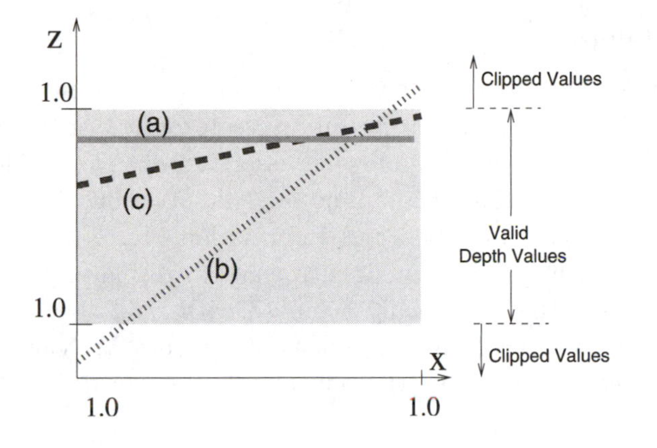

Figure 5.6. The plot shows depth buffer values along a scan line for points along constant depth: (a) using P_T; (b) after homography, $A_{4\times4}P_T$, the depth values range beyond $[-1,1]$; (c) with an approximation of depth buffer, $A'_{4\times4}P_T$, traditional graphics pipeline can be used to render perspectively correct images for a tracked moving user.

The rendering framework described the choice of view frustum for P_T based on a rectangle S. This chosen rectangle bounds the projected keystoned (quadrilateral) area (Figure 5.5(b)). Since S is larger than the displayed imagery, the normalized x and y coordinates due to P_T, m_{Tx}/m_{Tw} and $m_{Ty}/m_{Tw} \in [-1,1]$ for points displayed inside S. Hence,

$$\frac{(1-|a_{31}|-|a_{32}|)m_{Tz}}{(a_{31}m_{Tx}+a_{32}m_{Ty}+m_{Tw})} \in [-1,1].$$

This scale factor makes sure that the resultant depth values after warping with homography are $\in [-1,1]$. This will avoid undesired clipping with the near and far plane. Further, by construction of P_T, the angle between the projector's optical axis and the normal of the planar surface is the same as the angle between the optical axis and the normal of the retinal plane of the view frustum for P_T. To summarize, the modified projection matrix is $A'_{4\times4}P_T$, where

$$A'_{4\times4} = \begin{bmatrix} a_{11} & a_{12} & 0 & a_{13} \\ a_{21} & a_{22} & 0 & a_{23} \\ 0 & 0 & 1-|a_{31}|-|a_{32}| & 0 \\ a_{31} & a_{32} & 0 & 1 \end{bmatrix}. \quad (5.1)$$

5.1.5 Multiple Projectors

We can extend the same single-pass rendering and warping technique to register multiple overlapping projectors to create larger displays on a planar wall. In this section, we describe a necessary modification. More details are also available in [156]. Some popular systems using $m \times n$ arrays of projectors are PowerWall [143] and Information Mural [71]. The major challenge is matching the two images so that the displayed image appears seamless. Again, this is typically achieved with rigid structure which is difficult to maintain and needs frequent alignment. We can instead use roughly aligned projectors or even relatively oblique projectors.

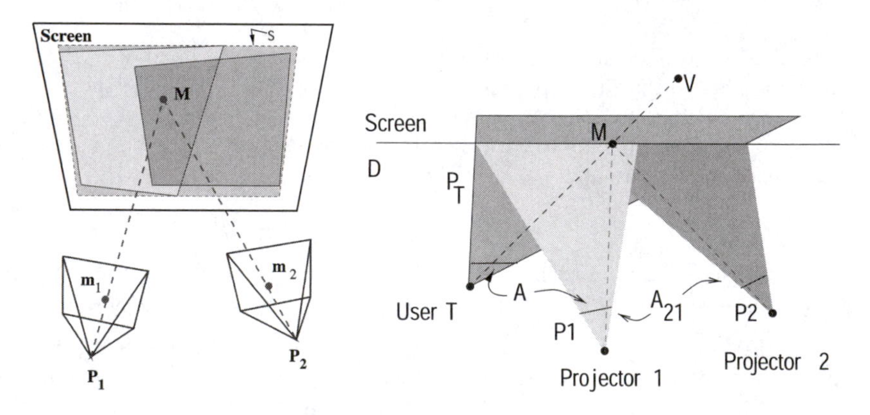

Figure 5.7. Homography between overlapping projectors on planar display surface.

Consider two overlapping projectors creating seamless images of virtual three-dimensional scenes on a shared planar surface (Figure 5.7). Let's say the projection matrices for the two projectors are P_1 and P_2. We exploit the homography between the images in the two projector frame buffers. If the 3×3 homography matrix mapping pixel coordinates from projector 1 to those in projector 2 is A_{21}, then the 3×4 projection matrix $\tilde{P_2}$ can be replaced with $\tilde{P_2'} = A_{21}\tilde{P_1}$.

Note that, as we have seen earlier, although the corresponding 4×4 matrix P_2' will create correct images of three-dimensional virtual points, we cannot use the traditional depth buffer for visibility and are forced to use approximated depth values. If P_1 itself is a result of oblique projection, so that $P_1 = A'_{4\times4}P_T$, the corresponding homographies $A_{3\times3}$, and A_{21} are used together. We will say $A_{21T} = A_{21}A_{3\times3}$. We first create the corresponding 4×4 matrix $A'_{21T-4\times4}$ (similar to Equation (5.1)). We then replace P_2 with $P_2' = A_{21T-4\times4}P_T$ to achieve correct rendering, warping, and depth buffer transformation. In practice, since it involves only eight unknown parameters, it is easier to compute the homography A_{21} than to calibrate the projector and compute the projection matrix P_2.

It is also necessary to achieve intensity blending of overlapping projectors. The complete problem of intensity feathering was addressed in the last section under generalized panoramic displays. In this case, however, due to the exact mapping defined by A_{21}, it is very easy to calculate the actual quadrilateral (or triangle in some cases) of overlap on the screen as well as in the projector image space. The intensities of pixels lying in this quadrilateral in both projectors are weighted to achieve intensity roll-off [97, 27, 185] and necessary blending. The technique for intensity blending in the case of two-pass rendering are described with more detail in Raskar[149] and Surati[181] and are applicable here for a single-pass rendering with minor modifications. The undesired parts of the projector image, for example the corners of projected regions outside a defined display region, can be truncated to create large and rectangular imagery with multiple projectors. The region in the projector frame buffer is masked off by rendering a black polygon at the near plane.

Implementation. We describe a case study of a multi-projector planar screen display [151]. We implemented a system consisting of three overlapping projectors displaying images on a 12×8 foot planar screen. The user is tracked with an optical tracking system. The homography between the desired image for the user and the first projector's frame buffer is calculated using approximately eight points. The mapping between overlapping projectors' pixels (again described by a homography) are actually computed by observing individual pixels of each of the projectors with a camera. The camera field of view covers the projection of all three projectors.

During preprocessing, one needs to compute the extent of projector illumination in world coordinates (WC), the transformation between the tracker and WC, and the homography $A_{3\times3}$. There are many ways of computing the homography and tracker-WC transformation. For example, one can have predefined locations on screen with known three-dimensional coordinates (e.g., corners of the cube in CAVE or a ruled surface on a workbench). In this case, one can manually, or using a camera-feedback, find which projector pixels illuminate these markers and then compute the unknown parameters. The origin in world coordinates is defined to be near the center of projection of the first projector.

Usually predefined markers are not available, for example, when one wants to simply aim a projector at a screen and render head-tracked images. We describe a technique that computes the homography and transformation simultaneously.

Assume the screen defines the WC: the plane $\Pi \equiv z = 0$ and origin (preferably) near the center of projected quadrilateral. We take tracker readings of locations on the screen illuminated by projector pixels. We will compute the transformation between the tracker and WC. Let us denote the tracker reading for a three-dimensional point M by M^{trk}.

- Assign one of the measured locations illuminated by a projected pixel as the origin of WC, O, (the tracker measurement effectively giving the translation between origins in both coordinate systems).

- To find the relative rotation, find the best-fit plane in tracker coordinates for the screen, the normal to the screen denoted the by vector $r3$.

- Assign one illuminated pixel in an approximate horizontal direction, M_x, as a point on the WC x-axis.

- Find the vector $M_x^{trk} - O^{trk}$ in tracker coordinates, normalize it to get vector $r1$. The rotation is $R = [r1;\ r3 \times r1;\ r3]$. The transformation is $[R\ \ (-R * O^{trk});\ 0\ 0\ 0\ 1]$.

- Find the three-dimensional coordinates of points on the screen illuminated by projector pixels

- Finally, Compute the homography $A_{3\times3}$. The homography allows the computation of extents of projector illumination in WC.

Wei-Chao Chen at the University of North Carolina (UNC) at Chapel Hill has implemented a stereo display using two completely overlapping projectors. One projector is used to display images for the left eye and the second for the right eye. The two images are distinguished by using polarizing optical filters on the projectors and eyeglasses. He computed P_T for each projector and the common homography matrix $A_{3\times3}$ using the UNC HiBall tracker. The same tracker is later used for tracking the user's head-motion [30].

For multiple overlapping projectors, we actually used a camera to find the homography between the projected images, and we also use the same information for computing weighting functions for intensity blending. The use of camera feedback for multiprojector displays is described in more detail later.

Summary of techniques. The sequence of steps suggested for using a single oblique projector to create a planar immersive display are given here.

During preprocessing:

- Turn on (four or more) projector pixels, $m_{P1}...m_{Pn}$ and find, in tracker coordinates, the position of the illuminated three-dimensional points on the plane, $M_1...M_n$.
- Compute the equation of the plane and transform the coordinate system so that it is coplanar with $\Pi \equiv z = 0$. We will denote the transformed points by M_i^{Π}.
- Find the homography, $A_{\Pi P}$, between M_i^{Π} and m_{Pi}. Using $A_{\Pi P}$, compute three-dimensional points on Π illuminated by corner pixels of the projector. Then find the axis-aligned rectangle, S, that bounds the four corners of the illuminated quadrilateral (the min and max of x and y coordinates).
- For a random user location T, compute P_T for a frustum created with S. Find normalized image coordinates $m_{Ti} = P_T M_i^{\Pi}$. Compute the homography $A_{3\times3}$, between m_{Ti} and projector pixels m_{Pi}. Create $A'_{4\times4}$ from $A_{3\times3}$ using Equation (5.1).

During run-time at every frame:

- Update T for the new user location and compute P_T (using a graphics API function such as `glFrustum`).
- Use $A'_{4\times4}P_T$ as the new projection matrix during rendering.

As one can see, the only human intervention involved is finding the three-dimensional tracker readings M_i for screen points illuminated by four or more projector pixels m_{Pi}.

5.1.6 Applications

The new techniques can easily be used in current immersive display systems. For example, in immersive workbenches, it is much easier to take tracker readings at a few points (illuminated by the projector), compute the homography off-line, and then use the modified projection matrix for the rest of the session. There is no need to worry about achieving some predefined alignment of the projector.

In CAVE, one can roughly align projectors so that they actually create images larger than the usable size of the screen. Since the usable screen sections are already marked (we can assume that the markers are stable unlike the configuration of the projectors and mirror), one needs to only find the projector pixels that illuminate those markers. The corners of the cube also allow geometric continuity across the screens that meet at right angles. The part of the imagery being projected outside the usable screen area can be masked during rendering.

Finally, this technique is useful for quickly creating a simple immersive display by setting up a screen, a projector, and a tracking system. For stereo displays, the off-axis projection matrix P_T is different for the left and right eyes, but the homography matrix $A_{3\times 3}$ remains constant.

5.1.7 Issues

The techniques are valid only when the assumed pinhole projection model (dual of the pinhole camera model) is valid. However, as seen in Figure 5.8, this assumption is valid for even low-cost single-lens commercial projectors. However, projectors typically have a limited depth of field and hence, when they are oblique, all pixels may not be in focus. The imagery is in focus for only a limited range of angles between the screen plane normal and the projector optical axis. The intensities across the screen are also not uniform when the projector image plane is not parallel to the screen. However, this can be compensated for by changing the intensity weights during rendering (using, for example alpha blending). In the case of a multi-projector tiled

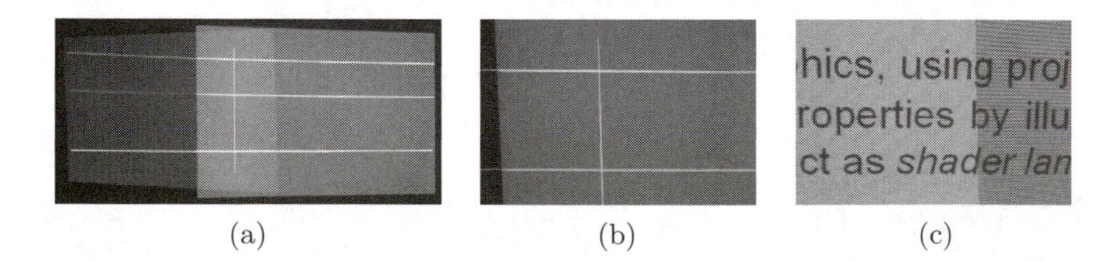

(a) (b) (c)

Figure 5.8. For a given plane of focus, a projector can be treated as a pin-hole device. The pictures here show experimental verification of linear perspective projection: (a) straight line segments in image plane appear straight on a planar display surface; (b) further, line segments projected by overlapping projectors can be aligned to sub-pixel accuracy using homography; (c) example of overlapping text in a warped image.

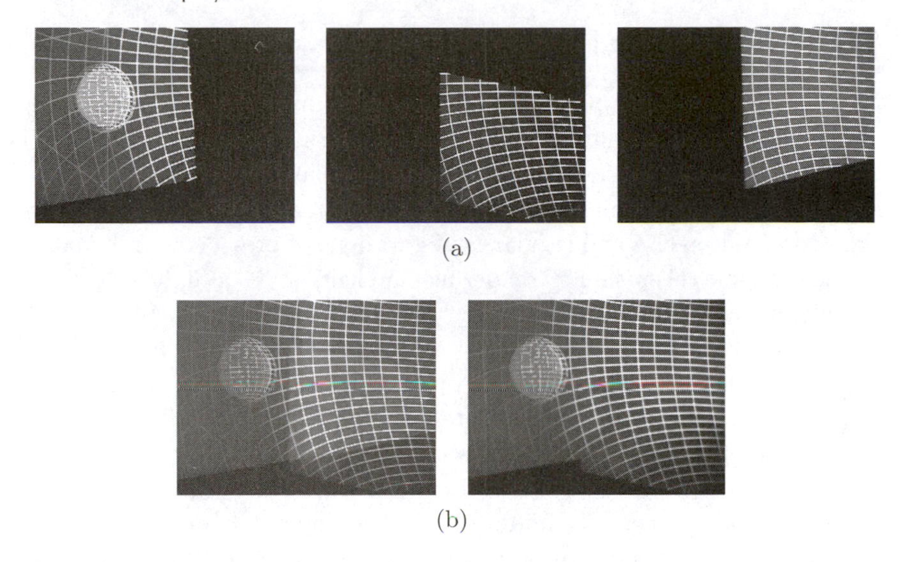

Figure 5.9. (a) Images displayed by an oblique projector which created keystoned imagery corrected using homography; (b) displayed image without and with intensity blending. (*Images reprinted from [151] © IEEE.*)

display, the alpha blending is performed for the overlap region, and hence it can be extended to include intensity correction for an oblique projector without any additional cost. The transformation of depth-buffer values essentially reduces the usable range of depth values, and hence the depth resolution is also decreased. In terms of resolution, rendering speed, and visibility computation, the displayed images were visually equivalent to those generated by orthogonal projection systems.

5.1.8 Discussion

The technique presented in this section allows rendering correct images even when the projectors are positioned without any precise alignment. Techniques for prewarping the images using a two-pass rendering are already available. However, the two major advantages of this technique are that it does not increase the cost of rendering and does not introduce resampling artifacts such as aliasing. The possibility of rendering images with oblique projectors that are visually equivalent can eliminate the need for cumbersome electro-mechanical adjustment of projectors, screens and the supporting structure.

5.2 Non-Planar Display

Let us consider the complete procedure for creating a display using a single projector on an non-planar display surface. This forms the basis of discussion for the multi-projector seamless display like the one shown in Figure 5.1. Conventional projector-based display systems are typically designed around precise and regular configurations of projectors and display surfaces. While this results in rendering simplicity and speed, it also means painstaking construction and ongoing maintenance.

The rendering framework provides a procedure for creating images when all three geometric components—projection model, representation of the display surface, and user location—are known. Here, we describe a camera-based method to estimate those parameters and use them in the rendering framework after some processing. In a limited fashion, Dorsey [38] and Surati [181] use cameras to find a simple two-dimensional image warping function for the purpose of two-dimensional image display. But we present a unified approach for camera-assisted detection of three-dimensional display configuration and projector-assisted active stereo vision.

The calibration procedures are based on using a stereo camera pair that is positioned on a wide baseline with each camera oriented to observe the entire projector-illuminated surface. Figure 5.10 illustrates this layout. Step 1 of the procedure involves calibration of the camera pair using

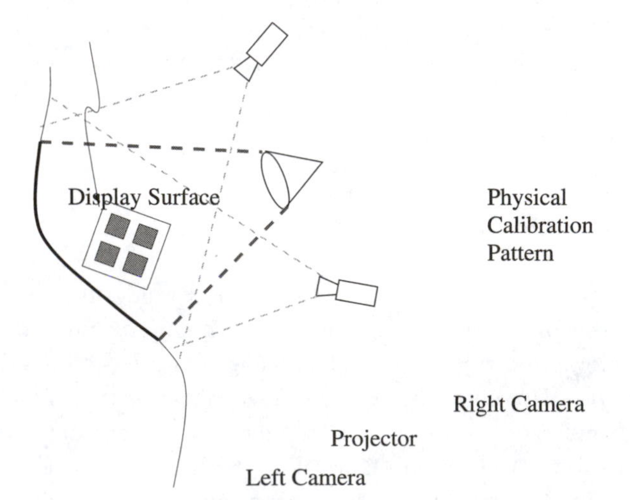

Figure 5.10. Configuration for single projector. (*Image reprinted from [151] © IEEE.*)

a physical calibration pattern. Calibration Step 2 involves estimation of display surface geometry and Step 3 evaluates projector intrinsic and extrinsic parameters. These methods are based on standard computer vision techniques and are relatively straightforward. However, they systematically build on each other. The calibration pattern can be retired once the calibration procedures are complete.

Camera calibration. To calibrate the camera pair, let us position a three-dimensional calibration pattern with spatially-known feature points within the intersection of their view frusta. By extracting feature points in the two-dimensional camera images corresponding to known three-dimensional points on the calibration pattern, we can determine the 3×4 projection matrix for each camera based on the perspective projection:

$$[s\,u \quad s\,v \quad s]^T = \tilde{C}\,[X\ Y\ Z\ 1]^T .$$

The perspective equation maps a three-dimensional point (X, Y, Z) in object space to a two-dimensional pixel (u, v) in camera image space, s is an arbitrary scale factor. The projection matrix $\tilde{C}$, determined up to a scale factor, represents a concatenated definition of the camera's intrinsic and extrinsic parameters assuming a pinhole optics model. With six or more correspondences between three-dimensional points on the calibration pattern and their mapping in the camera image, the 11 unknown parameters of $\tilde{C}$ can be solved using a least-squares method [44]. The intrinsic and extrinsic parameters of the camera can be obtained by a decomposition of $\tilde{C}$.

Display surface estimation. After the independent calibration of each camera, we can evaluate the geometry of the display surface using triangulation techniques based on correspondences extracted from the stereo image pair. Correspondences in the images are easily determined since we can use the projector to sequentially illuminate point after point until we have built a three-dimensional point cloud representing the display surface. By binary-coding the projector-illuminated pixels, we can efficiently determine n stereo correspondences in $\log_2(n)$ frames. The process of projecting patterns so that they can be uniquely identified by a camera is also known as an *active structured light technique*.

This three-dimensional surface representation, which is in the coordinate frame of the physical calibration pattern established in Step 1, is then reduced into a mesh structure in projector image space using two-dimensional Deluanay triangulation techniques.

Projector calibration. As a result of the surface extraction process, for each projector pixel (u, v), we now have a corresponding illuminated three-dimensional surface point (X, Y, Z). Using these correspondences, we can solve for the projector's projection matrix P as we did for the cameras.

A problem arises when the three-dimensional points of the display surface are co-planar. In this case, the least-square's method is degenerate due to the depth-scale ambiguity of viewing planar points; this means there exists a family of solutions. To develop a unique solution in this case, we can add surfaces into the scene and repeat the surface extraction procedure solely for the purpose of eliminating this ambiguity. Once this solution is achieved, the introduced surfaces are removed from the display environment.

Two-pass rendering algorithm. The calibration steps give the geometric definition of the two components of the rendering framework: the projection model of the projector and polygonal approximation of the display surface. The user location is continuously obtained using a head-tracking system. We are now ready to use the rendering process based on the rendering framework. To render perspectively correct imagery on irregular surfaces, we use the same two-pass rendering method and update the images as the user moves.

5.2.1 Multiple Projector Display

The remainder of this section will address the issues in scaling the calibration and rendering techniques from a single projector system to a multi-

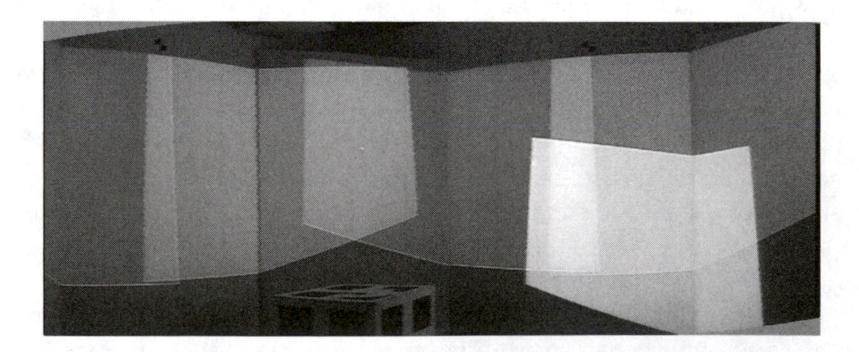

Figure 5.11. A panoramic image (100 degree FOV) of the panoramic five projector display system. (*Image reprinted from [151] © IEEE.*)

projector system. Let us assume we have accurately estimated the parameters. The effect of errors is addressed in the work of Raskar[151].

First, to calibrate multiple projectors, we repeat the procedures discussed in the previous section, but then we must re-register the display surface definitions and projector parameters for the entire system to a common world coordinate space.

Second, display surface regions where multiple projectors overlap are noticeably brighter because of multiple illumination. This is corrected by attenuating projector pixel intensities in the overlapped regions. Let us look at a versatile intensity feathering technique using camera-based feedback.

5.3 Projector Overlap Intensity Blending

Regions of the display surface that are illuminated by multiple projectors appear brighter, making the overlap regions very noticeable to the user. To make the overlap appear seamless, we assign an intensity weight $[0.0 - 1.0]$ for every pixel in the projector. The weights are assigned following these guidelines:

1. The sum of the intensity weights of the corresponding projector pixels is 1 so that the intensities are normalized.

2. The weights for pixels of a projector along a physical surface change smoothly in and near overlaps so that the interprojector color differences do not create visible discontinuity in displayed images.

Note that these two conditions cannot always be satisfied and hence they remain as guidelines. In the next chapter, we discuss the intensity blending algorithm in more detail, where presence of depth discontinuities in an overlap region makes a third guideline necessary. Here, we explain the algorithm assuming no depth discontinuities in the overlap region. We use an *alpha blending* technique to attenuate the projector pixel intensities. We essentially create an *alpha mask* for each projector, which assigns an intensity weight $[0.0 - 1.0]$ for every pixel in the projector. The weight is additionally modified through a gamma look-up table to correct for projector non-linearities.

Although various methods can be used to find the intensity weights, we use a feathering technique influenced by common techniques used in blending multiple images for panoramic photomosaics [170, 185]. To find the alpha mask, we use a camera to view the overlapped region of several

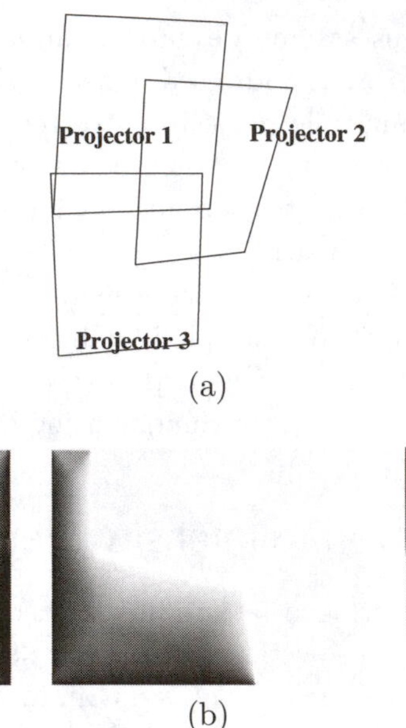

(a)

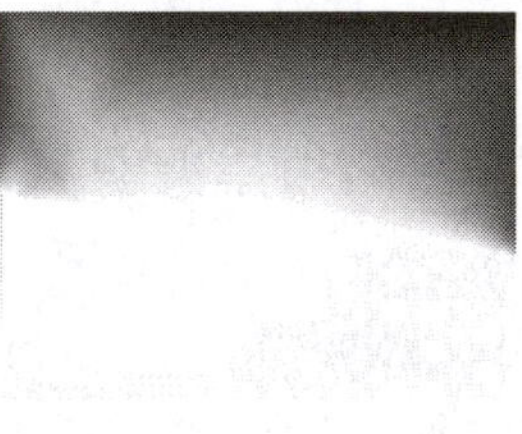

(b)

Figure 5.12. (a) The overlap position of three projectors; (b) the alpha masks created for projectors 1, 2, and 3 using feathering algorithm. (*Images reprinted from [151] © IEEE.*)

projectors. We form a convex hull H_i in the camera's image plane of the observed projector P_i's pixels. The alpha-weight $A_m(u,v)$ associated with projector P_m's pixel (u,v) is evaluated as follows:

$$A_m(u,v) = \frac{\alpha_m(m,u,v)}{\sum_i \alpha_i(m,u,v)}, \tag{5.2}$$

where $\alpha_i(m,u,v) = w_i(m,u,v) * d_i(m,u,v)$ and i is the index of the projectors observed by the camera (including projector m).

In Equation (5.2), $w_i(m,u,v) = 1$ if the camera's observed pixel of projector P_m's pixel (u,v) is inside the convex hull H_i; otherwise $w_i(m,u,v) = 0$. The term $d_i(m,u,v)$ is the distance of the camera's observed pixel of projector P_m's pixel (u,v) to the nearest edge of H_i. Figure 5.12 shows the alpha masks created for three overlapping projectors.

The feathering techniques used in photomosaics warp the source image to a single destination image. However, such a target two-dimensional panoramic image does not exist when projector images are blended on (possibly non-planar) display surfaces. Hence, we use the two-dimensional image of the display surface taken from a calibration camera and compute the intensity weights in this two-dimensional (camera) image space. Then, we transform this alpha mask in camera space into projector image space by using the same two-pass rendering techniques. With a fixed alpha mask for each projector, we simply render a textured rectangle with appropriate transparency as the last stage of the real-time rendering process.

5.3.1 Implementation

Let us look at a system that was built using the ideas described previously. The system setup includes five 1024×768 resolution SHARP LCD projectors and multiple JVC and Pulnix 640×480 resolution cameras. The projectors are ceiling mounted approximately three to four meters from the display surfaces. These projectors are casually positioned with multiple overlapping regions to produce a 180 degree field of view when the user is in the display center.

The *calibration* of the system (i.e., evaluation of camera and projector parameters and display surface estimation) is done once as a prerendering step. This is accomplished using a 0.6 meter cube that was constructed as the physical target pattern and a Dell NT workstation equipped with OpenGL graphics, Matrox Meteor II frame grabbers, and MATLAB software. The equipment is first used to capture the images of the physical target pattern and calibrate the cameras. Next, the workstation performs the structured-light projection and analysis, controlling one projector and a stereo camera pair at a time. The stereo correspondences acquired by projecting structured light form the data set needed for projector calibration, display surface reconstruction and unification, mesh generation, and alpha mask generation. The actual processing for these steps is done offline using MATLAB.

The required sampling density of the structure-light patterns depends on the complexity of the display surfaces and the need to accurately locate the edges of overlapping projectors for alpha mask generation. For our purposes, we used sampling density of every 8th and every 32nd display pixel. By binary encoding the structure-light, this process can be parallelized, and we are able to project and recover 16×12 correspondence points simultaneously. The complete operation for display surface recovery and light projector parameter estimation takes approximately 15 minutes per projector at the highest sampling density and less than one minute for the lower sampling density.

A moving user is tracked using an Origin Instruments' DynaSight infrared tracker [126]. The user wears a set of infrared LED beacons provided

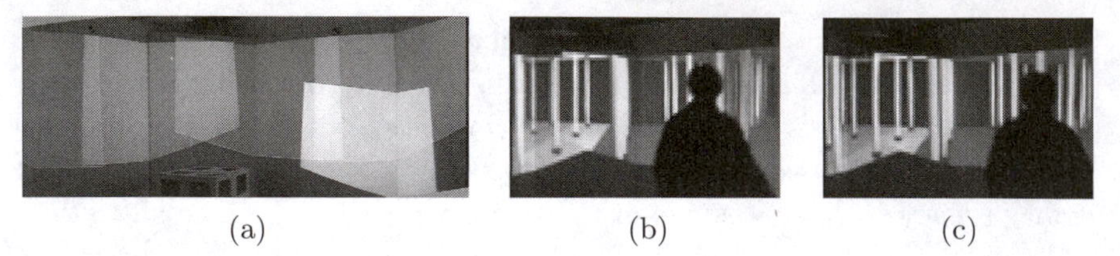

(a) (b) (c)

Figure 5.13. (a) Display environment with five projectors; (b) and (c) The images appear seamless and are updated as the user moves. (*Images reprinted from [151] © IEEE.*)

with the tracker. Tracker readings are acquired and processed (low-pass filtered and transformed into the world coordinates) by a Dell NT workstation before being dispatched in a network packet to the SGI image generation host.

The graphics rendering is done on an SGI InfiniteReality2 for each projector using the OpenGL API. While our rendering pipeline has additional computational cost due to the image warping steps, this cost is fixed and is independent of the rendered scene complexity.

Figure 5.13 shows our setup with five projectors forming a seamless panoramic image.

5.3.2 Special Cases

We presented a general solution for creating a multi-projector large area display where the display surface can be irregular and the images are correct for a head-tracked user. The techniques for this most general case can be simplified under certain conditions such as (1) when the viewer is static rather than moving, (2) when the display surface is known to be planar, or (3) the geometry of the display surface is the same as the virtual object being displayed. Let us look at the issues in a system built for a static user. Case (2) was described in the earlier part of this chapter and Case (3) is described later in the chapter.

Static user. Display systems with only a single "sweet spot" are commonly used because either the application guarantees that the user will always stay in a single location (e.g., flight simulator) or that many people will view the images simultaneously from or near the correct position, as in domed displays such as the Omnimax [105]. The relationship between desired image and projected image for each projector (i.e., the viewer-to-display mapping function) needs to be computed only once and subsequently remains fixed for that location.

This mapping function can be obtained directly by using a camera to imitate the viewer at a given location. The camera observes points illuminated by projectors in the display environment, establishing viewer-to-display correspondences. A detailed implementation of this method is described in Raskar [149]. Using this technique, the rendering process implements the two stages of the rendering framework: (1) compute the desired image and load it into texture memory and (2) warp the texture via the viewer-to-display mapping to produce the correct imagery. Intensity blending of overlapping projectors is handled as in Section 5.2.1. This special case avoids explicitly solving for three-dimensional parameters, but limits the user to one position in the display environment.

5.4 Quadric Curved Displays

Curved screens are increasingly being used for high-resolution visualization environments. We describe a parametric technique to display images on curved quadric surfaces such as those with spherical or cylindrical shape. We exploit a quadric image transfer function and show how it can be used to achieve sub-pixel registration while interactively displaying two- or three-dimensional data sets for a head-tracked user.

5.4.1 Basics

We present the notation and basics of quadric transfer. Mapping between two arbitrary perspective views of an opaque quadric surface Q in three-dimensions can be expressed using a quadric transfer function Ψ. While planar homography transfers can be computed from four or more pixel correspondences, quadric transfer requires nine or more correspondences. The quadric transfer can be defined in a closed form using the three-dimensional quadric surface Q and additional parameters that relate perspective projections of the two views. The quadric transfer in our case means image transfer from first view to the second view.

The quadric Q is a surface represented by a 4×4 symmetric matrix, such that three-dimensional homogeneous points X (expressed as a 4×1 vector) that lie on the surface satisfy the quadratic constraint,

$$X^T Q X = 0.$$

The quadric Q has 9 degrees of freedom corresponding to the independent elements of the matrix. The matrix is symmetric and defined up to an overall scale.

The homogeneous coordinates of the corresponding pixels x in the first view and x' in the second view are related by

$$x' \cong Bx - \left(q^T x \pm \sqrt{(q^T x)^2 - x^T Q_{33} x} \right) e.$$

From pixel correspondences (x, x'), we compute the 21 unknowns: 10 for the unknown 3D quadric Q, eight for a 3×3 homography matrix B, and three more for the epipole e in homogeneous coordinates. The epipole e is the image of the center of projection of the first view in the second view. The sign $\cong$ denotes equality up to scale for the homogeneous coordinates. Matrix Q is decomposed as follows:

$$Q = \begin{bmatrix} Q_{33} & q \\ q^T & d \end{bmatrix}.$$

Thus, Q_{33} is the top 3×3 symmetric submatrix of Q and q is a 3-vector. $Q(4,4)$, or d, is non-zero if the quadric does not pass through the origin (i.e., the center of projection of the first view). Hence, it can be safely assigned to be 1.0 for most display surfaces. The final two-dimensional pixel coordinate for homogeneous pixel x' is $(x'(1)/x'(3),\ x'(2)/x'(3))$.

The transfer described by 21 parameters has four dependent parameters [169, 203]. This ambiguity is removed [158] by defining

$$A = B - eq^T, \qquad E = qq^T - Q_{33},$$

so that,

$$x' \cong Ax \pm \left(\sqrt{x^T E x} \right) e.$$

The equation $x^T E x = 0$ defines the outline conic of the quadric in the first view. (The outline conic can be geometrically visualized as the image of the silhouette or the points on the surface where the view rays are locally tangent to the surface (e.g., the elliptical silhouette of a sphere viewed from outside the sphere.)) A is the homography via the polar plane between the first and the second view. The ambiguity in relative scaling between E and e is removed by introducing a normalization constraint, $E(3,3) = 1$. The sign in front of the square root is fixed within the outline conic in the image. The sign is easily determined by testing the equation above by plugging in coordinates for one pair of corresponding pixels.

Note that the parameters of the quadric transfer $\Psi = \{A, E, e\}$ can be directly computed from nine or more pixel correspondences in a projective coordinate system. So it is tempting to follow an approach similar to estimating planar homography for planar displays without computing any Euclidean parameters. However, in practice, it is difficult to robustly estimate the epipolar relationship in many cases. The discussion below is limited to a linear estimation but non-linear refinement is highly recommended [158].

Approach. All registration information is calculated relative to a stereo camera pair. We assume that the stereo camera pair can see the entire three-dimensional surface. One of the cameras is arbitrarily chosen as the origin. The cameras here are used to determine only the three-dimensional points on the display surface and not for any color sampling. Hence, any suitable three-dimensional acquisition system can be used. The following is an outline of the technique.

During preprocessing, the following steps are performed:

- For each projector i
  - Project structured light pattern with projector
  - Detect features in stereo camera pair and reconstruct three-dimensional points on the display surface
- Fit a quadric Q to all the three-dimensional points detected
- For each projector i
  - Find its pose with respect to the camera using the correspondence between projector pixels and three-dimensional coordinates of points they illuminate
  - Find the quadric transfer, Ψ_{0i}, between the camera and projector i
  - Find intensity blending weights, Φ_i, in overlap regions

At run time, the rendering for two-dimensional images or three-dimensional scenes follows these steps.

- Read head tracker and update quadric transfer Ψ_{h0} between virtual view and camera
- Read the input image in the two-dimensional video or compute the input image by rendering a three-dimensional scene from the virtual viewpoint

- For each projector i
  - Pre-warp input image into projector's frame buffer using quadric transfer, $\Psi_{0i} \circ \Psi_{h0}$
  - Attenuate pixel intensities with blending weights, Φ_i

5.4.2 Calibration

The goal is to compute the parameters of quadric transfer $\Psi_{0i} = \{A_i, E_i, e_i\}$ so that the projected images are geometrically registered on the display surface. The method to calculate quadric transfer parameters directly from pixel correspondences involves estimating the 4×4 quadric matrix Q in 3D [169, 34] using a triangulation of corresponding pixels and a linear method. If the internal parameters of the two views are not known, all the calculations are done in projective space after computing the epipolar geometry (i.e., the epipoles and the fundamental matrix). We describe an approach based on Euclidean parameters.

Quadric surface. We use a rigid stereo camera pair C_0 and C'_0 for computing all the geometric relationships. We arbitrarily choose one of the cameras to define the origin and coordinate system. We calibrate the small baseline stereo pair with a small checkerboard pattern [211]. Note that the cameras do not need to be near the sweet-spot in this setup which is an important difference with respect to some of the non-parametric approaches.

The stereo pair observes the structured patterns projected by each projector (Figure 5.14) and using triangulation computes a set of N three-dimensional points $\{X_j\}$ on the display surface. The quadric Q passing through each X_j is computed by solving a set of linear equations $X_j^T Q X_j = 0$ for each three-dimensional point. This equation can be written in the form

$$\chi_j \, V = 0,$$

where χ_j is a 1×10 matrix which is a function of X_j only and V is a homogeneous vector containing the distinct independent unknown variables of Q. With $N \geq 9$, we construct a $N \times 10$ matrix X and solve the linear matrix equation

$$\mathsf{X}V = 0.$$

Given points in general position, the elements of V (and hence Q) are the one-dimensional null-space of X.

Figure 5.14. Images captured by the 640×480 resolution camera during calibration. The resolution of each projector is significantly higher at 1024×768 and yet is captured in only a small part of the camera view.

Projector view. In addition to the quadric Q, we need to estimate the internal and external parameters of each projector with respect to the camera origin. We use the correspondence between the projector pixels and coordinates of the three-dimensional points they illuminate to compute the pose and internal parameters.

5.4.3 Camera to Projector Transfer

The idea is to use the perspective projection parameters of the camera along with the approximate projection matrix of the projector to find the camera to projector quadric transfer using linear methods. We then refine the solution using non-linear optimization.

The quadric transfer parameters $\Psi_{0i} = \{A_i, E_i, e_i\}$ between camera and projector i are easy to calculate from Q, camera projection matrix $[\, I \mid 0 \,]$ and projector projection matrix $[\, P_i \; | e_i]$:

$$A_i = P_i - e_i q^T, \qquad E_i = qq^T - Q_{33}.$$

The parameters found by the linear method can be used as an initial estimate for nonlinear minimization to refine the results.

5.4.4 Rendering

The rendering involves a two-step approach. For two-dimensional data, we extract the appropriate input image. For three-dimensional scenes, we first render the three-dimensional models from the head-tracked viewer's viewpoint. In the second step, the resultant image is then warped using the quadric image transfer into the projector image space.

5.4.5 Virtual View

When three-dimensional scenes are displayed on a curved screen, the images are perspectively correct from only a single point in space. We track this three-dimensional location, the virtual viewpoint, with a head-tracking system and update the view as the user moves.

We recalculate the quadric transfer Ψ_i between the virtual view image space and projector i frame buffer by cascading two transfers $\Psi_{0i} \circ \Psi_{h0}$, where Ψ_{0i} is calculated during preprocessing and Ψ_{h0} is updated as the user moves. Ψ_{h0} is calculated using a linear method from Q and the projection matrices of the camera and the virtual view.

5.4.6 Display Region

The view frustum for the virtual view is defined using the head-tracked position and the extents of an oriented bounding box (OBB) around the display surface. The look-at vector is from the virtual viewpoint toward the center of the OBB. (Note, the view frustum is only used for computing the VirtualViewProjection matrix and Ψ_{h0}, and not for rendering.)

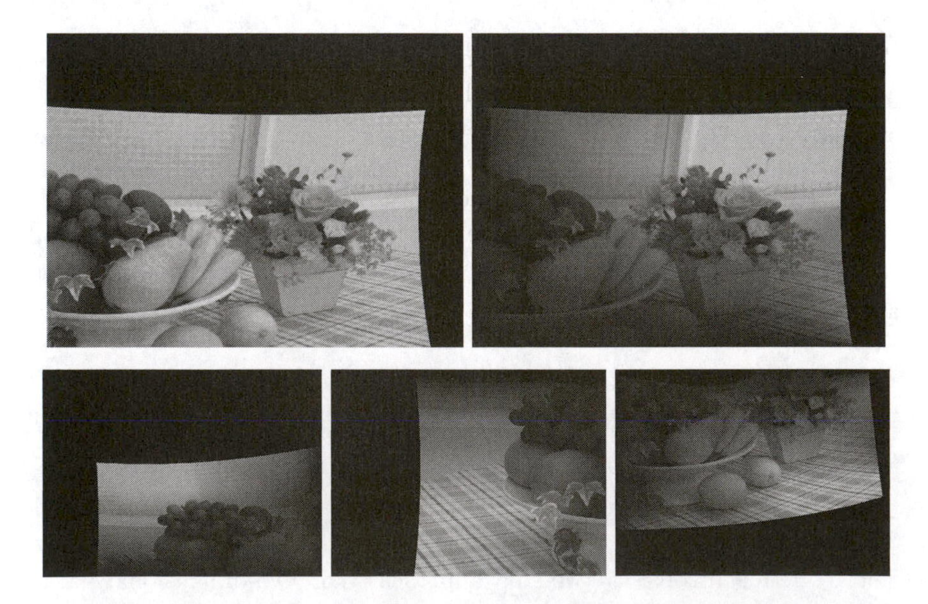

Figure 5.15. Top row: image in top-right projector's frame buffer, before and after attenuation with alpha map Φ_i; Bottom row: Three other projectors with intensity correction. Note the outer black areas which are automatically generated after quadric transfer. (*Images reprinted from [158] © ACM.*)

We crop the view frustum to an aesthetic shape such as a disk or a screen space rectangle. For three-dimensional applications, we draw a set of black quadrilaterals to cut out the areas outside the desired display region. For example, for a rectangular view, the viewport is made by four large quadrilaterals near the outer edge of the viewport in the frame buffer. The black quadrilaterals along with rest of the three-dimensional models get rendered and warped as described below (see also Figure 5.15). For two-dimensional applications, areas outside the input image to be displayed are considered black.

5.4.7 Image Transfer Using a Single-Pass Rendering

We present a single-pass rendering solution to prewarp rendered images of a three-dimensional scene before projection on a curved screen. A single-pass solution avoids the costly post-rendering warp, and it also eliminates the aliasing artifacts common in texture mapping of limited resolution input images. A single-pass solution is possible due to the parametric approach that does not require a look-up table and involves only a small number of parameters.

Given a three-dimensional vertex M in the scene to be rendered, we find its screen space coordinates m in the virtual view. Then, we find the transferred pixel coordinate m_i in the frame buffer of projector i, using the quadric transfer $\Psi_i = \{A_i, E_i, e_i\}$. The polygons in the scene are then rendered with vertices M replaced with vertices m_i. Thus, the rendering process at each projector is the same. Each projector's frame buffer automatically picks up the appropriate part of the virtual view image, and there is no need to explicitly figure out the extents of the projector.

- At each projector, i
  - For each vertex, M
    - Compute pixel m via VirtualViewProjection(M)
    - Compute warped pixel m_i via quadric transfer $\Psi_i(m)$
  - For each triangle T with vertices $\{M^j\}$
    - Render triangle with two-dimensional vertices $\{m_i^j\}$

There are two issues with this approach. First, only the vertices in the scene, but not the polygon interiors, are accurately prewarped. Second, visibility sorting of polygons needs special treatment.

After quadric transfer, the edges between vertices of the polygon, theoretically, should map to second-degree curves in the projector frame buffer. But scan conversion converts them to straight line segments between the warped vertex locations. This problem will not be discernible if only a single projector is displaying the image. But, overlapping projectors will create individually different deviations from the original curve and, hence, the edges will appear misregistered on the display screen. Therefore, it is necessary to use a sufficiently fine tessellation of triangles. Commercial systems are already available that tessellate and predistort the input models on the fly [43, 72, 77] so that they appear straight in a perspectively correct rendering on the curved screen. So our method is compatible with fine tessellation provided by such systems. Predistortion of the scene geometry in commercial systems is used to avoid the two-pass rendering, which involves a texture-mapping result of the first pass. In our case, instead of predistorting the geometry, we predistort the image space projection. Our approach, arguably, is more practical thanks to the programmable vertex shaders now available (see Appendix A).

Scan conversion issues. When pixel locations in the projection of a triangle are warped, information needs to be passed along so that the depth buffer will create appropriate visibility computations. In addition, for perspectively correct color and texture coordinate interpolation, the appropriate "w" values need to be passed. Our solution is to post-multiply the pixel coordinates with w:

$$\begin{aligned} m(x, y, z, w) &= \text{VirtualViewProjection}(M(X)), \\ m'_i(x'_i, y'_i, w'_i) &= \Psi_i(m(x/w, y/w), 1), \\ m_i(x_i, y_i, z_i, w_i) &= [wx'_i/w'_i, wy'_i/w'_i, z, w]. \end{aligned}$$

Thus, the z and w values are preserved using explicit assignment. Homogeneous coordinates of m_i have the appropriate final pixel coordinate $(x_i/w_i, y_i/w_i) = (x'_i/w'_i, y'_i/w'_i)$ due to quadric transfer along with original depth, $(z_i/w_i) = z/w$, and $w_i = w$ values. The corresponding code is in Appendix A and Figure 5.16 shows a photo of the dome displaying a sample interactive animation that is synchronized and displayed on the dome with four overlapping projectors.

For rendering two-dimensional input images, we densely tessellate the virtual view image space into triangles and map the image as a texture on these triangles. Vertex m of each triangle is warped using the quadric transfer into vertex (and pixel) m_i, as above. Scan conversion automati-

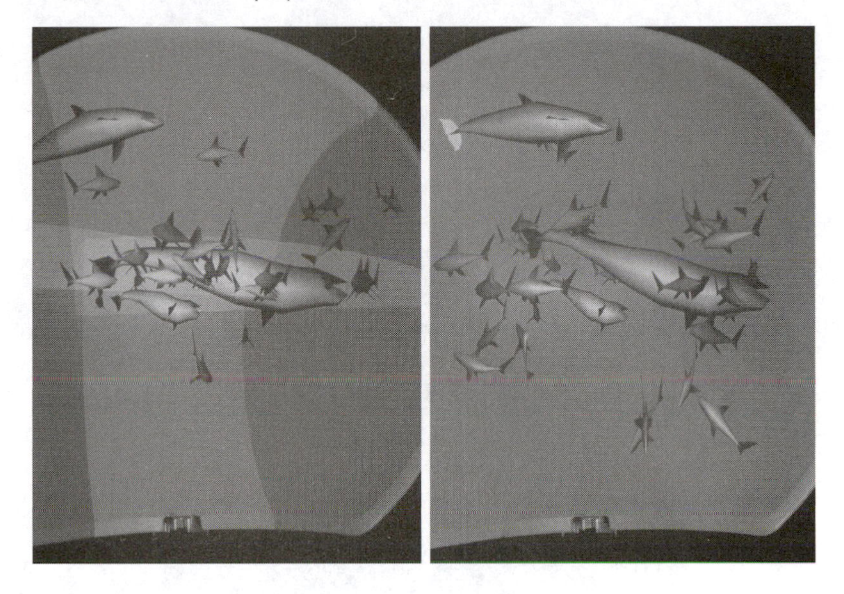

Figure 5.16. An unmodified 3D rendering application displayed with correct depth sort and texture mapping without and with intensity blending. (*Images reprinted from [158] © ACM.*)

cally transfers colors and texture attributes at m to m_i and interpolates in between. If required, two-pass rendering of three-dimensional scenes can also be achieved in this manner, by first rendering the input image due to VirtualViewProjection.

5.4.8 Intensity Blending

Pixel intensities in the areas of overlapping projectors are attenuated using alpha blending of the graphics hardware. Using the parametric equations of quadric transfer, the alpha maps are calculated robustly and efficiently.

For every projector pixel x_i in projector i, we find the corresponding pixels in projector k using the equation

$$x_k \cong \Psi_{0k}\left(\Psi_{0i}^{-1}\left(x_i\right)\right).$$

For cross-fading, pixels at the boundary of the projector frame buffer are attenuated. Hence, the weights are proportional to the shortest distance from the frame boundary. The weight assigned to pixel x_i, expressed in normalized window pixel coordinates (u_i, v_i) which are in the range $[0, 1]$, is

$$\Phi_i(x_i) \cong d(x_i) \Big/ \sum\nolimits_k d(x_k),$$

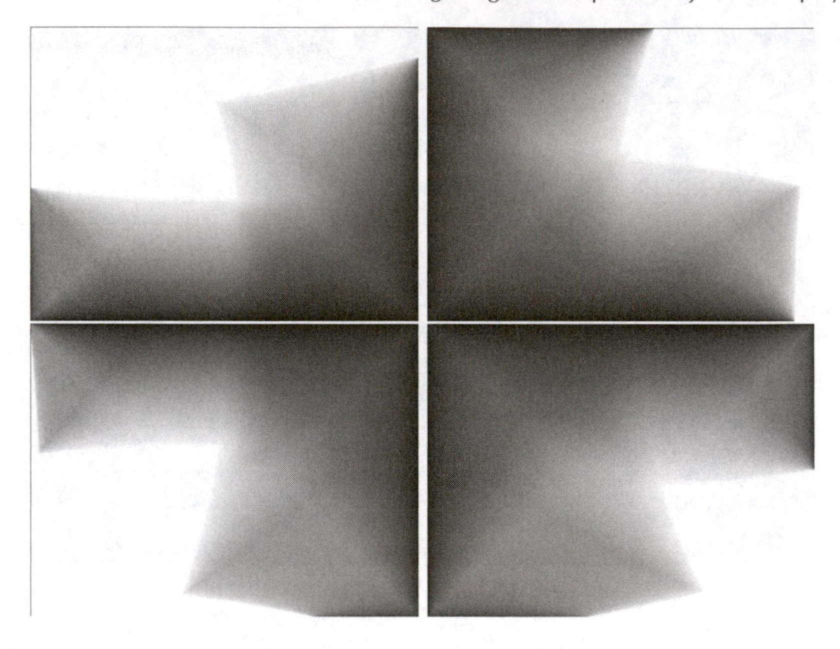

Figure 5.17. Calculated alpha maps Φ_i for intensity correction of four overlapping projectors. (*Images reprinted from [158] © ACM.*)

where $d(x)$ is $\min(u, v, 1-u, 1-v)$ if $0 \leq u, v \leq 1$, else $d(x) = 0$. Thanks to a parametric approach, we are able to compute corresponding projector pixels x_k and, hence, the weights at those locations at subpixel registration accuracy. The sum of weights at corresponding projector pixels accurately adds to 1.0.

At each projector, the corresponding alpha map Φ_i is loaded as a texture map and rendered as screen-aligned quadrilaterals during the last stage of the rendering (Figure 5.17).

5.5 Illuminating Objects

The idea of projector-based augmentation is best used when virtual objects are close to the physical objects on which they are displayed. For example, one can replace a physical object with its inherent color, texture, and material properties with a neutral object and projected imagery, reproducing the original appearance directly on the object. Furthermore the projected imagery can be used to reproduce alternative appearances, including alternate shading, lighting, and, even, animation. The approach is to effectively

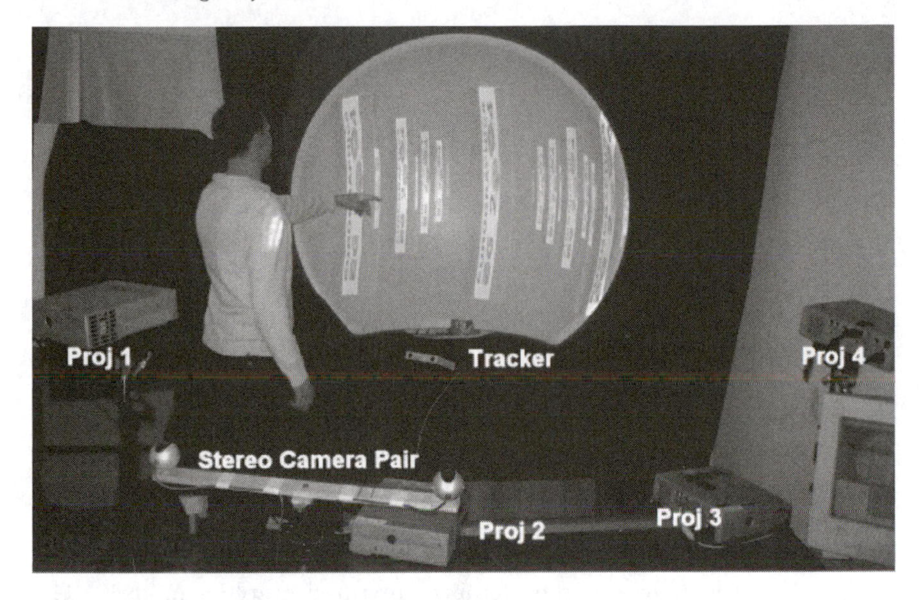

Figure 5.18. The setup of four casually installed projectors, stereo camera pair, and tracker with a concave dome. The camera pair is not near the sweet-spot. (*Images reprinted from [158]* © *ACM.*)

lift the visual properties of the object into the projector and then re-project onto a neutral surface (Figure 4.1). We use the term shader lamps to describe this mode of operation for projectors. The image-based illumination of physical objects has been explored by many. But, we believe, two main challenges have kept the previous efforts to only expensive, large scale, or one-off implementations.

1. The first problem is geometric registration which is cast as matching the projection of a single two-dimensional image with an object. The projection of a perspective device has up to 11 degrees of freedom (six external and five internal) [44]; therefore, any effort to manually achieve the registration is likely to be extremely tedious.

2. The second problem is the complete illumination of non-trivial physical objects in the presence of shadows due to self-occlusion. We illuminate the shadowed parts of the object by adding more projectors and then address the issue of merging overlapped images from multiple projectors.

With the advent of digitally-fed projectors and real-time three-dimensional graphics rendering, a new approach for image-based illumination is now possible. We approach these problems by exploiting the three-dimensional geometric relationship between the projector, display surface, and the user.

5.5.1 Authoring and Alignment

One of the important tasks in achieving compelling visualization is to create the association between the physical objects and the graphics primitives that will enhance those objects when projected. For example, how do we specify which texture image should be used for the face of a tabletop building model, or what color distribution will look better for a physical object? We need the physical object as well as its geometric three-dimensional representation and real or desired surface attributes. Many hardware and software solutions are now available to scan/print three-dimensional objects and capture/create highly detailed, textured graphics models. The authoring can also be performed interactively by "painting" directly on top of the physical objects. The result of the user interaction can be projected on the objects and also stored on the computer [152]. Ideally, a more sophisticated user interface would be used to create and edit graphics primitives of different shape, color, and texture.

To align a projector, first we position the projector approximately and then adapt to its geometric relationship with respect to the physical object. That relationship is computed by finding the projector's intrinsic parameters and the rigid transformation between the two coordinate systems. This is a classical computer vision problem [44] addressed earlier in the chapter which is solved using camera-based feedback. Here we use a different approach. We take a set of fiducials with known three-dimensional locations on the physical object and find the corresponding projector pixels that illuminate them. This allows us to compute a 3×4 perspective projection matrix up to scale, which is decomposed to find the intrinsic and extrinsic parameters of the projector. The rendering process uses the same internal and external parameters, so that the projected images are registered with the physical objects.

5.5.2 Intensity Correction

The intensity of the rendered image is modified on a per-pixel basis to take into account the reflectance of the neutral surface, the local orientation, and the distance with respect to the projector using Equation (7.2). Since the surface normals used to compute the $1/cos(\theta_P)$ correction are available only at the vertices in polygonal graphics models, we exploit the rendering pipeline for approximate interpolation. We illuminate a white diffuse version of the graphics model (or a model matching appropriate $k_u(x)$ of the physical model) with a virtual white light placed at the location of the projector lamp and render it with black fog for squared distance attenuation. The resultant intensities are smooth across curved surfaces due to shading interpolation and inversely proportional to $(d(x)^2/k_u(x)cos(\theta_P))$. To use the limited dynamic range of the projectors more efficiently, we do not illuminate surfaces with $|\theta_P| > 60$ (since, for $|\theta| \in [60, 90]$ the range $1/cos(\theta) \in [2, \infty])$. This avoids the low sampling rate of the projected pixels on oblique surfaces and also minimizes the misregistration artifacts due to any errors in geometric calibration. During the calculations to find the overlap regions, highly oblique surfaces are considered not to be illuminated by that projector.

5.5.3 Steps

During preprocessing

- Create three-dimensional graphics model, G, of the physical object
- Create three-dimensional graphics model, B, of the background
- Position the projector approximately
- Find perspective pose, P, of the projector with respect to the physical object

During run-time

- Get user location, U
- Get animation transformation, T
- Modify G's surface attributes
- Render G using the pose, P, and user location, U
- Transform B using T-1, B
- Render B using the pose, P, and user location, U
- Modify image intensity to compensate for surface orientation

5.5.4 Occlusions and Overlaps

For complete illumination, using additional projectors is an obvious choice. This leads to the more difficult problem of seamlessly merging images from multiple projectors. A naive solution may involve letting only a single projector illuminate any given surface patch. But, there are two main issues when dealing with overlapping CRT, LCD, or DLP projectors, which compel the use of feathering (or cross-fading) of intensities. The first is the lack of color equivalence between neighboring projectors [99], due to the manufacturing process and temperature color drift during their use. The second is the desire to minimize the sensitivity to small errors in the estimated geometric calibration parameters or mechanical variations.

Feathering, as described earlier, is commonly used to generate seamless panoramic photomosaics by combining several views from a single location [186]. Similar techniques are exploited in multi-projector wide field-of-view

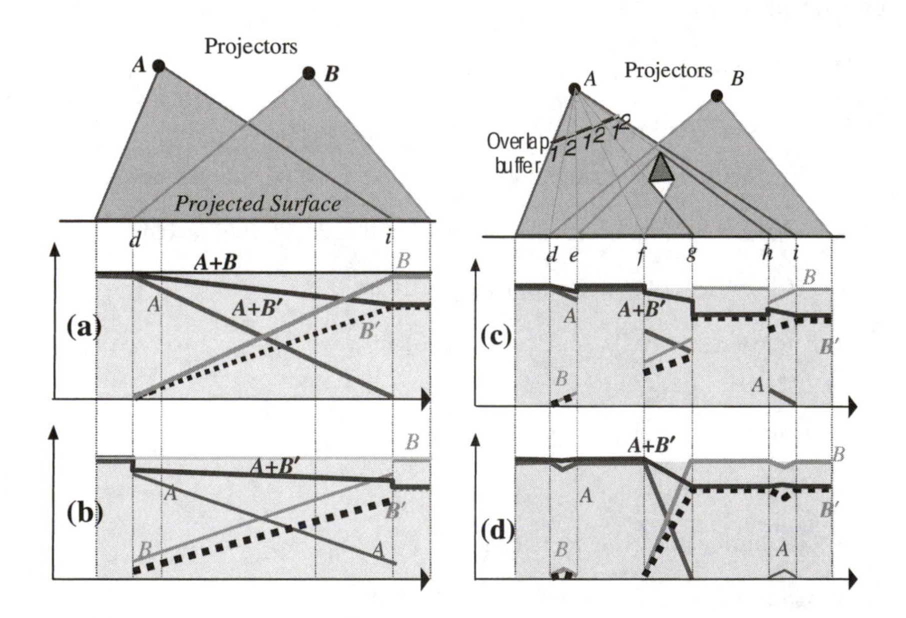

Figure 5.19. Intensity weights using feathering methods. The plots show the contribution of projectors A, B and B' and the resultant accumulation $A+B$ and $A+B'$ along the lit planar surface:(a) simple intensity ramps on planar overlap create smooth transitions; (b) weights computed using surface normals are not sufficient; (c) occluders create sharp transitions; (d) our solution, which considers depth continuity, maintains the local smoothness of intensity weights. (*Images reprinted from [155] © Springer-Verlag.*)

displays [128, 191, 151], and two-dimensional arrays of flat projections. In such cases, the overlap region is typically a (well-defined) contiguous region on the display surface as well as in each projector's frame buffer. In the algorithm used in [186, 151], the intensity of a pixel is weighted proportionally to the Euclidean distance to the nearest boundary (zero contribution) pixel of the (projected) image. The per-pixel weights are in the range $[0, 1]$. They are multiplied with the pixel intensities in the final rendered image. The pixels weights near the boundary of a source image are near zero and the pixels contribute very little, so that there is a smooth transition to the next source image. This leads to the commonly seen intensity roll-off as shown in Figure 5.19(a). Under ideal conditions and assuming color equivalence, the weight contribution of both projectors $A + B$ adds up to 1. Even when projector B's color response is different than that of A (say, attenuated-shown as B'), the resultant $A + B'$ (shown as dark solid line) transitions smoothly in the overlap region.

5.6 Summary and Discussion

We discussed four different methods for creating images for spatial projection. They include exploiting homography for planar surfaces, two-pass rendering for non-planar surfaces, quadric image transfer for quadric curved surfaces, and Shader Lamps for directly augmenting objects. In the next chapter, we take the concept of the projector as the dual of a camera further and describe a new class of visualization methods by combining optical see-through elements.

6

Generating Optical Overlays

Spatial optical see-through displays overlay the real environment with computer graphics in such a way that the graphical images and the image of the real environment are visible at the same time. In contrast to head-attached or body-attached optical see-through displays, spatial displays generate images that are aligned with the physical environment. They do not follow the users' movements but rather support moving around them. They are comparable to spatial projection displays—but do not share the opaque characteristic of such displays.

An essential component of an optical see-through display is the *optical combiner*—an optical element that mixes the light emitted by the illuminated real environment with the light produced with an image source that displays the rendered graphics. Creating graphical overlays with spatial optical see-through displays is similar to rendering images for spatial projection screens for some optical combiners. For others, however, it is more complex and requires additional steps before the rendered graphics is displayed and optically combined.

While monitors, diffuse projection screens, and video projectors usually serve as light emitting image sources, two different types of optical combiners are normally used for such displays—*transparent screens* and *half-silvered mirror beam combiners*. Rendering techniques that support creating correct graphical overlays with both types of optical combiners and with different images sources will be discussed in this chapter. We make the convention that all the following elements, such as optical combiners, image sources, observers, virtual and real environments, etc. are defined within the same *Cartesian coordinate system*. This keeps the explanations particularly simple and allows a straightforward implementation of the techniques.

6.1 Transparent Screens

Transparent screens have two main properties: they are transparent to a certain degree to allow the transmission of the image of the real environment, and they also emit the light of the rendered graphics. Observers can see directly through the screen (and through the image displayed on it) to the real environment.

In some cases, such screens are *active* and contain light-emitting elements. Examples are liquid crystal displays that are modified (by removing the opaque back light source) to enable their see-through capabilities and flexible polymers with light-emitting *transparent semi-conductors*. In other cases, external image sources, such as video or laser projectors are used to generate light that is diffused by a *transparent projection screen*. Examples include *holographic projection screens* that diffuse the light in a narrow angle to achieve an enhanced brightness for restricted viewing angles and *transparent film screens* that diffuse the light in a wide angle to support a more flexible viewing and a higher degree of transparency.

On the one hand, the use of video projectors allows building large screens that do not necessarily have to be planar. Active screens, on the

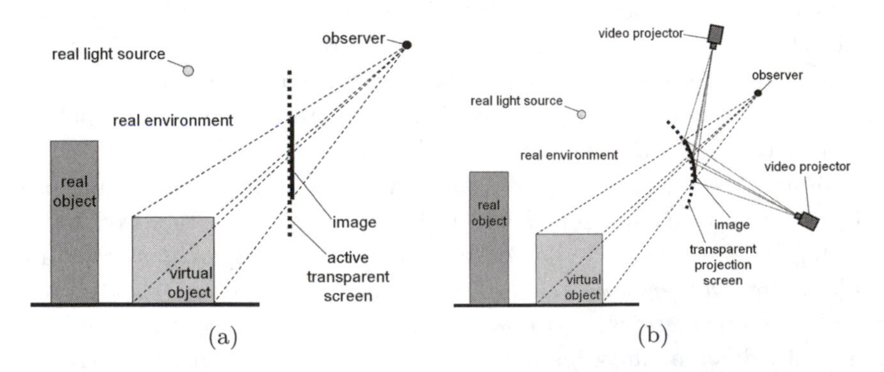

Figure 6.1. (a) Planar active transparent screen; (b) curved transparent projection screen with multiple projectors.

other hand, provide more flexibility since they are not constrained to a stationary configuration. In the future new materials, such as light emitting polymers may allow building active transparent screens that are both flexible and scalable.

Rendering for transparent screens is essentially equivalent to rendering for regular projection screens or monitors. While for planar screens an affine off-axis projection transformation is sufficient, *curvilinear image warping* is required for curved screens. For large or extremely curved screens, the image has to be composed of the contribution displayed by multiple projectors. The different pieces have to be geometrically aligned and blended to result in a consistent final image. All these techniques are explained in detail for opaque screens in Chapters 3 and 5. They are the same for transparent screens.

Sections 6.3 through 6.5 describe a rendering framework for spatial optical see-through displays that use mirror beam combiners. Most of this framework, such as refraction correction and multi-screen and multi-plane beam combiner transformations, is exactly the same for transparent screens. The main difference to the mirror beam combiner is that reflection transformations do not apply in this case. They have to be ignored if the framework is used for displays with transparent screens. If projection displays are used to generate images, the outlined screen transformation is equivalent to the rendering techniques explained in Chapters 3 and 5.

An important fact that needs to be mentioned is that the rendered image appears directly on the surface of the transparent screen. Similar to head-attached displays, this causes focus problems if the real environment to be augmented is not located in the same place as the image. For transparent screens, this can never be the case since the screen itself (and therefore the image) can never take up exactly the same space within the environment. It is not possible for our eyes to focus on multiple distances simultaneously. Thus, in extreme cases, we can only continuously shift focus between the image and the real environment, or perceive either one unfocused.

If stereoscopic graphics need to be displayed on transparent projection screens, we must pay attention to the materials used. Not all materials preserve the polarization of light that is produced by *passive stereoscopic projection* setups—especially when materials are bent to form curved screens. Transparent film screens, for instance, have the property to preserve polarization in any case—even if bent. Active stereo projection systems do not cause such problems.

6.2 Mirror Beam Combiners

Mirror beam combiners are the most common optical combiners because of their availability and low cost. If they cannot be obtained from an optics store, they can be homemade in almost any size and shape. For instance, regular float glass or Plexiglas can be coated with a *half-silvered film*, such as 3M's Scotchtint sun protection film. These materials are easily available in hardware stores. The advantage of half-silvered film is that it can be coated onto a flexible carrier, such as thin Plexiglas, that can easily be bent to build curved displays. The drawback of such a material is its non-optimal optical properties. Instead of having transmission and reflection factors of 50%, these materials normally provide factors of approximately 70% reflection and 30% transmission, due to their sun-blocking functionality. An alternative material is the so-called *spyglass* that offers better and varying transmission/reflection factors without sun-blocking layers. However, the reflective film is usually impregnated into float glass—thus it is not well suited for building curved mirror displays.

Spatial optical see-through displays that apply mirror beam combiners as optical combiners have to display the rendered graphics on a *secondary screen* that is reflected by the mirror optics. The observer sees through the optical combiner and through the reflected image at the real environment. The image that appears on the secondary screen usually should not be visible. To hide it, the secondary screen is often coated with *light directing film*, such as 3M's Light Control Film. Such a film can be used to direct the displayed image only towards the mirror optics and not to the observer. Thus, only the reflection of the displayed image can be seen and not the image itself.

In contrast to transparent screens, mirror beam combiners create an optical reflection of the secondary screen and the displayed image. There are no physical constraints in aligning the image closer to the real environment. The reflected image can even intersect with the real environment. Consequently, the focus problem of transparent screens can be improved although not completely solved.

6.3 Planar Mirror Beam Combiners

To render a three-dimensional graphical scene on spatial optical see-through displays applying planar mirror beam combiners requires neutralizing the optical transformations that are caused by half-silvered mirrors: reflection

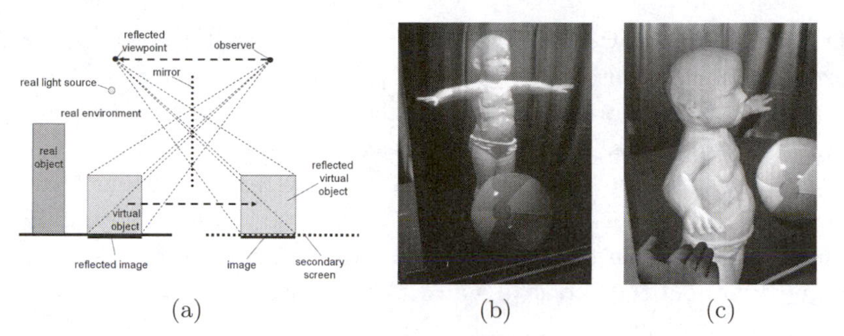

Figure 6.2. (a) Affine reflection transformation for planar mirror beam combiner; (b) a virtual baby observed through a large beam combiner reflecting a horizontal projection screen; (c) due to parallax effects, virtual objects (e.g., the baby's arm) can appear in front of the mirror. (*Images (b) and (c) reprinted from [10] © MIT Press.*)

and refraction. The goal is to render the graphical content and display it on the secondary screen in such a way that its reflection appears perspectively correct and aligned with the real environment.

6.3.1 Reflection

A planar mirror beam combiner divides the environment into two subspaces: the one that contains the observer and the secondary screen, and the one that contains the real environment to be augmented and the physical light sources that illuminate it.

Note that from a geometric optics point of view, the real environment behind the mirror equals the mirror's *image space* (i.e., the reflection that appears behind the mirror by looking at it).

Virtual objects that consist of graphical elements (such as geometry, normal vectors, textures, clipping planes, virtual light sources, etc.) are defined within the same global world coordinate system in which the real environment, the observer, the secondary screen, and the mirror beam combiner are located. In contrast to conventional virtual reality scenarios, this coordinate system actually exceeds the boundaries of the display screen and extends into the surrounding real environment (see Figure 6.2).

The virtual objects are either defined directly within the real environment or they are transformed to it during an *object registration process*.

We now consider the mirror and compute the reflection of the observer's physical eye locations (as well as possible virtual headlights). We then apply the inverse reflection to every graphical element that is located within the real environment. In this manner these graphical elements are transformed and can be projected and displayed on the secondary screen. The displayed image is optically reflected back by the mirror into the real environment (or optically, into the mirror's image space).

When the setup is sufficiently calibrated, the real environment and the mirror's image space overlay exactly. The virtual objects appear in the same position within the image space as they would within the real environment without the mirror (if a direct display possibility was given within the real environment).

Note that the *reflection transformation* of planar mirrors is a *rigid-body transformation*, and preserves all properties of the transformed geometry.

As discussed in Chapter 2, with known plane parameters of the mirror beam combiner within the world coordinate system, a point in three-dimensional space can be reflected by

$$p' = p - 2(np + d)n, \tag{6.1}$$

where p' is the reflection of p over the mirror plane $[n, d] = [a, b, c, d]$.

This is equivalent to multiplying p with the 4×4 *reflection matrix*

$$R = \begin{bmatrix} 1-2a^2 & -2ab & -2ac & -2ad \\ -2ab & 1-2b^2 & -2bc & -2bd \\ -2ac & -2bc & 1-2c^2 & -2cd \\ 0 & 0 & 0 & 1 \end{bmatrix}.$$

Note that $R = R^{-1}$.

The reflected viewpoint e' of the observer (which, for instance, is head-tracked) can be computed using Equation (6.1) or by multiplying the original viewpoint e with the reflection matrix.

The inverse reflection of the virtual scene that is located within the real environment is simply computed from its reflection with respect to the mirror plane. Since we assume that the real environment and the mirror's image space exactly overlay, we can also assume that the reflection of the graphical elements located within the real environment results in the inverse reflection of the image space; that is, they are transformed to their corresponding positions on the observer's side of the mirror and can be displayed on the secondary screen. Consequently, the additional model transformation (i.e., the inverse reflection of the scene) is achieved by multiplying the reflection matrix onto the current model-view matrix of the transformation pipeline (between scene transformation and view transformation).

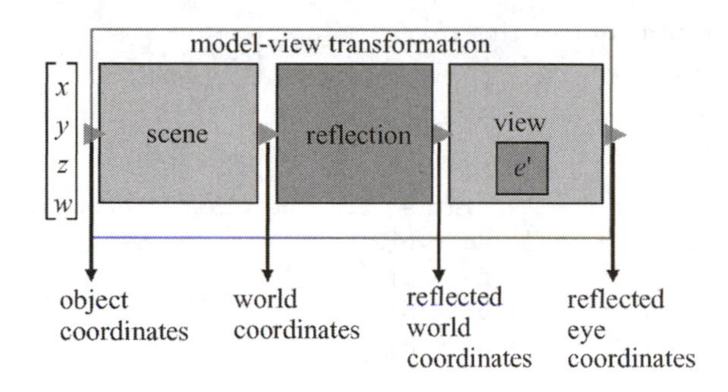

Figure 6.3. The integration of reflection transformations into the rendering pipeline.

Consequently, geometry that is registered to the real environment is first transformed from object coordinates into the coordinates of the world coordinate system, then into *reflected world coordinates*, and finally into *reflected eye coordinates*. After the model-view transformation has been applied, the rendering pipeline is continued in the normal way (see Chapter 2); the reflected eye coordinates are off-axis projected into clip coordinates, then, after the perspective division, transformed into normalized device coordinates, and finally (via the viewport transformation) mapped into window coordinates.

By applying the reflection matrix, every graphical element is reflected with respect to the mirror plane. A side effect of this is that the order of reflected polygons is also reversed (e.g., from counterclockwise to clockwise) which, due to the wrong front-face determination, results in a wrong rendering (e.g., lighting, culling, etc.). This can easily be solved by explicitly reversing the *polygon order*. Note that transformations and rendering have to be done for both viewpoints (left and right) if stereoscopic rendering is activated.

The reflection transformation can be entirely executed by accelerated graphics hardware that provides a fixed function rendering pipeline. The following OpenGL code fragment can be used to configure the rendering pipeline with reflection transformation:

```
...

// mirror plane = a,b,c,d
// viewpoint = e[0..2]
// world coordinate system = x/y-axes horizontal plane,
```

```
// z-axis points up
float e[3], e_[3];
float NP;
float R[16];

// compute reflected viewpoint e_ from original viewpoint e
NP = a * e[0] + b * e[1] + c * e[2];
e_[0]= e[0] - 2.0 * (NP + d) * a;
e_[1]= e[1] - 2.0 * (NP + d) * b;
e_[2]= e[2] - 2.0 * (NP + d) * c;

// set up reflection matrix
R[0]  = 1-2*a*a; R[1]  =-2*a*b;   R[2]  =-2*a*c;   R[3] =0;
R[4]  =-2*a*b;   R[5]  = 1-2*b*b; R[6]  =-2*b*c;   R[7] =0;
R[8]  =-2*a*c;   R[9]  =-2*b*c;   R[10] = 1-2*c*c; R[11]=0;
R[12] = -2*a*d;  R[13] =-2*b*d;   R[14] =-2*c*d;   R[15]=1;

// configure rendering pipeline
glViewport(...);
glMatrixMode(GL_PROJECTION);
glLoadIdentity();
glFrustum(...);
glMatrixMode(GL_MODELVIEW);
glLoadIdentity();
gluLookAt(e_[0], e_[1], e_[2], e_[0], e_[1], 0, 0, 1, 0);
glMultMatrixf(R);

// reverse polygon order
glFrontFace(GL_CW + GL_CCW - glGetIntegerv(GL_FRONT_FACE));

// draw scene with scene transformation

...
```

An alternative to a reflection of the entire geometry and viewpoint is to set up a viewing frustum that is defined by the reflected image plane instead of the image plane on the physical secondary screen. In Figure 6.2(a), the secondary screen is reflected from the right sides of the beam combiner to its left side. The unreflected viewing frustum (right side) can be used if defined relative to the reflected image plane (left side). For this, the exact position and orientation of the reflected image plane has to be known. They can also be derived from the parameters of the mirror plane.

In addition, the displayed image has to be reflected over the on-screen axis that is perpendicular to the two-dimensional intersection vector (i_x, i_y) of the mirror plane and the secondary screen. In a simple example, this in-

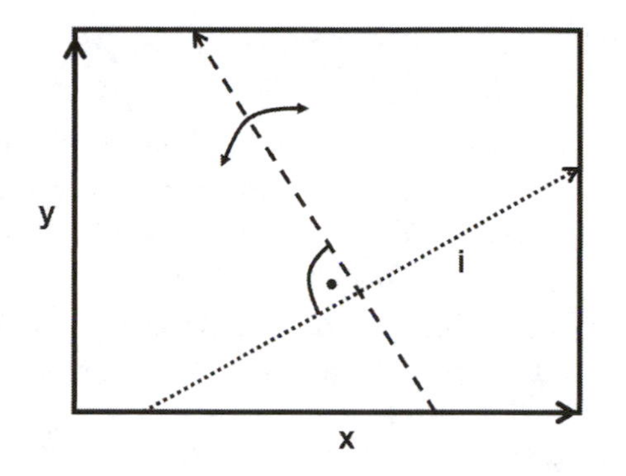

Figure 6.4. Simple example for reflection over on-screen axis.

tersection equals the x-axis (or the y-axis) of the screen coordinate system, as illustrated in Figure 6.4. In this case, an additional scaling transformation can be applied after the projection transformation that causes the fragments to be reflected within the normalized device space. For instance, `glScale(i_x, i_y, 0)` with $i_x = -1, i_y = 0$ causes a reflection over the y-axis for the case that the intersection of the mirror with the screen is on the x-axis (all in normalized screen/device coordinates). For an arbitrary intersection vector, however, a scaling transformation alone is not sufficient. An additional rotation transformation is required that first aligns the intersection vector with either the x- or the y-axis in the screen coordinate system. Then the scaling is transformed along this principle axis (for example over the x-axis). Finally the reverse rotation transformation has to be applied to produce the correct effect.

It is important to apply this reflection transformation after the projection (e.g., before `glFrustum()` in OpenGL's reversed matrix stack notation) since it has to be the final transformation, and it must not influence other computations, such as lighting and depth culling.

An interesting optical effect can be observed by applying mirrors in combination with stereoscopic secondary screens; convex or planar mirrors can optically only generate virtual images. However, in combination with stereoscopic graphics and the effects caused by stereopsis, virtual objects can appear in front of the mirror optics (Figure 6.2(c)). We refer to this effect as *pseudo real images*. In nature, real images of reflected real objects can only be generated with concave mirrors. Note that a restricted direct manipulative interaction with pseudo real images in front of the mirror optics is supported.

6.3.2 Refraction

From an optics point of view, the glass or Plexiglas carriers used for optical combiners (i.e., mirror beam combiner or transparent screens) are lenses that cause refraction distortion. A homogeneous medium that is bound by two plane-parallel panels is referred to as a planar lens. The refraction distortion is small and can be neglected for thin planar lenses, but has to be corrected for thick plates. Note that the following techniques are also applicable for transparent screens that suffer from refraction distortion.

All virtual objects that are registered to the real environment are virtual points that are not physically located behind the optical combiner. They are images that are created by the optical combiner. Mirror beam combiners are usually *front surface mirrors* (i.e., the mirror film coated on the side of the carrier that faces the image source and the observer) while transparent screens can be front projected causing the same registration problem: the displayed image is not physically refracted by the optical combiner. However, the transmitted light which is emitted by the real environment and perceived by the observer, is refracted. Consequently, the transmitted image of the real environment cannot be precisely registered to the reflected virtual objects, even if their geometry and alignment match exactly within our world coordinate system. Unfortunately refraction cannot be undistorted by a rigid-body transformation, but approximations exist that are sufficient for augmented reality display types.

All optical systems that use any kind of see-through element have to deal with similar problems. For head-mounted displays, aberrations (optical distortion) caused by refraction of the integrated lenses are mostly assumed to be static [4]. Due to the lack of eye-tracking technology as a component of head-mounted displays, the rotation and the exact position of the eyeballs as well as the movement of the optics in front of the observer's eyes is not taken into account. Thus, a static refraction distortion is precomputed for a *centered on-axis* optical constellation with methods of *paraxial analysis*. The result is stored in a *two-dimensional look-up table*. During rendering, this look-up table is referenced to transform rendered vertices on the image plane before they are displayed. Rolland and Hopkins [162], for instance, describe a polygon-warping technique that uses the look-up-table to map projected vertices of the virtual objects' polygons to their predistorted location on the image plane. This approach requires subdividing polygons that cover large areas on the image plane. Instead of predistorting the polygons of projected virtual objects, the projected image itself can be predistorted, as described by Watson and Hodges [200], to achieve a higher rendering performance.

For spatial see-through displays, however, aberrations caused by refraction are dynamic, since the optical distortion changes with a moving viewpoint that is normally off-axis and off-centered with respect to the optical combiner.

For some *near-field displays*, such as reach-in displays, the displacement caused by refraction can be estimated [204]. An estimation of a constant refraction might be sufficient for near-field displays with a fixed viewpoint that use a relatively thin beam combiner. For larger spatial optical see-through setups that consider a head-tracked observer and apply a relatively thick beam combiner, more precise methods are required. In the following explanation, we want to consider such a display constellation.

Since we cannot predistort the refracted transmitted image of the real environment, we artificially refract the virtual scene before it is displayed, in order to make both images match. In contrast to the static transformation that is assumed for head-mounted displays, refraction is dynamic for spatial optical see-through displays and cannot be precomputed. In general, refraction is a complex curvilinear transformation and does not yield a *stigmatic mapping* (the intersection of multiple light rays in a single image point) in any case; we can only approximate it.

In the case of planar lenses, light rays are refracted twice—at their entrance points and at their exit points. This is referred to as *in-out refraction*. In the case of planar lenses, the resulting out-refracted light rays have the same direction as the corresponding original rays, but they are shifted by the amount Δ parallel to their original counterparts. Due to refraction, an object that is physically located at the position p_o appears at the position p_i. To maintain registration between a virtual object (that will appear at position p_o) and the corresponding real object (that will appear at position p_i), the virtual object has to be repositioned to p_i.

The offset Δ can be computed as

$$\Delta = t\left(1 - \frac{\tan\alpha_t}{\tan\alpha_i}\right),$$

where t is the thickness of the planar lens, α_i is the angle between the plane normal of the plate and the line of sight, and α_t is given by *Snell's law of refraction*:

$$\eta_1 \sin\alpha_i = \eta_2 \sin\alpha_t.$$

It is constrained to the following boundaries:

$$\lim\left(\alpha_i \rightarrow \frac{\pi}{2}\right) \Rightarrow \Delta = t,$$

$$\lim\left(\alpha_i \rightarrow 0\right) \Rightarrow \Delta = t\left(1 - \frac{\sin\alpha_t}{\sin\alpha_i}\right) = t\left(1 - \frac{1}{\eta_2}\right) = \text{const.}$$

A special case of the previous transformation is an on-axis situation where the incidence angle is perpendicular to the lens and parallel to the optical axis (i.e., $\alpha_i = 0$). In this situation, the offset equation can be simplified to $\Delta = t\,(1 - 1/\eta_2)$.

In the case of an optical combiner that is located in air, the refraction index η_1 is equal to 1. The refraction index η_2 is that of the carrier's base material (e.g., glass or Plexiglas).

A simple solution to simulate refraction is the assumption that, despite its optical nature, it can be expressed as a rigid-body transformation. This is a common approximation used by the three-dimensional computer graphics community to render realistic-looking refraction effects in real time. In beam-tracing [61], for instance, it was assumed that, considering only *paraxial rays* (entering light rays that are exactly or nearly perpendicular to the refracting plane), objects seen through a polygon with a refraction index of η appear to be η times their actual distance. This is because light travels slower in denser materials by precisely this factor. For this approximation, the incidence angles of the optical line of sight are not taken into account but, instead, a constant incidence angle of $\alpha_i = 0$ to the optical axis is assumed. This, however, covers on-axis situations only. For off-axis situations that occur with spatial optical see-through displays, the incident

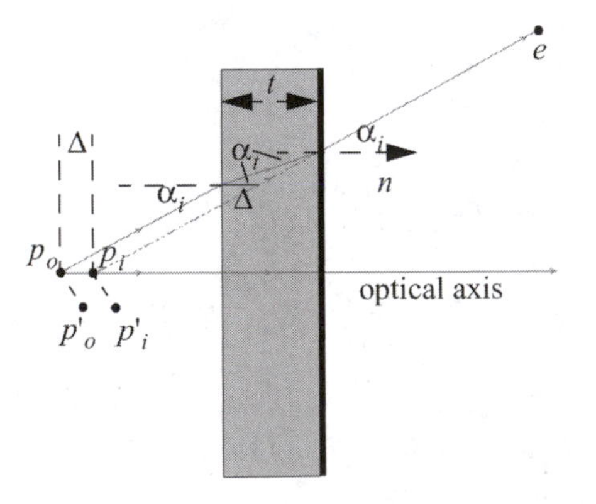

Figure 6.5. Off-axis refraction transformation for planar lenses.

angle has to be considered. As a rigid-body transformation, refraction can be expressed by a homogeneous 4×4 matrix:

$$F = \begin{bmatrix} 1 & 0 & 0 & \Delta a \\ 0 & 1 & 0 & \Delta b \\ 0 & 0 & 1 & \Delta c \\ 0 & 0 & 0 & 1 \end{bmatrix}.$$

The advantage of this simple translation along the plate's optical axis is that it can be integrated into a fixed function rendering pipeline and can be carried out by graphics hardware. The drawback, however, is that this is only a rough approximation for refraction distortion, since every vertex is translated by the same amount Δ which is computed from a common incident angle (such as the angle between the viewing direction and the optical axis, for instance). The curvilinear characteristics of optical refraction and the individual incident angles of the viewing vectors to each scene vertex are not taken into account.

Programmable rendering pipelines allow per-vertex transformations directly on the graphics hardware. Thus the correct *curvilinear refraction transformation* can be implemented as a vertex shader, rather than as a rigid-body transformation expressed by a homogeneous matrix.

The following Cg shader fragment can be used to configure a programmable rendering pipeline with the refraction transformation:

```
...

// viewpoint = e[0..2]
// vertex = v[0..3]
// plane normal = n[0..2]
// plane thickness = t
// refraction index of material = r
float3 n,d;
float alpha_i, alpha_t, delta;

// compute viewing vector to vertex
d = e - v;
d = normalize(d);
n = normalize(n);

// compute angle between normal and viewing vector
alpha_i = acos(dot(d,n));

// compute delta
alpha_t = asin(sin(alpha_i) / r);
```

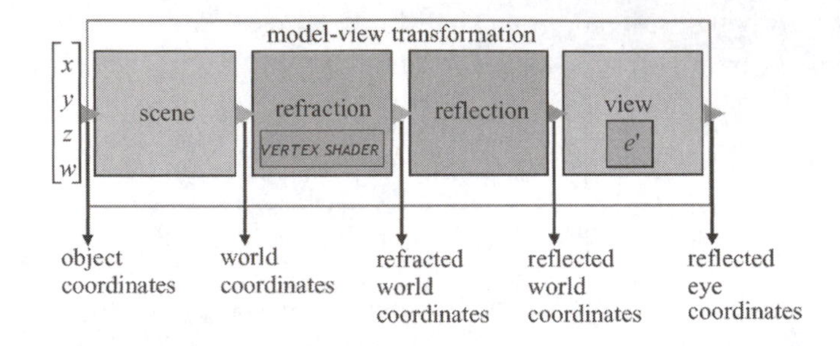

Figure 6.6. The integration of reflection transformations into the rendering pipeline.

```
delta = t * (1 - (tan(alpha_t)) / tan(alpha_i));

// compute refracted vertex
v.xyz = v.xyz + n * delta;

...
```

Whether refraction is implemented as a rigid-body transformation or as a per-vertex transformation, the sequence of the transformations carried out by the rendering pipeline is important for spatial optical see-through displays.

Figure 6.6 illustrates the extension of Figure 6.3. The refraction transformation has to be carried out after the scene transformation (either a rigid-body or a per-vertex transformation). Vertices are transformed from world coordinates to *refracted world coordinates* first. If the optical combiner is a mirror beam combiner, the refracted vertices are then reflected. If the optical combiner is a transparent screen, reflection transformation is not used. Finally, the vertices are mapped into reflected eye coordinates (either with the reflected or with the unreflected viewpoint, depending on the optical combiner), projected, converted into window coordinates, rasterized, and displayed.

The pseudocode below summarizes the rendering steps for spatial optical see-through displays that use one planar screen and one planar beam combiner. Only one user is supported.

```
for left and right viewpoints i
  initialize transformation pipeline and polygon order
  compute reflected viewpoint e'_i = Re_i
  compute refraction offset Δ_i for i
  set transformation pipeline: RMF_iV_iP
```

```
  reverse polygon order
  render scene from e'_i
endfor
```

First, the polygon order and the transformation pipeline have to be set to an initial state. Then the reflected viewpoint and the view-dependent refraction offset (Δ_i) are computed. The transformation pipeline is then set to the corresponding concatenation of transformation matrices: reflection transformation (R), model transformation (M), refraction transformation (F_i, with e_i), view transformation (V_i, with e'_i), and the projection transformation (P). Finally, the polygon order is reversed and the scene is rendered. Note that we assume that the screen coordinate system is equivalent to the world coordinate system. Note also that R might not be static but has to be recomputed continuously (e.g., if moving components have to be supported—see Section 6.5).

6.4 Screen Transformation and Curved Screens

If the coordinate system of the secondary screen does not match with the world coordinate system, an additional *screen transformation* has to be added to the rendering pipeline. It expresses the screen's transformation within the world coordinate system and is usually a composition of multiple translation and rotation transformations. Similar to the reflection transformation, not only the reflected geometry but also the reflected viewpoint has to be mapped into the *screen coordinate system*.

Once again the sequence in which all transformations are carried out is important. The screen transformation is applied after the reflection trans-

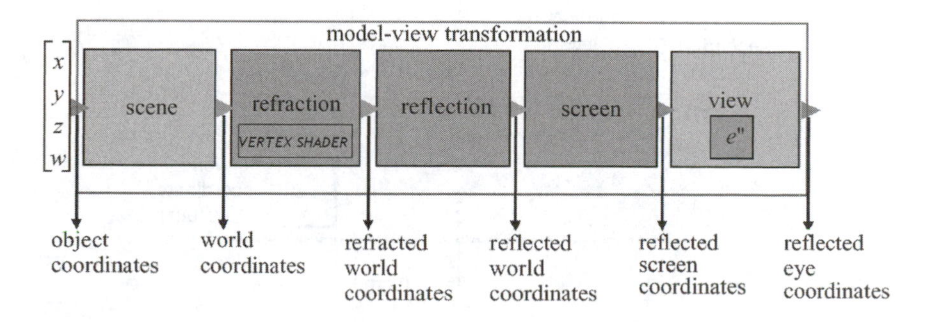

Figure 6.7. The integration of screen transformations into the rendering pipeline.

formation and before the view transformation—mapping reflected world coordinates into *reflected screen coordinates*. The view transformation has to be performed with the reflected viewpoint in screen coordinates e'', that can be computed by simply applying the screen transformation to the reflected viewpoint e'.

To give an example, consider a 30 cm × 40 cm screen that is not aligned with the world coordinate system. In particular, assume it is rotated by 180 degrees around the z-axis and translated in such a way that the screen's y edge is aligned with the edge of a mirror beam combiner, located 20 cm away from the world origin. In addition, the screen is tilted 30 degrees about the edge that faces the mirror to achieve a higher contrast and a better position of the reflected image.

The corresponding OpenGL screen transformation matrix can be calculated as follows:

```
...

glRotatef(180,0,0,1); //rotate around z-axis by 180 deg
glTranslatef(0,15,0); //translate screen edge to origin
glRotatef(30,1,0,0);  //rotate around screen edge by 30 deg
glTranslatef(0,20,0); //translate screen edge to mirror edge

...
```

Remember that the OpenGL notation has to be read bottom-up (but it has to be implemented this way). To understand the transformation,

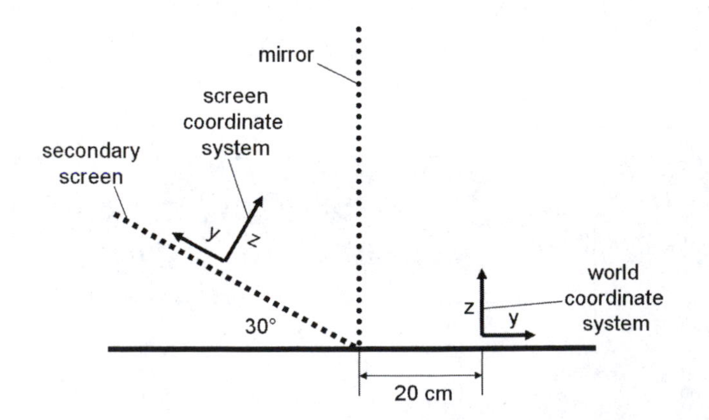

Figure 6.8. Example of a screen transformation.

it is helpful to think that the misaligned screen coordinate system has to be transformed back into the world coordinate system, since this corresponds to the transformation that needs to be applied to the rendered geometry. Reading the OpenGL fragment bottom to top, the screen is first translated by 20 cm along the positive y-axis of the world coordinate system aligning the adjacent edge with the world's origin. It is then rotated around this edge by 30 degrees fitting it to the x/y plane of the world coordinate system. Then, the screen's origin position is matched with the world's origin position by translating the screen further up the positive y-axis (by exactly 15 cm which is half of the screen height). Finally, the screen is rotated by 180 degrees to fully align both coordinate systems. A read-back from OpenGL's model-view stack (e.g., with `glGetFloatv(GL_MODELVIEW_MATRIX,S);`) allows us to obtain this matrix and multiply it with the reflected viewpoint before the view transformation is added.

As described in Chapter 5, projective texture mapping [168] can be used in combination with *two-pass rendering* to support single or multiple front/back projections onto a multi-plane or curved display surface. Projective texture mapping is explained in Chapter 2.

Projective textures utilize a *perspective texture matrix* to map projection-surface vertices into texture coordinates of those pixels that project onto these vertices. A first rendering pass generates an image of the virtual scene that will look perspectively correct to the user. During the second pass, this image is projected out from the user's current point of view onto a registered virtual model of the display surface using projective texture mapping. Finally, the textured model is rendered from the projector's point of view and the result is beamed onto the real display surface. If multiple projectors are used, the second pass has to be repeated for each projector individually. The generated images have to be geometrically aligned, and color and edge blended appropriately, to realize a seamless transition between them.

To support planar mirror beam combiners that require affine model and view transformations in combination with curved secondary screens, the method previously outlined can be slightly modified. Instead of rendering the original scene from the observer's actual viewpoint, the reflected scene has to be rendered from the reflected viewpoint. The reflection transformations that are applied to the scene and the viewpoint depend on the optics. All other rendering passes remain unchanged. Note that if multiple planar display surfaces are used, and if one projector is assigned to one projection plane, projective textures and two-pass rendering are unnecessary. Instead, regular multi-pipeline rendering can be used (as it is done for surround-screen projection systems, such as CAVEs or multi-sided workbenches).

6.5 Moving Components

Some spatial see-through displays do not require a static constellation of screens and optical combiners; they rather allow moving them freely within the world coordinate system during run time. With respect to rendering, it makes no difference whether the optical combiner and the screen are fixed or movable. The only modification to the transformations that are described in the previous section is that the plane parameters of the optical combiner (which influences reflection and refraction transformations), and/or the parameters of the screen transformation change continuously. This means that the rendering pipeline has to be updated every time a component has been moved; every frame in the worst case.

6.6 Multi-Plane Beam Combiners

More that one planar beam combiner can be utilized to build displays that provide views on an augmented environment from multiple—very different—angles, or to support multiple users simultaneously. This applies to mirror beam combiners as well as to transparent screens. If multiple planar beam combiners need to be supported by an optical see-through display simultaneously, the rendering pipeline that drives this display can be configured from the basic elements that have been discussed previously. Convex mirror assemblies unequivocally tessellate the surrounding space into individual mirror reflection zones which, for the same point of view, do not intersect or overlap and consequently provide a definite one-to-one mapping between the screen space and the reflection space. The constellation of basic transformations for these displays depends on the physical constellation of optical combiners (e.g., mirror beam combiners or transparent screens) and image sources (e.g., video projectors or secondary screens). Several scenarios are discussed in the following sections.

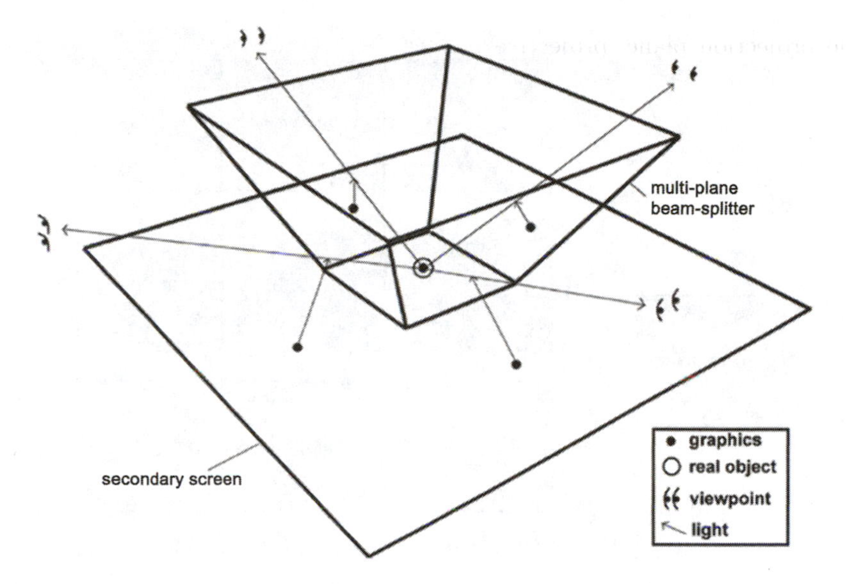

Figure 6.9. Example of a multi-plane beam combiner constellation: four mirror beam combiners and one secondary screen that support observing the augmented real environment from 360 degrees.

6.6.1 Single Viewer

For the following explanation, we assume a *multi-plane beam combiner* that provides different perspectives onto the same scene to a single observer. We also want to assume that mirror beam combiners are used as optical combiners, and a single secondary screen is used to display the rendered images. For example, four half-silvered mirrors can be assembled in the form of an upside-down pyramid frustum that reflects a large, horizontal projection screen. The real environment that needs to be augmented would be located inside the pyramid frustum.

The step from single-plane to multi-plane optical see-through displays is similar to the step from single-plane projection displays (like walls or workbenches) to multi-plane projection displays (like caves or two-sided workbenches).

The rendering pipeline is split into several (four in our example) *sub-pipelines*. Each sub-pipeline is responsible for rendering a single image onto the secondary screen. Since only one screen is used in this example, four separate images have to be rendered into the same frame buffer that is displayed on the screen on each frame.

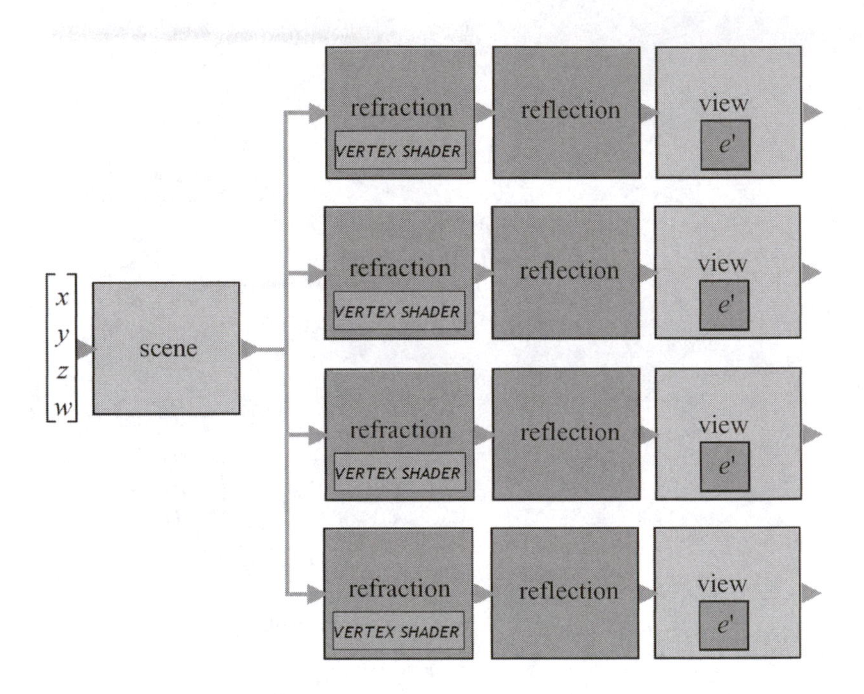

Figure 6.10. Dependent rendering pipelines that support a multi-plane beam combiner in combination with a single screen and a single observer.

The images differ in their individual mirror transformations (reflection and, optionally, refraction). The scene and the view transformation is the same for each sub-pipeline except that individual reflections of the same viewpoint have to be applied together with the view transformation.

Note that the number of sub-pipelines is doubled for stereoscopic viewing. In this case, the left and right stereo images may be rendered into two different frame buffers; the back-left and the back-right buffer if the graphics hardware provides a quad buffer.

If the display is calibrated precisely enough (i.e., the parameters of the pipeline, such as mirror-plane parameters, etc., have been determined correctly), the different images merge into a single, consistent reflection from all points of view. The perpendicular mirror constellation that was chosen in our example ensures that the different images never overlap in the frame buffer. Only two images and the corresponding two mirrors are visible from a single point of view at a time. The reflections of these images result in a single graphical scene (Figure 6.11).

Figure 6.11. Two images are reflected by two mirrors and merge into a single consistent graphical scene. (*Image reprinted from [9] © IEEE.*)

The sub-pipelines can be carried out sequentially on the same computer (i.e., by rendering each image into the frame buffer before swapping it). This divides the frame rate by a factor that equals the number of sub-pipelines.

Alternatively, hardware *image composition technology* (such as the Lightning-2 device [175]) can be applied to distribute the sub-pipelines to multiple computers and merge the resulting images within a single frame buffer before the outcome is displayed on the screen. In this case, multiple rendering nodes and the displays can be connected to a specific display subsystem. The subsystem allows the image data generated by the connected nodes to be mapped to any location of the connected displays without losing much time for transmission and composition.

To support configurations that apply multiple planar beam combiners and a single screen, the following algorithm can be used for a single user:

```
for left and right viewpoints i
  for each to i front-facing beam combiner j
    initialize transformation pipeline and polygon order
    compute reflected viewpoint e'_i = R_j e_i
    compute refraction offset Δ_ij for i and j
    set transformation pipeline: R_j M F_ij V_i P
    reverse polygon order
    render scene from e'_i
  endfor
endfor
```

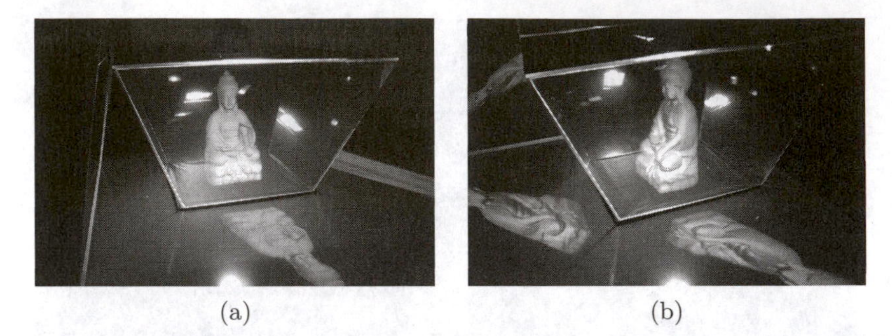

(a) (b)

Figure 6.12. Individual views of the same scene generated for two different observers. (*Images reprinted from [9] © IEEE.*)

The scene has to be rendered for each viewpoint and each beam combiner. If the configuration of the optics unequivocally tessellates the surrounding space into individual mirror reflection zones which do not intersect or overlap, a single object that is displayed within the screen space appears exactly once within the reflection space.

6.6.2 Multiple Viewers

To support *multiple viewers*, each observer can be assigned to an individual beam combiner (and corresponding screen portion and rendering sub-pipeline).

Instead of using the same viewpoint for the view transformation of each sub-pipeline, different viewpoints of the assigned observer have to be used. This means that the generated images are completely independent of each other and will no longer merge into a single, consistent three-dimensional scene. They represent the individual views of the augmented scene of each observer.

If we use the previous display example, the same scene is transformed and rendered for each observer individually, before all images are displayed on the same screen at the same time. In this case, each observer is restricted to a limited viewing area that allows observing the scene through the assigned beam combiner only.

The following algorithm will support such configurations:

```
for left and right viewpoints i of all viewers
  initialize transformation pipeline and polygon order
  compute reflected viewpoint e'_i = R_i e_i
  compute refraction offset Δ_i for i
  set transformation pipeline: R_i M F_i V_i P
```

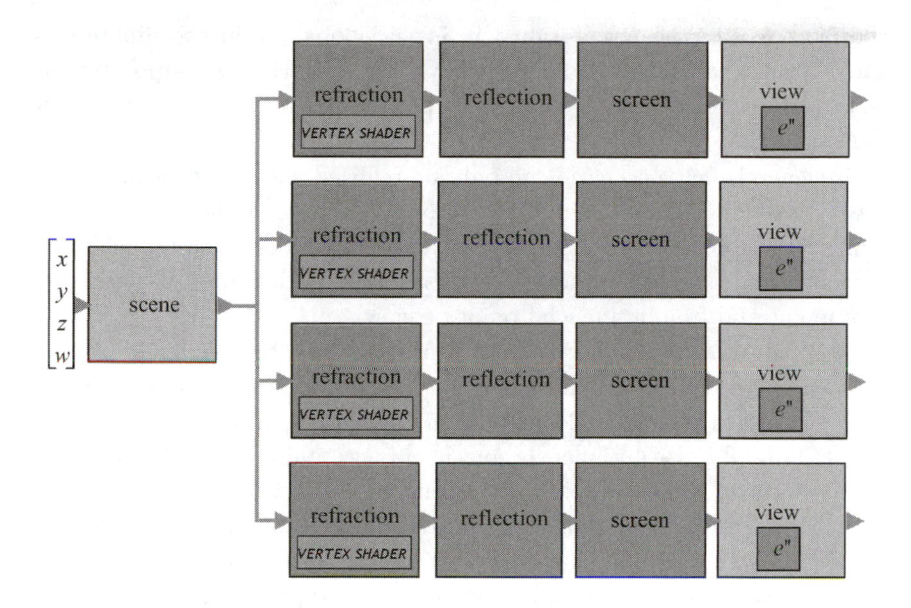

Figure 6.13. Dependent rendering pipelines that support a multi-plane beam combiner in combination with multiple screens and a single or multiple observer(s).

```
  reverse polygon order
  render scene from e'_i
endfor
```

Note that each user is restricted to the viewing zone that is given by the assigned beam combiner.

6.6.3 Multiple Screens

Using a single screen for multiple beam combiners has the disadvantage that each sub-pipeline renders its image into a portion of the common frame buffer. The resolution of a single image is only a fraction of the frame buffer's total resolution. To overcome this problem, an individual screen can be applied in combination with each beam combiner, sub-pipeline, and, optionally, with each viewer.

This implies that each image is rendered in full resolution into separate frame buffers that are displayed on different screens. Each sub-pipeline has to be extended by a screen transformation that copes with each screen's relative position and orientation within the global world coordinate system. Note, that single and multiple viewer scenarios are supported by this frame work depending on the viewpoints that are passed to the view transformations.

In general such a scenario can only be realized with *multi-channel rendering pipelines* that offer multiple frame buffers and the possibility to connect to more than one screen. Nowadays, *PC clusters* are much more efficient than monolithic graphics workstations, both from a performance standpoint and from a financial point of view.

If PC clusters are used, it is important that the entire rendering process is synchronized correctly. The generated images have to be displayed at exactly the same time. This means that not only scene information has to be updated on all PCs continuously and at an acceptable speed, but also the signal that causes the swapping of the rendered images from back buffers to front buffers might have to be distributed. This swapping synchronization is called *GenLocking* and is an essential feature for synchronizing multiple active stereoscopic screens. While some newer graphics cards can be connected to distribute the GenLock signal directly, older graphics cards do not provide this feature. However, open source software solutions (called *SoftGenLocking*) exist [2] that offer an acceptable workaround. The scene information is usually distributed over the network, using distributed scene graph frameworks or other distributed graphics concepts. This ensures that the same scene state is displayed on each screen at a time. This scene synchronization is referred to as *frame locking*.

Display configurations that assign an individual screen and beam combiner to each user benefit from a higher resolution than those configurations that split the screen (and its resolution) into different portions. To support such configurations, an additional screen transformation (S) has to be introduced (see Section 6.4). This matrix represents a concatenation of several translations, rotations, and scaling transformations, that map coordinates from the world coordinate system into an individual coordinate system of the corresponding screen. One can also think of the inverse transformation of the screen itself within the world coordinate system.

The following algorithm can be used in combination with such configurations:

```
for left and right viewpoints i of all viewers
  initialize transformation pipeline and polygon order
  compute reflected viewpoint e'_i = R_i e_i
  compute refraction offset Δ_i for i
```

```
  set transformation pipeline: RiMFiViSiP
  reverse polygon order
  render scene from e'i
endfor
```

Note, that by keeping track of individual model transformations (M) and by rendering individual scenes for each user, the display configuration acts as multiple individual-user displays. Otherwise, multiple users share and interact with the same scene. To use individual model transformations and to render the same scene for each user has been proven to be an effective way of precisely registering virtual objects to real ones cancelling out slight physical misalignments of the screens and mirrors (whose measured/calculated parameters that are stored in R_i and S_i reflect these errors) within each M. Thereby, the object transformation that is used to align real and virtual objects is stored in each M. This also applies to the algorithm that is presented in Section 6.6.2 although only one screen is used.

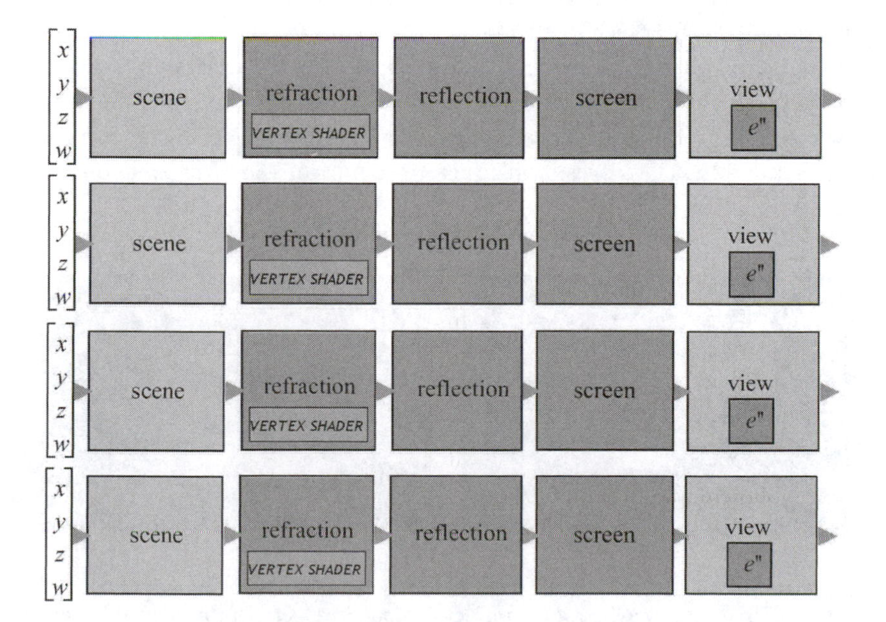

Figure 6.14. Independent rendering pipelines that support individual scenes.

6.6.4 Individual Scenes

For some multi-user scenarios, a common scene is not required. Rather, each user observes and interacts with his or her own scene representation on the same display. The actions that are carried out by a user do not effect the scene state of the other users.

In this case, the scene transformation becomes individual to each user and is carried out within each sub-pipeline. This makes all sub-pipelines completely independent from each other and allows distributing them (together with copies of the scene description) on separate rendering nodes without the need of any synchronization mechanism. Now every user is assigned to completely independent components that are integrated into a single spatial optical see-through display.

6.7 Curved Mirror Beam Combiners

Optical see-through displays can be assembled from many planar optical elements (mirror beam combiners or transparent screens). Every element requires its own rendering sub-pipeline. However, if many small optical elements are used to approximate curved surfaces, the above rendering concept quickly becomes inefficient.

Compared to calibrating multiple optical elements, the calibration effort for a single curved element can be reduced and optical aberrations that result from miscalibration can be decreased by several orders of magnitude. In addition, a curved optics can provide an *edge-free view* onto the augmented scene.

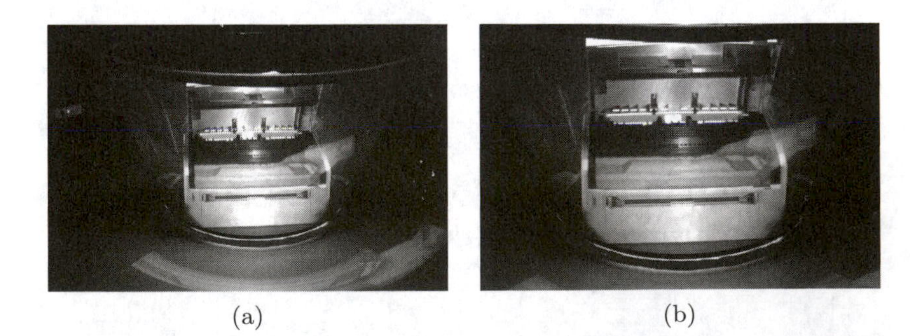

(a) (b)

Figure 6.15. (a) A warped and projected image reflected in a curved (conical) mirror beam combiner; (b) a virtual cartridge is being placed in a physical printer. (*Images reprinted from [9] © IEEE.*)

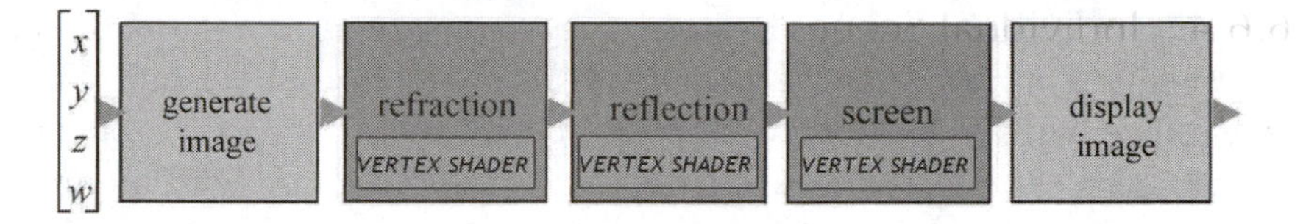

Figure 6.16. Two-pass image warping pipeline for neutralizing curvilinear optical transformations.

As for curved projection screens (see Chapter 5), the rendering for spatial optical see-through displays that use *curved optics* (mirror beam combiners or transparent screens) has to be warped before being displayed. This transformation is view-dependent and curvilinear rather than a rigid-body transformation and requires a *per-vertex viewpoint* and *model transformation*. A single accumulation of simple homogeneous transformation matrices (as described for planar optics) cannot express such a complex deformation for the entire scene. Consequently, fixed function rendering pipelines cannot be fully used for rendering graphics on curved optical see-through displays.

Note that since convex mirrors map a larger portion of the screen space into a smaller portion within the image space inside the optics, a high density of pixels can be compressed into a small reflection (Figure 6.15). Consequently, the *spatial resolution* within the reflection is regionally higher than the spatial resolution of the display device!

Warping the scene geometry itself requires highly tessellated graphical models to approximate the *curvilinear optical transformation* well enough. Instead, multi-pass rendering is frequently used. Rather than warping geometry, images of the geometry are warped and displayed. This causes the same effect, but the image transformation does not depend on the scene geometry (e.g., its type, its complexity, its shading, etc.).

The *first rendering pass* generates an image of the graphical overlay perspectively correct from the viewpoint of the observer. This image contains all visual information that is expected to appear within the real environment (e.g., shading, shadows, occlusion effects, etc.), and can be rendered completely in hardware. It is geometrically approximated by a *two-dimensional tessellated grid* that is transformed into the current viewing frustum in such a way that it is positioned perpendicular to the optical axis.

The grid vertices and the image's texture coordinates are transformed (warped) with respect to the viewpoint, the optics, and the image source. These transformations are applied on a *per-vertex level* and neutralize the image deformations that are caused by the optics (such as reflection and refraction of mirror beam combiners and transparent screens) and the image source (such as lens distortion of video projectors, or projections onto curved screens). As indicated, per-vertex transformations can be implemented as vertex shaders on programmable rendering pipelines.

Finally, the image that was generated during the first pass is mapped onto the warped grid using texture mapping and *bi-* or *tri-linear texture filtering*, and is displayed during the second pass. This process is repeated for multiple individual viewpoints (e.g., for stereoscopic viewing and for multi-user scenarios).

6.7.1 Generating Images

The first rendering pass generates the undeformed graphical overlay with all the information that is expected to be visible within the real environment. Conventional graphics techniques that are provided by fixed function rendering pipelines are sufficient for this step.

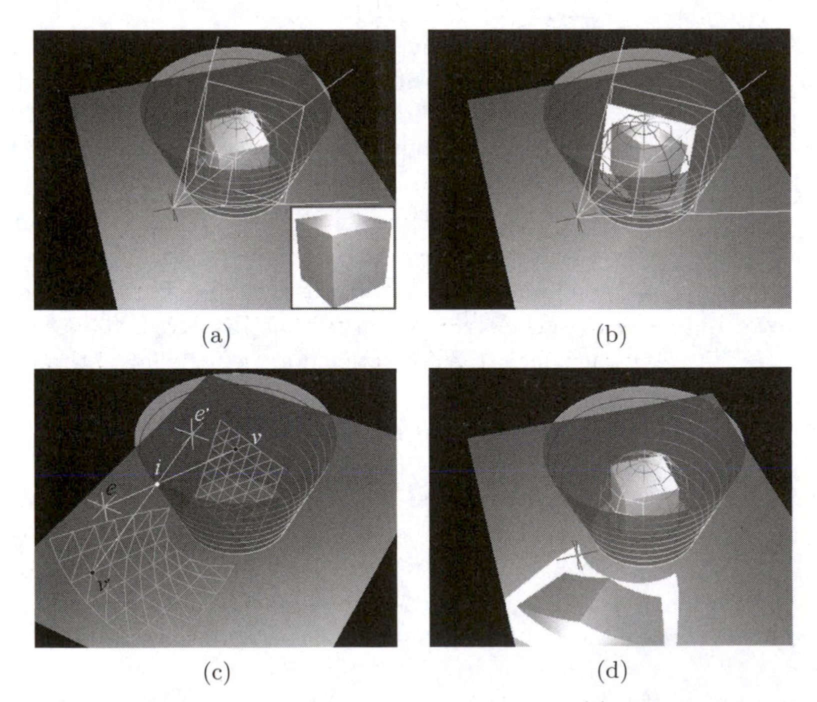

Figure 6.17. Different steps of image warping process: (a) image generation; (b) transformation of image grid; (c) texturing; (d) displaying deformed image grid. (*Images reprinted from [9] © IEEE.*)

To capture the entire virtual scene in the image, the *scene's bounding sphere* is used to define an *on-axis viewing frustum* from the viewpoint of the observer. The *apex* of the frustum is chosen in such a way that it encloses the bounding sphere exactly. This ensures that the complete scene is visible in the image and that it covers a large portion of the image. Finally the scene is rendered through this frustum into a texture memory block rather than directly into the frame buffer. Older graphics cards do not allow drawing an image into the on-board texture memory directly. In this case the scene has to be rendered into the back frame buffer, and then copied into the texture memory of the graphics card. This step is referred to as *read-back*.

New graphics cards do allow rendering the image directly into an auxiliary on-board memory block (usually called P-buffer) that can be referenced during texture mapping. This step is referred to as render-to-texture and avoids the time-consuming operation of transferring the image information from the frame buffer into the texture memory.

The following OpenGL code fragment can be used to configure the correct on-axis frustum for generating the undeformed image of the virtual scene:

```
...

// scene's center = p[0..2]
// scene's bounding sphere radius = r
// viewpoint = e[0..2], up-vector = u[0..2]
// texture height, texture width = th, tw
float l, d;
float left, right, bottom, top, near, far;

// compute parameters of on-axis frustum
l = sqrt((p[0] - e[0]) * (p[0] - e[0]) +
         (p[1] - e[1]) * (p[1] - e[1]) +
         (p[2] - e[2]) * (p[2] - e[2]));
d = r * (l - r) / l;
left = -d; right = d;
bottom = -d; top = d;
near = l - r; far = l + r;

// configure rendering pipeline
glViewport(0, 0, tw, th);
glMatrixMode(GL_PROJECTION);
glLoadIdentity();
glFrustum(left, right, bottom, top, near, far);
```

```
glMatrixMode(GL_MODELVIEW);
glLoadIdentity();
gluLookAt(e[0], e[1], e[2],
          p[0], p[1], p[2],
          u[0], u[1], u[2]);

// draw scene with scene transformation into texture memory

...
```

The big advantage of this image-based method is that it is completely independent of the scene's content and the way this content is rendered. Instead of using a geometric renderer, other techniques (such as image-based and non-photorealistic rendering, interactive ray tracing, volume rendering, or point-based rendering, etc.) can be employed to generate the image.

The image that has been generated during the first rendering pass has to be transformed in such as way that it is perceived undistorted while observing it through or with the display's combiner optics. To support these image deformations, a geometric representation of the image plane is generated initially. This *image geometry* consists of a *tessellated grid* (e.g., represented by an *indexed triangle mesh*) which is transformed into the current viewing frustum in such a way that, if the image is mapped onto the grid, each line of sight intersects its corresponding pixel. Thus, the image grid is perpendicular to the optical axis and centered with the scene geometry. The following deformation steps transform the positions and the texture coordinates of every grid vertex individually. Note that a pixel-individual warping would be possible with pixel shaders. However, holes caused by the stretching deformation would have to be filled accordingly. Rendering a dense and textured image grid through a vertex shader ensures a correct interpolation of missing image portions.

6.7.2 Reflection on Convex Mirrors

As discussed in Section 6.6, convex multi-plane mirror assemblies unequivocally tessellate the surrounding space into individual mirror reflection zones which, for the same point of view, do not intersect or overlap. Consequently, they provide a definite one-to-one mapping between screen space and reflection space. This is also true for curved convex mirrors. Curved convex mirrors cannot provide true stigmatism between all object-image pairs, but rather a close approximation which introduces small optical aberrations. Due to the limited resolution of the eyes, human observers can usually not perceive these aberrations.

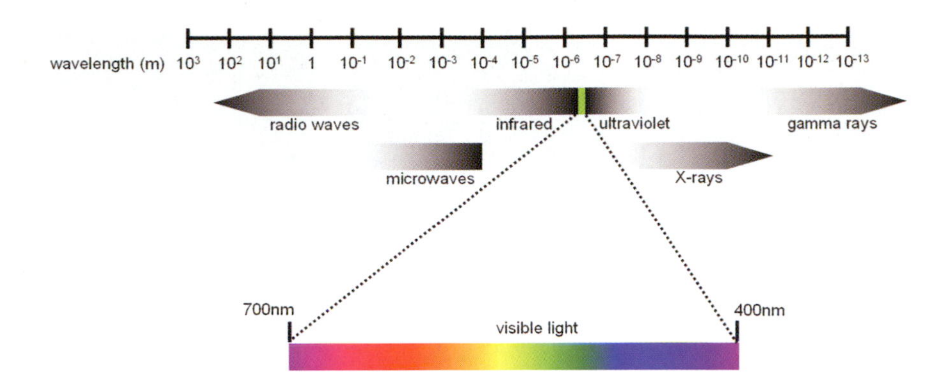

Plate I. EM spectrum and spectrum of visible light. (See Figure 2.2.)

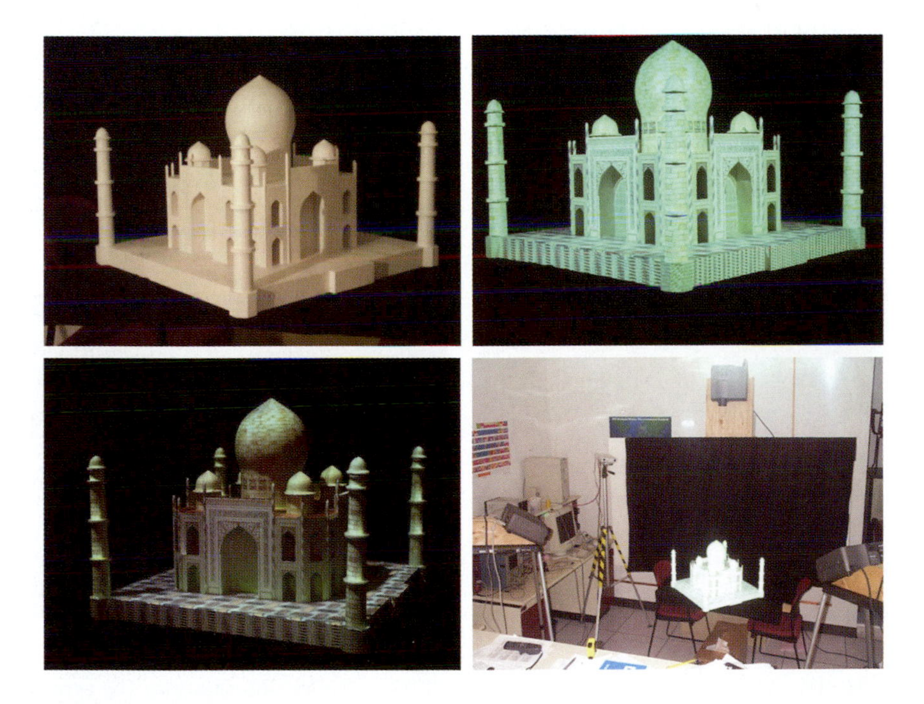

Plate II. ShaderLamps with Taj Mahal: The wooden white model is illuminated; the scanned geometry of the Taj Mahal is augmented to add texture and material properties; The geometry is then registered to the real Taj Mahal and displayed from the projector's viewpoint. (*Images reprinted from [155] © Springer-Verlag; see Figure 3.16.*)

Plate III. The surface appearance of a neutral colored object is changed by 'painting' it with projected light. (*Images reprinted from [155] © Springer-Verlag; see Figure 7.1.*)

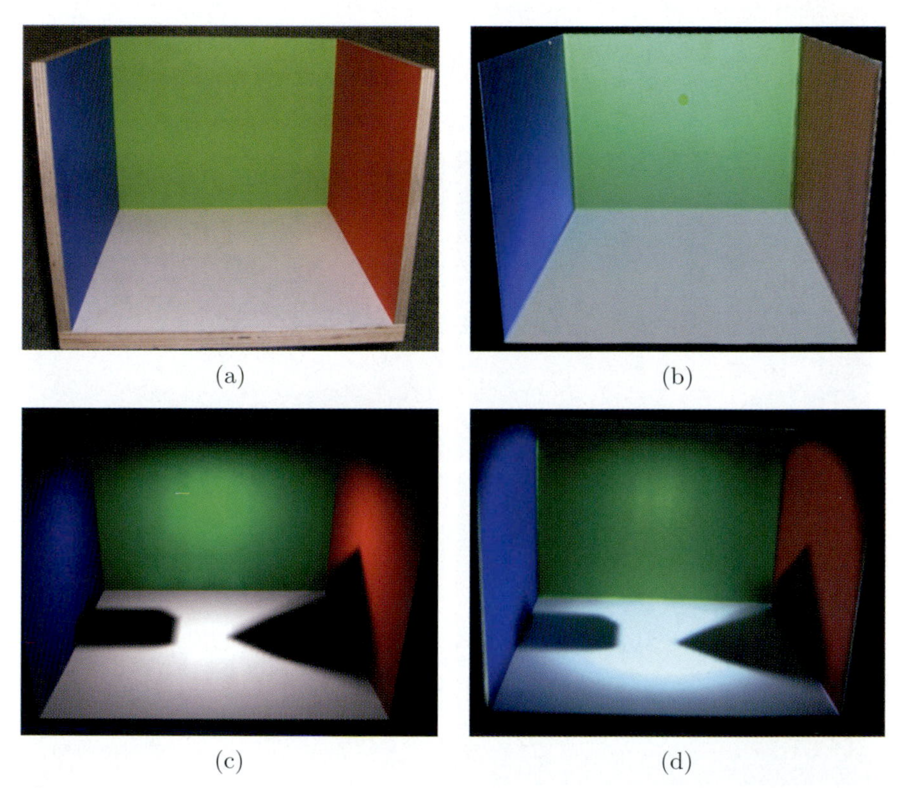

Plate IV. (a) Photograph of original object under room illumination; (b) screen shot of captured reflectance relit with virtual point light source and Phong shading; (c) screen shot of simulated radiosity solution with captured reflectance, virtual surface light source (shown in Figure 7.11), and two virtual objects (shown in Figure 7.11); (d) photograph of original object illuminated with the computed irradiance. (*Images reprinted from [15] © IEEE; see Figure 7.12.*)

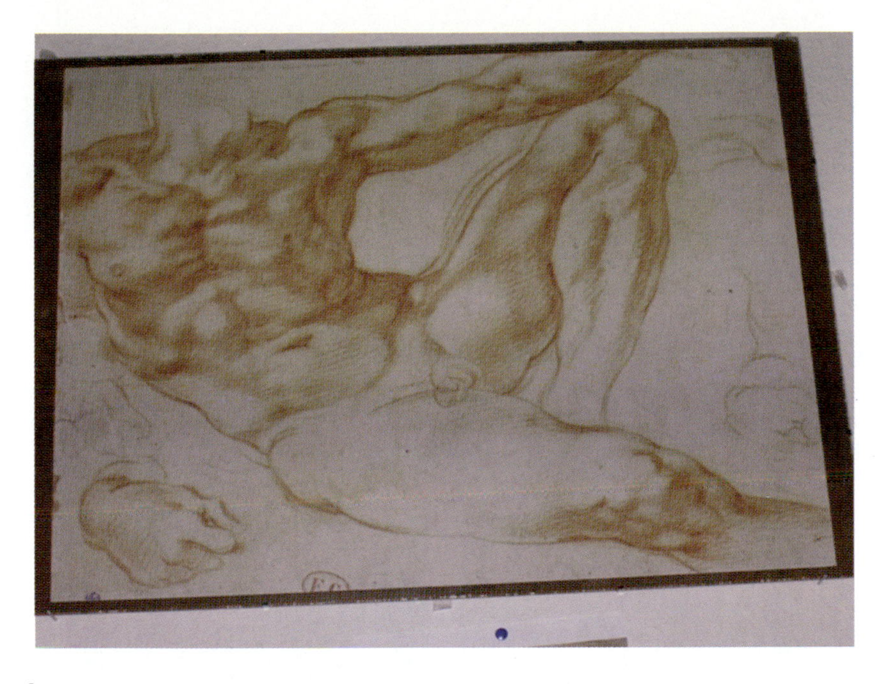

Plate V. Results of a color correction process with a single projector on a real drawing: (a) real drawing (64 × 48 cm) under environment light; (b) output image emitted onto drawing; (c) partially augmented drawing; (d) output image on a white piece of paper. (*Displayed artwork courtesy of the British Museum, London; images reprinted from [19] © IEEE; see Figure 7.19.*)

(a) (b)

Plate VI. Results of color correction process with two projectors: (a) the limited intensity capabilities of a single projector result in visible artifacts; (b) the contribution of a second projector reduces these effects. (*Displayed artwork courtesy of the Vatican Museum, Rome and the British Museum, London; images reprinted from [19] © IEEE; see Figure 7.24.*)

(a) (b)

Plate VII. (a) A green paper illuminated with white light; (b) the white diffuse surface on the right is illuminated with green light. In this special case, the secondary scattering off the white surface below is similar for both parts. (*Images reprinted from [155] © Springer-Verlag; see Figure 7.3.*)

(a) (b)

(c) (d)

Plate VIII. Rembrandt's self-portrait: (a) copy of original painting as it looks today (illuminated under environment light); (b)–(d) various cleaning stages to remove the overpainted layers form 1935(d), 1950(c) and 1980(b) are projected onto (a). Only black and white photographs of these stages are available. The high black-level of the video projectors prevents the creation of a totally black color on the canvas. Extreme regions, such as overlaid hair and hat cannot appear completely black for this reason. (*Displayed artwork courtesy of the Museum het Rembrandthuis, Amsterdam; images reprinted from [19] © IEEE; see Figure 7.25.*)

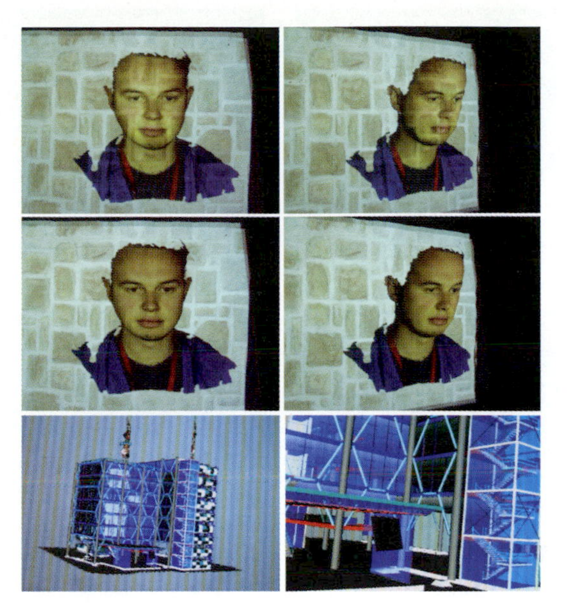

Plate IX. Geometrically corrected stereo-pairs of 3D scene projected onto a natural stone wall: (top) without radiometric compensation; (middle) with radiometric compensation; (bottom) radiometrically corrected stereoscopic walk-through projected onto a papered wall. (*Prototype realized by the Bauhaus-University Weimar; see Figure 7.28.*)

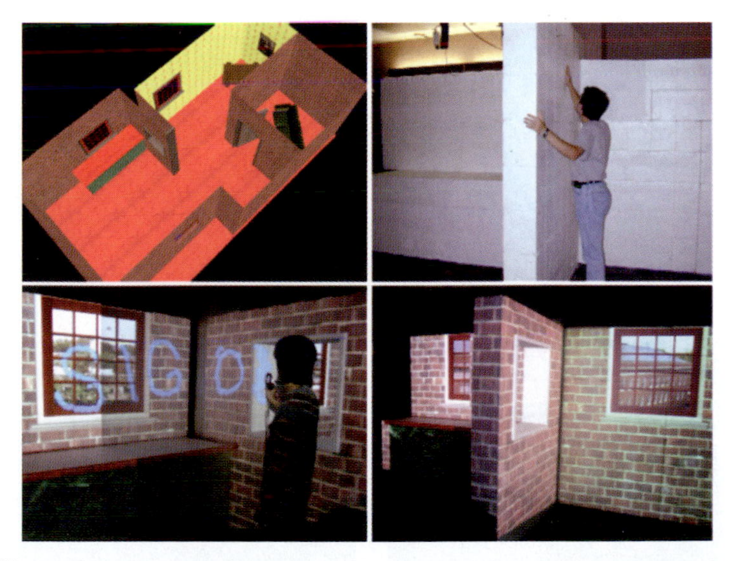

Plate X. Spatially augmenting large environments: (Top left) virtual model; (Top right) physical display environment constructed using styrofoam blocks; (Bottom row) augmented display—note the view dependent nature of the display, the perspectively correct view through the hole in the wall and the windows. (*Images courtesy of Kok-Lim Low [93]; see Figure 8.6.*)

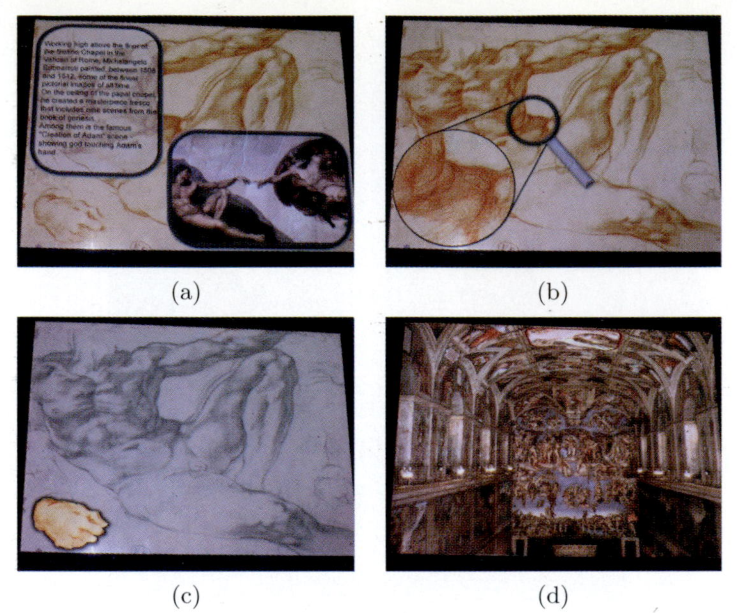

(a) (b) (c) (d)

Plate XI. Examples of presentation techniques: (a) inlay text and image; (b) magnification; (c) focus through highlighting and decolorization; (d) three-dimensional flythrough through the Sistine chapel. Note, that in images (a)–(c) the Adam drawing itself is not projected. (*Displayed artwork courtesy of the Vatican Museum, Rome and the British Museum, London; images reprinted from [19] © IEEE; see Figure 8.23.*)

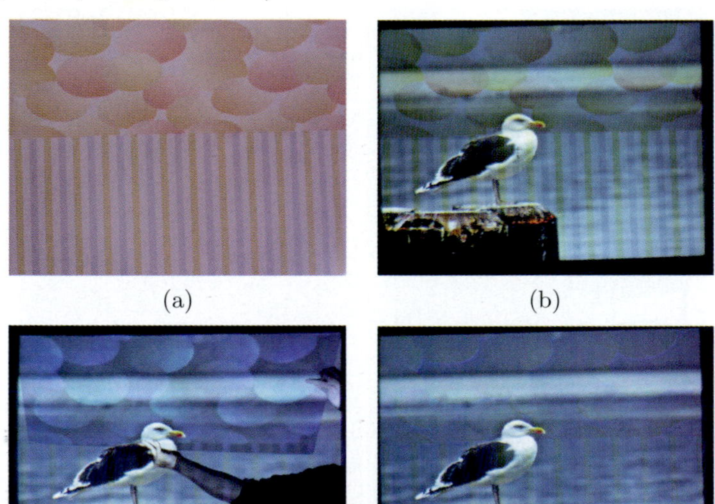

(a) (b) (c) (d)

Plate XII. (a) Projection onto a wallpapered pitched roof area; (b) projection with uncorrected colors; (c) color correction projected onto white piece of paper; (d) color corrected image on wallpaper. All projections are geometry corrected. (*Images reprinted from [20] © IEEE; see Figure 7.27.*)

(a) (b)

(c) (d)

(e) (f)

Plate XIII. (a) Projection onto a scruffy corner; (b) normal uncorrected projection; (c)–(d) line-strip scanning of surface; (e) geometry-corrected projection on virtual canvas; (f) final geometry and color-corrected image (*Movie footage: The Jackal, © 1997 Universal Pictures; images reprinted from [20] © IEEE; see Figure 7.26.*)

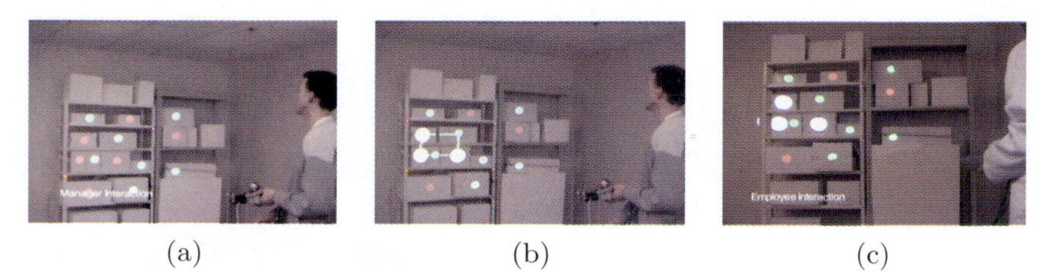

(a) (b) (c)

Plate XIV. Warehouse scenario implementation: (a) manager locates items about to expire (marked in white circles); (b) annotates some of those items (marked in larger white circles); (c) employee retrieves and views the same annotations from a different projector view. (*Images reprinted from [159] © ACM; see Figure 9.3.*)

Plate XV. Visualizing registered structures of (top) an adjacent room; (bottom) stairways and lower building level for architectural applications in a realistic environment. (*Prototype realized by the Bauhaus-University Weimar, 2005; see Figure 8.29.*)

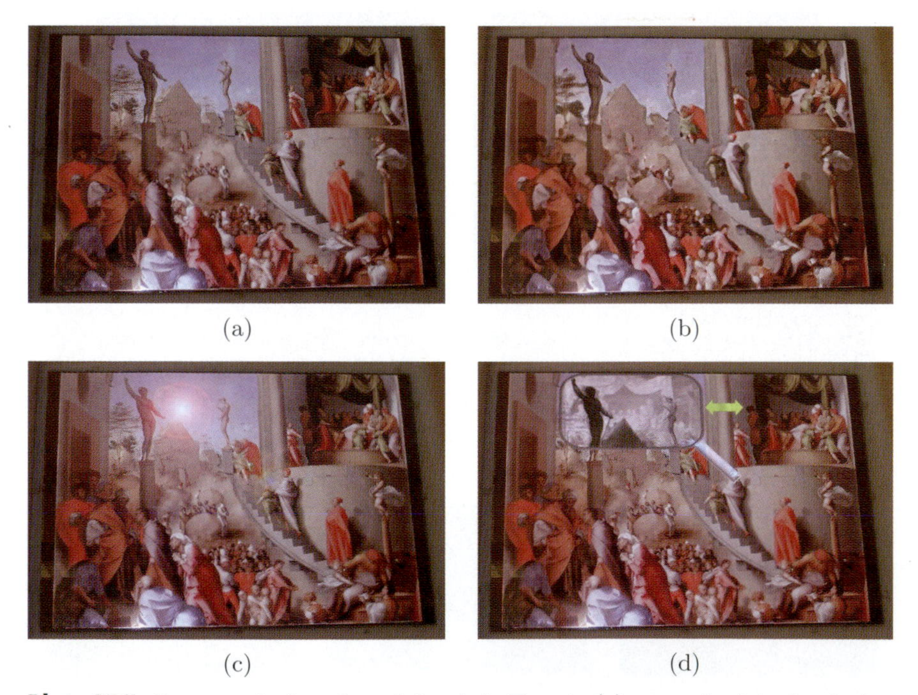

(a) (b)

(c) (d)

Plate XVI. Pontormo's Joseph and Jacob in Egypt: (a) copy of original painting illuminated under environment light; (b) modification of painting style from oil on wood to watercolor on paper via a 2D artistic filter; (c) reillumination and lens flare; (d) registered visualization of underdrawings (infrared recordings are black and white) and explanation. (*Displayed artwork courtesy of the National Gallery, London; images reprinted from [19] © IEEE; see Figure 8.24.*)

Images formed by convex mirrors appear to be reduced and deformed versions of the reflected objects. Hence, the image of the scene that results from the first rendering pass has to be stretched before displaying it on the secondary display. This results in the original (unscaled) image after the physical reflection.

Several curved mirror displays exist that generally don't predistort the graphics before they are displayed. Yet, some systems apply additional optics (such as lenses) to stretch or undistort the reflected image (e.g. [106, 107]). But these devices constrain the observer to a single point of view or to very restricted viewing zones. However, if a view-dependent rendering is required to support freely moving observers, interactive rendering and real-time image warping techniques are needed which provide appropriate error metrics.

The reflection transformation for displays that use convexly curved mirror beam combiners will be explained based on an example of the conical mirror display illustrated in the figures shown previously. Any other surface type can be used in combination with the described techniques.

Each grid point v of the image geometry has to be transformed with respect to the mirror surface M, the current viewpoint e, and the secondary screen S and the texture of each grid point in the image that we generated during the first rendering pass. For all grid vertices v, the intersection i of the geometric line of sight with the mirror surface has to be computed (that is, the ray r that is spanned by the eye e and the vertex v). Next, the normal vector n at the intersection i needs to be determined. The intersection point, together with the normal vector, gives the tangential plane at the intersection. Thus, they deliver the individual plane parameters for the per-vertex reflection. This information is then used to compute the reflection ray r' and the intersection i' of r' with the secondary screen surface S.

Having a geometric representation (e.g., a triangle mesh) to approximate the mirror's and the screen's surfaces, M and S, supports a flexible way of describing the dimensions of arbitrary shapes. However, the computational cost of the per-vertex transformations increases with a higher resolution surface geometry. A high resolution surface description, however, is required to neutralize the optical reflection distortion effectively. For triangle meshes, a fast ray-triangle intersection method (such as [113]) is required that automatically delivers the barycentric coordinates of the intersection within a triangle. The barycentric coordinates can then be used to interpolate between the three vertex normals of a triangle to approximate the normal vector at the intersection.

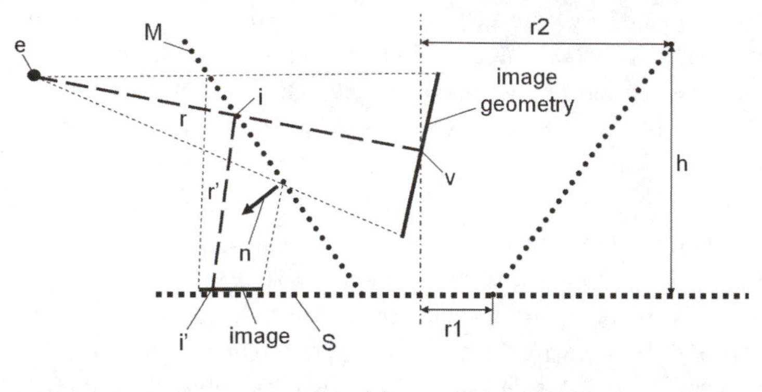

(a)

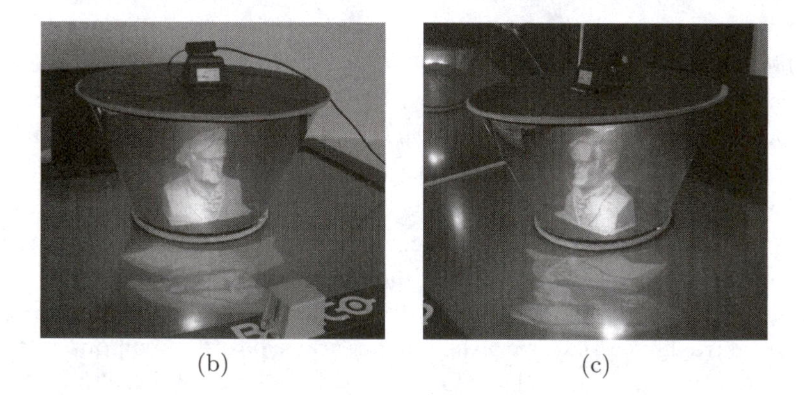

(b) (c)

Figure 6.18. Reflection transformation on a conical mirror. (*Images (b) and (c) reprinted from [9] © IEEE.*)

As explained for projection surfaces in Section 5.4.1, a more efficient way of describing surface shapes for this purpose is to use *parametric functions* in the form of $F(x, y, z) = 0$. Parametric functions to describe mirror and screen surfaces ($M(x, y, z) = 0$ and $S(x, y, z) = 0$) can be used to calculate the intersections and the normal vectors (using its first order derivatives) with an unlimited resolution by solving simple equations. However, not all shapes can be expressed by parametric functions.

Obviously, we have the choice between a numerical and an analytical approach for describing mirror and screen surfaces. If an analytical solution is given, it should be preferred over the numerical variant. Higher-order curved surfaces, however, require the application of numerical approximations.

The parametric function for the conical mirror example and its first-order derivatives are

$$M(x, y, z) = \frac{x^2}{r_1^2} + \frac{y^2}{r_2^2} - \frac{z^2}{h^2} = 0$$

and

$$n = \left[\frac{M}{\delta x}, \frac{M}{\delta y}, \frac{M}{\delta z}\right] = 2\left[\frac{x}{r_1^2}, \frac{y}{r_2^2}, -\frac{z}{h^2}\right].$$

where r_1 and r_2 are the cone's radii with its center located at the world coordinate system's origin, and h is the cone's height along the z-axis.

To intersect the ray r with the mirror surface, it has to be transformed from the world coordinate system into the coordinate system that is used by the parametric function. Then it can be intersected easily with the surface by solving a (in our example, quadratic) equation created by inserting a parametric ray representation of $r(x, y, z)$ into the mirror equation $M(x, y, z)$.

Given the surface intersection i and the normal n at i, the specular reflection ray can be computed by

$$r' = r - 2n(nr).$$

Finally, r' has to be transformed from the coordinate system of M into the coordinate system of S to compute the intersection i' with S by solving another equation created by inserting r' into S.

The following Cg vertex shader fragment can be used to compute the reflection transformation for conical mirror surfaces (this example applies to the truncated conical mirror surface illustrated in Figure 6.18):

```
...

// viewpoint = e[0..2]
// vertex = v[0..3]
// truncated cone parameters: lower radius = r1,
upper radius = r2,
height = h
float3 i, i_, n, r, r_;
float a, b, c, r, s, t, u, l1, l2, l3;

// compute cone parameters
a = r2; b = r2; c = (r2 * h) / (r2 - r1);

// transformation to cone-coordinate system (r = v - e)
```

```
v.z = v.z + (c - h); e.z = e.z + (c - h);

// compute viewing direction (r)
r = v - e; r = normalize(r);

// compute cone intersection
s = (e.x * e.x) / (a * a) +
    (e.y * e.y) / (b * b) -
    (e.z * e.z) / (c * c);
t = 2.0 * ((e.x * r.x) / (a * a) +
           (e.y * r.y) / (b * b) -
           (e.z * r.z) / (c * c));
u = (r.x * r.x) / (a * a) +
    (r.y * r.y) / (b * b) -
    (r.z * r.z) / (c * c);
l1 = (-t + sqrt(t * t - 4.0 * u * s)) / (2.0 * u);
l2 = (-t - sqrt(t * t - 4.0 * u * s)) / (2.0 * u);
i = e + r * l2;

// back-transformation to world coordinate system
i.z = i.z - (c - h);

// compute cone normal
n.x = i.x / (a * a);
n.y = i.y / (b * b);
n.z = i.z / (c * c);
n = normalize(n);

// compute incident ray (i - e)
r = i - e; r = normalize(r);

// compute reflector
r_ = reflect(r,n);

// compute intersection with screen
// (x/y plane in this example)
l1 = -dot(float3(0, 0, 1), i);
l2 =  dot(float3(0, 0, 1), r_);
l3 = l1 / l2;
i_ = i + l3 * r_;

// compute projected vertex
v.xyz = i_.xyz; v.w = 1.0;

...
```

Note, that the same shader can be used in combination with any other parametric functions that describe the mirror and the screen surfaces. Only the surface equation specific parts have to be replaced. Also note the rendering analogy between parametric optical surfaces and parametric projection surfaces (see Chapter 5).

6.7.3 Refraction

Refraction distortion is discussed for planar lenses in Section 6.3.2. A vertex transformation is applied to the rendered geometric model between scene and reflection transformation. To cope with the curvilinearity of refraction, vertex shaders are preferred over a rigid-body approximation if programmable function pipelines are available.

For the image-based method, a per-pixel transformation on the image plane would be more efficient than a per-vertex transformation of the image grid. Pixel shaders of current programmable function pipelines do support a per-pixel displacement mapping. However, holes that are caused by stretching deformations have to be filled. Discrete pixel transformations and the interpolation of empty image portions can be approximated with vertex shaders. The computation of new texture coordinates for each vertex allows offsetting of the corresponding pixel position on the image plane. The positions of pixels in between vertices (i.e., within patches of the image grid) are then approximated via a linear interpolation of the texture mapping. As for the reflection transformation, this introduces optical errors that can be large if the resolution of the image grid is too low.

As shown in Figure 6.19, two surface descriptions (M_o and M_i) are required to compute in- and out-refracted rays for such optical combiners since they represent curved, surface-parallel lenses. As for the reflection transformation, explicit functions in the form of $F(x, y, z) = 0$ are most efficient for computing the required intersections i,i' and the corresponding parameters of the tangent planes.

Similar to the reflection case, each ray from the viewpoint e to all grid vertices v has to be processed. Each ray has to be transformed from the world coordinate system into the coordinate system that is used by the parametric function and then intersected with the outer in-refracting surface M_i. The computed intersection i and its tangential plane parameters are used to determine the in-refractor, which is intersected again with the inner out-refracting surface M_o. The second intersection i' and its tangential plane parameters allow computing the out-refractor. Any point on this ray (x) can be projected onto the image plane from i'. This results in posi-

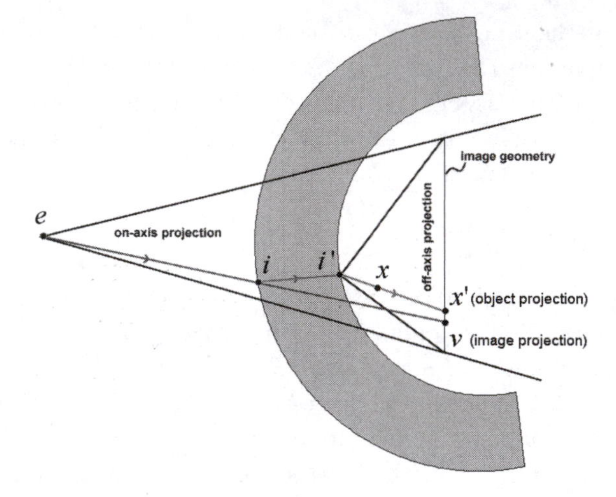

Figure 6.19. Image-based in-/out-refraction transformation on curved lens.

tion x' in projected and normalized device coordinates (e.g., [-1,1]). After a conversion from device coordinates into normalized texture coordinates (e.g., [0,1]), the texture coordinates of the pixel that needs to appear at grid vertex v have been determined. To simulate refraction, these texture coordinates are simply assigned to v. Note, that x' can also be computed by intersecting the out-refractor directly with the image plane. This may require a different conversion from world coordinate systems into normalized texture coordinates.

The composition of an appropriate texture matrix that computes new texture coordinates for the image vertex is summarized in the algorithm below:[1]

```
compute texture normalization correction:
```

$$S = scale(0.5, 0.5, 0.5), s = s \cdot translate(1, 1, 0)$$

```
compute off-axis projection transformation:
```

$$\phi = \frac{(i'_z - 1)}{i'_z}, \quad left = -\phi(1 + i'_x), \quad right = \phi(1 - i'_x)$$
$$bottom = -\phi(1 + i'_y), \quad top = \phi(1 + i'_y)$$
$$near = i'_z - 1, \quad far = i'_z + 1$$

[1]This applies for OpenGL-like definitions of the texture and normalized device coordinates.

$P = frustum(left, right, bottom, top, near, far)$

```
compute view transformation:
```

$V = lookat(i'_x, i'_y, i'_z, i'_x, i'_y, 0, 0, 1, 0)$

`compute new texture coordinate` x' `for particular` $x(v)$`,`
`including perspective division`

$x' = S \cdot \frac{(P \cdot V \cdot x)}{w}$

As illustrated in Figure 6.19, an off-axis projection transformation is used, where the center of projection is i'. Multiplying x by the resulting texture matrix and performing the perspective division projects x to the correct location within the normalized texture space of the image. Finally, the resulting texture coordinate x' has to be assigned to v'.

Note that if the image size is smaller than the size of the allocated texture memory, this difference has to be considered for the normalization correction. In this case, the image's texture coordinates are not bound by the value range of [0,1].

As mentioned previously, the matrix operations can be replaced by an explicit ray casting.

For plane parallel lenses, the in-refracting and out-refracting surfaces are the same but offset by the thickness t of the lens. Explicit in- and out-refractors do not have to be computed in this case. Rather, the offset Δ can be computed for every geometric line of sight $(v - e)$ as described in Section 6.3.2. The new projection on the image plane can be computed by translating the viewpoint e and the corresponding grid vertex v by the amount Δ along the normal of the lens. The new texture coordinates can be computed by projecting v' onto the image plane from e' and converting the result into the normalized texture space.

Nevertheless, both refraction methods face the following problems for outer areas on the image:

- Given a geometric line of sight to an outer grid vertex, its corresponding optical line of sight does not intersect the image. Thus, a grid vertex exists but its new texture coordinate cannot be computed. This results in vertices with no, or wrong, texture information.

- Given an optical line of sight to an outer pixel on the image, its corresponding geometric line of sight does not intersect the image. Thus,

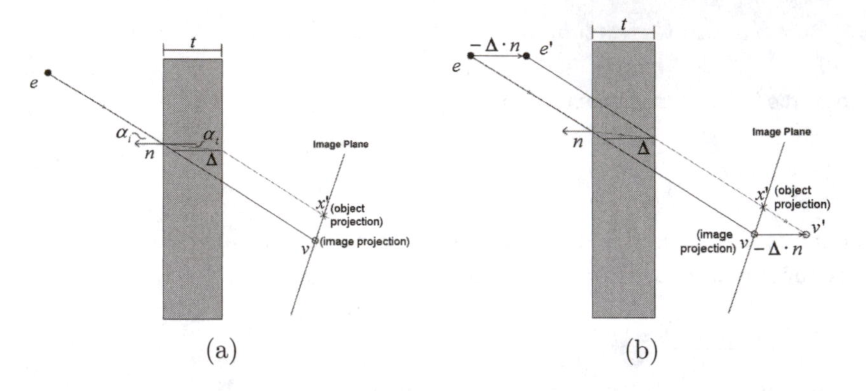

Figure 6.20. Image-based in-/out-refraction transformation on planar lens.

a texture coordinate can be found but an assignable grid vertex does not exist. Consequently, the portion surrounding this pixel cannot be transformed. This results in image portions that aren't mapped onto the image geometry.

A simple solution to address these problems does not avoid them, but ensures that they do not occur for image portions which contain visible information. As described in Section 6.7.1, the image size depends on the radius of the scene's bounding sphere. Its radius can simply be increased by a constant amount before carrying out the first rendering pass. An enlarged image does not affect the image content, but subjoins additional outer image space that does not contain any visible information (i.e., just black pixels). In this way, we ensure that the previously mentioned problems emerge only at the new (black) regions. Yet, these regions will not be visible in the optical combiner.

Note, that in contrast to the image-based reflection transformation (Section 6.7.2) which transforms grid vertices, the refracted image transform remaps texture coordinates. However, all image transformations have to be applied before the final image is displayed during the second rendering pass.

6.7.4 Screen Transformation and Non-Planar Screens

In cases where the screen coordinate system is not equivalent to the world coordinate system in which beam combiners, user(s), and scene(s) are defined, an additional screen transformation has to be used to project the reflected/refracted image-geometry vertices to their correct places. This can be achieved with additional affine transformation operations that are applied to the image-geometry itself, as described in Section 6.4, or by intersecting the traced rays with the geometry of the transformed screen plane (e.g., if explicit ray casting is used inside a vertex shader).

If the secondary screen is not planar, an explicit ray-casting approach still leads to a correct result. The traced rays simply have to be intersected with a geometric representation of the curved screen.

Note that an approach that uses projective texture-mapping [168] for planar mirror beam combiners fails for a combination of curved mirror optics and curved screens. Although the first rendering pass is similar to the first pass that generates the image (see Section 6.7.1), the second pass cannot be used, because multiple centers of projection exist (one for each transformed image vertex).

6.7.5 Displaying Images

During the second rendering pass, the transformed image geometry is finally displayed within the screen space, mapping the outcome of the first rendering pass as texture onto its surface.

Note, that if the reflection/refraction transformations of the previous steps deliver device coordinates, but the secondary screen and the mirror optics are defined within the world coordinate system, a second projection transformation (such as `glFrustum`) and the corresponding perspective divisions and viewpoint transformation (such as `gluLookAt`) aren't required. If a plane secondary screen is used, a simple scale transformation suffices to normalize the device coordinates for example, `glScale(1/device_width/2, 1/device_height/2, 1)`. A subsequent view-port transformation finally scales them into the window-coordinate system for example, `glViewport(0, 0, window_width, window_height)`.

Time-consuming rendering operations that aren't required to display the two-dimensional image (such as illumination computations, back-face culling, depth buffering, and so on) should be disabled to increase the rendering performance. In this case, the polygon order doesn't need to be reversed before rendering, as we noted previously.

Obviously, one can choose between numerical and analytical approaches to represent curved mirror surfaces. Simple shapes can be expressed as parametric functions, but higher-order curved mirrors require numerical approximations. In addition, the grid resolution that is required for the image geometry also depends on the mirror's shape. Pixels between the triangles of the deformed image mesh are linearly approximated during rasterization (that is, after the second rendering pass). Thus, some image portions stretch the texture while others compress it. This results in different regional image resolutions. However, because of the symmetry of simple mirror setups (such as cones and cylinders), a regular grid resolution and a uniform image resolution achieve acceptable image quality. In Section 6.7.8, a selective refinement method is described that generates a non-uniform image geometry to minimize the displacement error of the image portions, the complexity of the image geometry, and, consequently, the number of vertex transformations and triangles to be rendered.

Since primitive-based (or fragment-based) *antialiasing* doesn't apply in deformed texture cases, bilinear or trilinear texture filters can be used instead. As with antialiasing, texture filtering is usually supported by the graphics hardware.

Note that the image's background and the empty area on the secondary screen must be rendered in black, because black doesn't emit light and therefore won't be reflected into the reflection space.

6.7.6 Multiple Viewers

To support multi-user applications, the individual viewer images must be composed and the black background must be *color-blended* appropriately. For convex beam combiners the images are stretched within the screen

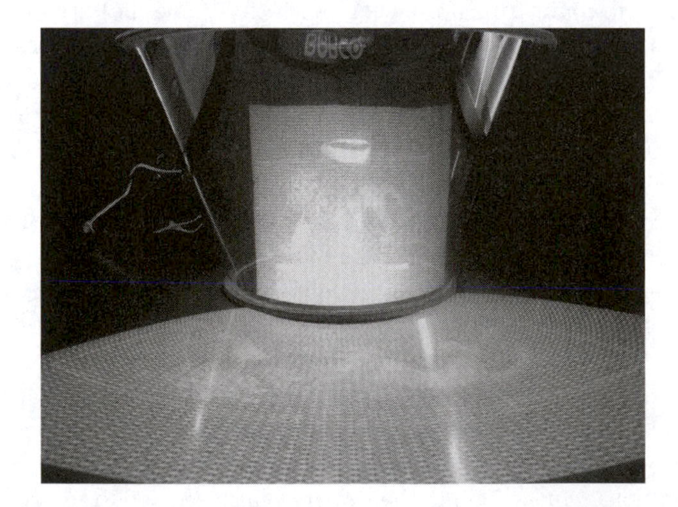

Figure 6.21. Stretched image geometry projected on secondary screen and reflected in conical mirror beam combiner. (*Image reprinted from [16] © Elsevier.*)

space to appear correctly within the reflection space (Figure 6.21), therefore multiple images for different observers might intersect.

In these cases, individual observers can perceive the (perspectively wrong) images of other users in addition to their own images. The amount of intersection depends on the size of the graphical scene, the positions of the observers, and the parameters of the mirror optics and the secondary screen. The larger the graphical scene, for instance, the more space on the secondary screen is required, and conflicting situations become more likely. Such conflicts also arise for multi-plane beam combiner configurations if multiple users are allowed to move freely around the display (as it has been described for a single user in Section 6.6.1).

6.7.7 Concave Beam Splitters

Concave mirrors can generate both real images and virtual images. Light rays which are reflected off convex mirror assemblies do not intersect. In contrast, light rays that are reflected off a *parabolic concave mirror* do intersect exactly within its focal point. For concave mirrors that deviate from a parabolic shape (e.g., *spherical mirrors*) the light rays do not intersect within a single point, but bundle within an *area of rejuvenation*.

The rendering techniques for curved mirror beam combiners are described on the basis of convex mirrors. However, they can be used for concave mirrors without or with only minor modifications to the algorithms.

Figure 6.22 illustrates how, given the viewpoint and the location of the mirror, the geometry of a cube is transformed by the algorithm described

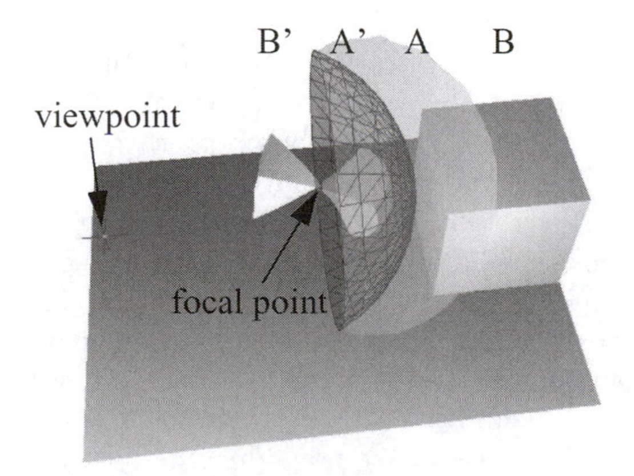

Figure 6.22. Reflected geometry at a concave mirror.

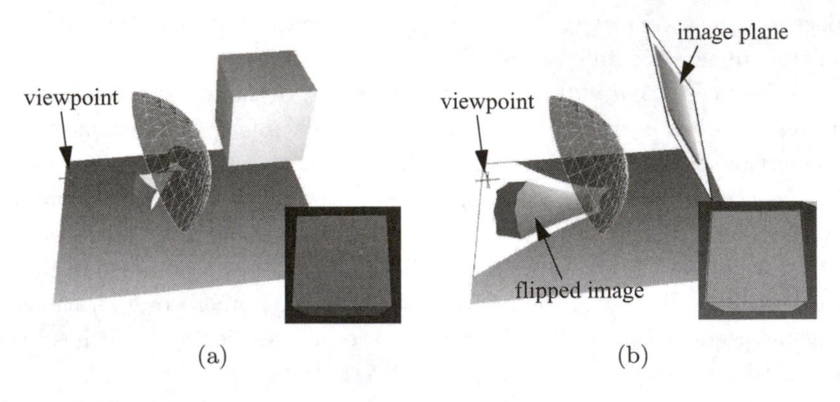

Figure 6.23. (a) Reflecting scene geometry; (b) reflecting image geometry at a concave mirror.

in Section 6.6.2 (without projection onto the screen plane). Note, that the vertices of a highly tessellated surface approximation of the cube are reflected, instead of the vertices of the image geometry.

The portion of the geometry that is located within area A is mapped to area A' (behind the mirror's focal point as seen from the viewpoint). This mapping has a similar behavior as for convex mirrors. The portion of the geometry that is located within area B is mapped to area B' (in front of the mirror's focal point as seen from the viewpoint). In this case, the mapped geometry is flipped and the polygon order is changed. The geometry that is located on the intersecting surface of A and B is mapped to the focal point.

Figure 6.23 shows that concave mirrors can be treated just like convex mirrors. The left image indicates the reflection of the scene geometry itself (again, without projection onto the screen plane). The transformed geometry and the mirror model have been exported to a ray tracer, and the image that is seen from the viewpoint has been ray traced. The result is outlined at the lower right of Figure 6.23(a) (dark cube).

It can be seen, that the transformed geometry appears untransformed as a reflection in the mirror. Note that due to the different orders of polygons that are located in front of and behind the focal point, polygon order related options offered by the applied rendering package should be disabled. Otherwise, additional actions have to be taken to determine the location of each transformed polygon and its order.

The same experiment has been carried out reflecting the image geometry (including the projection onto the screen plane). The image geometry was transformed and the image that was generated during the first rendering pass was mapped onto the transformed image grid. Figure 6.23(b) illustrates that the projected image is a flipped version of the original image. If ray-traced from the given viewpoint, this deformed image appears as a correct reflection in the mirror (see lower right of the right image). Note that the contours of the ray-traced result have been highlighted, since the planar projection of the deformed image does not provide sufficient normal or depth information to generate correct shading effects with the ray tracer. Note also, that reflecting the scene geometry and projecting the result onto the screen plane yields the same results as reflecting and projecting the image geometry.

Mirrors with a *mixed convexity* (i.e., simultaneously convex and concave mirrors) can cause multiple images of the same object, or they can reflect multiple objects to the same location within the image space. In such cases, the transformations from screen space to reflection space and vice versa are not definite and are represented by many-to-many mappings. Such types of mirrors should be decomposed into convex and concave parts (as done by Ofek [122, 123]) to ensure a correct functioning of the algorithms. Although for many surfaces this can be done fully automatically [172], spatial optical see-through configurations do not make use of mirrors with mixed convexities. This is because one of the initial goals of such displays, namely to unequivocally overlay real environments with computer graphics in an AR manner, is physically not supported by such mirrors. Consequently, mirrors with mixed convexities are not considered for spatial augmented reality.

6.7.8 Non-Uniform Image Geometry

If image warping is used for a *non-linear predistortion*, the required grid resolution of the underlying image geometry depends on the *degree of curvature* that is introduced by the display (e.g., caused by the properties of a mirror, a lens, or a projection surface).

Pixels within the triangles of the warped image mesh are linearly approximated during rasterization (i.e., after the second rendering pass). Thus, some image portions stretch the texture, while others compress it. This results in different local image resolutions. If, on the one hand, the geometric resolution of an applied uniform grid is too coarse, texture artifacts are generated during rasterization (Figure 6.24(a)). This happens because a piecewise *bi-* or *tri-linear interpolation* within large triangles

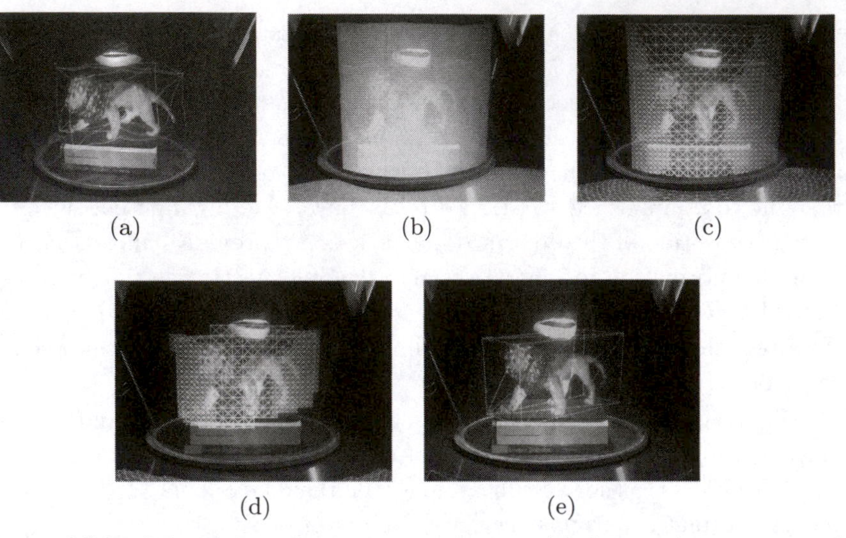

Figure 6.24. Selectively refined image geometry: (a) distorted image created with an undersampled uniform image grid; (b) oversampled uniform image grid; (c) undistorted image; (d) selectively refined image grid; (e) grid portion that projects onto bounding container. (*Images reprinted from [16] © Elsevier.*)

is only a crude approximation to neutralize the curvilinear optical transformations.

If, on the other hand, the grid is oversampled to make the interpolation sections smaller (Figure 6.24(b)), interactive frame rates cannot be achieved. In addition to the speed of the first rendering pass, the performance of view-dependent multi-pass methods depends mainly on the complexity of the image geometry that influences geometric transformation times and on the image resolution that influences *texture transfer* and *filter times*.

This section describes a selective refinement method [16] that generates image grids with appropriate local grid resolutions on the fly. This method avoids oversampling and the occurrence of artifacts within the final image.

The main difficulty for a selective refinement method that supports curved mirror beam combiners is that in contrast to simpler screens, a *displacement error* that defines the refinement criterion has to be computed within the image space of the mirror optics, rather than in the screen space of the display. For convexly curved mirror beam combiners, this requires fast and precise numerical methods. For displays that do not contain view-dependent optical components (e.g., curved screens), these computations are much simpler because analytical methods or look-up tables can be used to determine the error within the screen space.

Recent advances in *level-of-detail* (LOD) *rendering* take advantage of *temporal coherence* to *adaptively refine* geometry between subsequent frames. Terrain-rendering algorithms locally enhance terrain models by considering viewing parameters.

Hoppe introduced *progressive meshes* [65] and later developed a view-dependent refinement algorithm for progressive meshes [66, 67]. Given a complex triangle mesh, Hoppe first pregenerates a coarse representation called *base mesh* by applying a series of *edge collapse operations*. A sequence of precomputed *vertex split operations* that are inverse to the corresponding edge collapse operations can then be applied to the base mesh's regions of interest to successively refine them.

The selection of the appropriate vertex split operations is based on his refinement criteria. Lindstrom [90] describes a method that generates a view-dependent and continuous LOD of height fields dynamically in real time, instead of precomputing a coarse base mesh and a sequence of refinement steps. He hierarchically subdivides a regular height field into a *quadtree* of *discrete grid blocks* with individual LODs. Beginning with the highest LOD, Lindstrom locally applies a two-step surface simplification method. He first determines which discrete LOD is needed for a particular region by applying a coarse *block-based simplification*, and then performs a *fine-grained retriangulation* of each LOD model in which vertices can be removed. To satisfy continuity among the different LODs, Lindstrom considers vertex dependencies at the block boundaries.

The main difference between both methods is that Lindstrom performs a dynamic simplification of high-resolution height fields for domains in R^2 during rendering. Lindstrom's mesh definition provides an *implicit hierarchical LOD structure*. Hoppe uses refinement steps to low-resolution LODs of arbitrary meshes during rendering. His mesh definition is more general and does not require an implicit hierarchical LOD structure. Consequently, the refinement steps and the low resolution base mesh have to be precomputed. In addition, he uses *triangulated irregular networks* (TINs) for triangulation, rather than regular grids. Note that these two types of refinement methods may be representative for related techniques. Since the image geometry for spatial optical see-through displays can also be parameterized in R^2 and provides an implicit hierarchical LOD structure, multiple LODs or appropriate refinement steps do not need to be precomputed but can be efficiently determined on the fly. This is similar to Lindstrom's approach. However, simplifying a high-resolution mesh instead of refining a low-resolution mesh would require a retransformation of all grid vertices of the highest LOD after a change in the viewpoint occurred. This is very inefficient, since for the type of displays that we consider, viewpoint changes normally happen at each frame.

In contrast to the static geometry that is assumed in Lindstrom's and Hoppe's cases, the image geometry generated for spatial optical see-through displays is not constant, but dynamically deforms with a moving viewpoint. Consequently, the geometry within all LODs dynamically changes as well.

This section describes a method that dynamically deforms the image geometry within the required LODs while *selectively refining* the lowest-resolution base mesh during rendering. The method aims at minimizing the *displacement error* of the image portions, the complexity of the image geometry, and, consequently, the number of vertex transformations and triangles to be rendered.

Note that this method cannot be implemented with the limited capabilities of vertex shaders, yet. Consequently it has to be executed on the CPU rather than on the GPU. This implies that rendering a high-resolution uniform image grid on a GPU might be faster than creating a selectively refined image grid on the CPU. Nevertheless, the LOD method is explained to impart knowledge about displacement errors that are generated by approximating a curvilinear optical transformation with discretized image grids. Future generations of vertex shaders will offer more possibilities. Recursions that enable the implementation of the following method completely in a shader might become feasible in the future.

Image triangulation. Instead of transforming and rendering a uniform high-resolution mesh, one can start from the coarsest geometry representation and successively refine it locally until certain *refinement criteria* are satisfied. Due to the well-defined structure of the image grid, all possible *global or discrete LODs* can be computed at runtime including the highest, which is the uniform high-resolution representation of the mesh itself.

Figure 6.25 illustrates a quadtree-based image triangulation, which is similar to Lindstrom's triangulation method for height fields [90]. While Figure 6.25(a) shows an unrefined patch at LOD n, Figure 6.25(b) shows the same patch at LOD $n+1$, with lower LOD neighbors. Given a highest LOD of m, an indexed $(2^m+1) \times (2^m+1)$ matrix structure can be chosen to store the grid vertices.

For illustration purposes the following types of patch vertices can be differentiated:

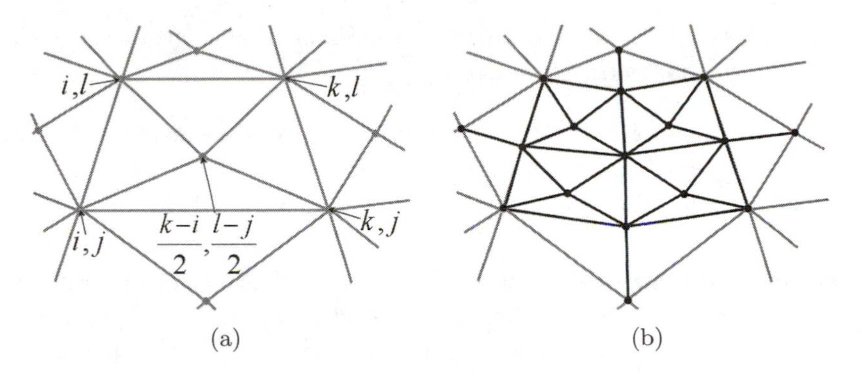

Figure 6.25. (a) Triangulation of unrefined patch at LOD n; (b) triangulation of refined patch at LOD $n+1$ with resolution transitions. (*Images reprinted from [16] © Elsevier.*)

- *L-vertices* are vertices at the corners of a patch (e.g., at indices $[i,j]$,$[i,l]$,$[k,l]$ and $[k,j]$ in Figure 6.25(a));
- *X-vertices* are vertices at the center of a patch (e.g., at index $[(k-i)/2,(l-j)/2]$ in Figure 6.25(a));
- *T-vertices* are vertices that split the patch edges after refinement (e.g., at indices $[i,(l-j)/2]$, $[k,(l-j)/2]$, $[(k-i)/2,l]$, and $[(k-i)/2,j]$ in Figure 6.25(b))

To refine a patch, it is divided into four sub-patches by computing the corresponding four T-vertices, as well as the four X-vertices that lie inside the sub-patches. Note that the matrix structure in this particular case is equivalent to a *quadtree data structure*. To ensure consistency during rasterization, the T-vertices have to be connected to their neighboring X-vertices wherever a LOD transition occurs (e.g., at all four neighbors of the refined patch, as shown in Figure 6.25(b)).

Due to the well-defined matrix structure that contains the image grid, the following conditions are given:

1. A clear relationship between the X-vertices and T-vertices exists. X-vertices can never be T-vertices and vice versa.
2. Each patch has definite L-vertices, T-vertices, and X-vertices, whose indices can always be computed.
3. Each X-vertex can be explicitly assigned to a single patch at a specific LOD.

4. Each T-vertex can be explicitly assigned to exactly one or two adjacent patches at the same LOD.

The triangulation methods described require *continuous level-of-detail transitions* [111]. This implies that neighboring patches do not differ by more than one LOD.

Recursive grid refinement. The objective of this step is to generate an image grid that provides a sufficient local grid resolution (i.e., appropriate discrete LODs) to avoid artifacts within the rasterized texture that would result from undersampling, as well as oversampling.

The following pseudocode illustrates an approach to recursively refine a grid patch, which initially is equivalent to the lowest LOD (i.e., the patch at the lowest LOD is outlined by the L-vertices at the four corners of the image geometry):

```
RecursiveGridRefinement(i, j, k, l)
1: begin
2: a = (k − i)/2, b = (l − i)/2
3: if GeneratePatch([i, j],[i, l],[k, l],[k, j],[a, b])
4: begin
5:    TransformPatchVertices([i, j],[i, l],[k, l],[k, j],[a, b])
6:    P = P ∪ {[i, j, k, l]}
7:    if RefineFurther([i, j],[i, l],[k, l],[k, j],[a, b])
8:    begin
9:      RecursiveGridRefinement(i, j, a, b)
10:     RecursiveGridRefinement(a, j, k, b)
11:     RecursiveGridRefinement(i, b, a, l)
12:     RecursiveGridRefinement(a, b, k, l)
13:     if j < 2^m + 1 TC[a, j] += 1
14:     if k < 2^m + 1 TC[k, b] += 1
15:     if l > 1 TC[a, l] += 1
16:     if i > 1 TC[i, b] += 1
17:     end
18:   end
19: end
```

The patch that has to be refined is stored at indices $[i, j, k, l]$ within a matrix structure. Condition (2) allows us to locate the position of the patch (individual X-vertex) at indices $[a, b]$ (line 2). First, it is evaluated whether or not a patch has to be generated at all (line 3). The four L-vertices and the X-vertex are transformed from the image plane to the display surface (line 5), as described in detail in Sections 6.7.2 and 6.7.3. Normally the image plane is located within the image space of the mirror optics. In this case, these mappings composite the individual vertex model-view transformations to neutralize reflection and refraction, as well as the projection transformation that maps a vertex onto the display surface. Note that vertices are only transformed once even if the recursive refinement function addresses them multiple times. This is realized by attaching a marker flag to each vertex. A reference to the transformed patch is stored in patch set P by adding the patch's indices to P (line 6). In line 7, a function is called that evaluates the transformed patch based on predefined refinement criteria and decides whether or not this patch has to be further refined. The main refinement criterion is described below. If this decision is positive, the patch is divided into four equal sub-patches, and the refinement function is recursively called for all of these sub-patches (lines 9–12). Note that Condition (2) also allows us to determine the indices of the patch's four T-vertices, which become L-vertices of the sub-patches in the next LOD. Consequently, the `GeneratePatch` and the `RefineFurther` functions represent the exit conditions for the recursion.

It was mentioned that T-vertices have to be connected to their neighboring X-vertices whenever an LOD transition occurs to ensure consistency during rasterization. To detect LOD transitions, a counter (TC) is attached to each T-vertex. This counter is incremented by 1, each time the corresponding T-vertex is addressed during the recursive refinement (lines 13–16). Note that the if-conditions ensure a correct behavior of the counter at the image geometry's boundaries. Due to Condition (4), each counter can have one of the following three values:

- 0—indicates that the T-vertex is located at a boundary edge of the image geometry or it is contained by a discrete LOD that is higher than the required one for the corresponding region.

- 1—indicates a LOD transition between the two neighboring patches that, with respect to Condition (4), belong to the T-vertex.

- 2—indicates no resolution transition between the two neighboring patches that belong to the T-vertex.

After the image grid has been completely generated, all patches that are referred to in P are rendered with appropriate texture coordinates during the second rendering pass. Thereby, the counters of the patch's four

T-vertices are evaluated. Depending on their values, either one or two triangles are rendered for each counter. These triangles form the final patch. Counter values of 0 or 2 indicate no LOD transition between adjacent patches. Consequently, a single triangle can be rendered which is spanned by the T-vertex's neighboring two L-vertices and the patch's X-vertex (this is illustrated in Figure 6.25(b)). A counter value of 1, however, indicates a LOD transition. Two triangles have to be rendered that are spanned by the T-vertex itself, the two neighboring L-vertices, and the X-vertex of the adjacent patch (this is illustrated in Figure 6.25(a)). A selectively refined image geometry is shown in Figure 6.24(d).

Generation and refinement criteria. This section discusses the patch generation and refinement criteria that are implemented within the `GeneratePatch` and `RefineFurther` functions. The input for these functions is the four L-vertices, as well as the X-vertex of a patch. They deliver the Boolean value *true* if the patch has to be generated, transformed, rendered, or further refined, or *false* if this is not the case.

In general, the `RefineFurther` function can represent a Boolean concatenation of multiple refinement criteria (such as maximal patch size, angular deformation, etc.). An important refinement criterion for LOD methods is the *screen space error*. Since the computations of this displacement error are display specific, an important variation of the screen space error that can be applied for convex mirror beam combiners is described—the *image space error*.

Spatial limits. The multi-pass rendering method that is described in Sections 6.6.1–6.7.7 uses the scene's *bounding sphere* to determine the parameters of the symmetric viewing frustum and the image size (Figure 6.17). Since all image generation methods assume a rectangular image shape that is adapted to today's screen shapes, the bounding sphere provides enough information to determine the rectangular image size.

Bounding spheres, however, are only rough approximations of the scene's extensions and consequently cover a fairly large amount of void space. This void space results in grid patches on the image plane whose texture does not contain visible color information.

To achieve a speed-up, these patches are avoided while creating the image grid. This implies that they are not transformed and refined during the `RecursiveGridRefinement` algorithm and that they are not rendered during the second rendering pass. As a result, an image grid is generated and rendered that is not rectangular, but dynamically approximates the silhouette of the scene as perceived from the observer's perspective (Figure 6.24(e)). A condition that causes the recursion to exit in these cases is implemented within the `GeneratePatch` function.

A tighter convex container (such as an oriented convex hull or a bounding box) of the scene is evaluated that is generated either in a preprocess for static objects, or at run-time for animated scenes. For each untransformed patch that is passed recursively into the `RecursiveGridRefinement` algorithm, it has to be determined whether the container is visible on that patch, partially or as a whole. The approximation is twofold. First, the geometric lines of sight from e to all four L-vertices of the patch are intersected with the front-facing portion of the container. Second, the geometric lines of sight from e to the front-facing container vertices are intersected with the patch. If at least one of the resulting rays causes an intersection, the patch might contain visible color information, and it will be further processed. If, however, none of the rays cause intersections, the patch is not treated further (i.e., it won't be transformed, refined, or rendered). If the convex container is represented as a triangle mesh, a fast *ray-triangle intersection* method [113] is used together with the front-facing container triangles. Note, that as for vertex transformations, the intersection information is buffered and looked up in the memory, rather than recomputing it multiple times while evaluating adjacent patches.

Oriented convex hulls can be used as containers. It is obvious that the complexity of the container influences the performance of this method. Although a precise container can eliminate a maximum number of patches, the number of intersection tests increases with the container's number of vertices and polygons. Experiments have shown that the highest speed-ups are reached if the container is as simple as an oriented bounding box or a very coarse, but tighter, convex hull (Figure 6.26). However, the complexity of the convex hull that maximizes the speed-up and balances intersection tests with patch generations depends on the scene and the required rendering precision.

Image space error. The consideration of the screen space error that is determined relative to a display surface is a common measure for many computer graphics methods (e.g., [66, 67, 90]). In contrast to traditional displays, curved mirror beam combiners create a view-dependent, non-planar image surface which is located inside the mirror optics. Consequently, an appropriate error function has to consider this optical transformation. The error is determined by first warping the curved image surface into an image plane and then computing the screen space error on this plane. This error is referred to as *image space error* (δ_{is}).

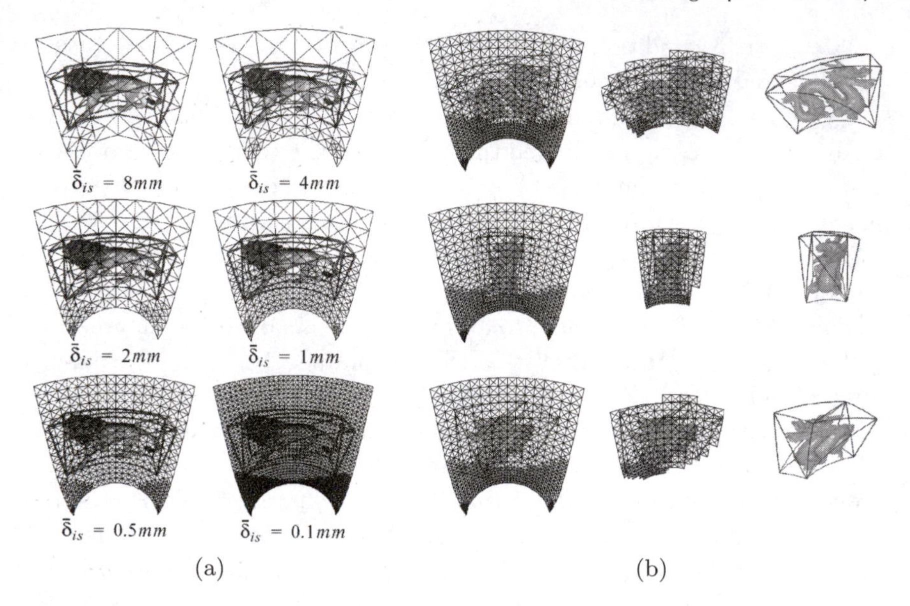

Figure 6.26. (a) Selective grid refinement for different image space error thresholds; (b) spatially limited grids for different perspectives. (*Images reprinted from [16] © Elsevier.*)

The image space error is a variation of a *screen space error* that can be computed for convex mirror beam combiners which present the image plane within the image space of their optics, rather than on a display surface. The image space error is defined as the geometric distance between the desired position (v_d) and the actual appearance (v_a) of a point on the image plane. Consequently, the image space error is given by

$$\delta_{is} = |v_d - v_a| .$$

It delivers results in image space coordinates (e.g., mm in this case).

In case of convexly curved mirror optics, the image space is the reflection of the screen space (i.e., the secondary screen in front of the mirror optics) that optically overlays the real environment inside the mirror optics. In addition, the optically deformed pixels do not maintain a uniform size within the image space. They are deformed in exactly the same way as the entire image after being reflected from the screen space into the image space, although on a smaller scale. Consequently, the Euclidean distance between geometric points can be chosen as an error metric, rather than expressing the image space error with a uniform pixel size.

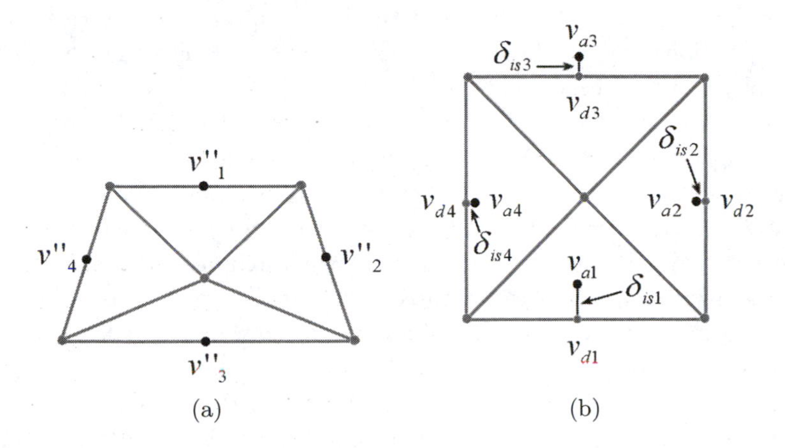

Figure 6.27. (a) Samples on the transformed patch; (b) the distance between desired and actual appearance of samples near the untransformed patch results in the image space error. (*Images reprinted from [16] © Elsevier.*)

For any given pixel on the transformed patch with texture coordinates u and v, δ_{is} can be computed as follows (Figure 6.27):

First, the pixel's world coordinate v'' at u,v within the screen space (i.e., on the secondary display surface) is determined. Note that the pixels, which are mapped onto the patch's transformed vertices, optically appear at their correct locations on the image plane inside the image space. This is because their exact mappings have been determined during the patch's transformation. This transformation considers the laws of geometric optics (i.e., reflection and refraction laws). Note also that the pixels that are displayed anywhere else (i.e., inside one of a patch's triangles) do not necessarily appear at their correct locations on the image plane. This is because their positions on the display surface are approximated by a *linear texture interpolation*, rather than by optical laws.

The second step is to determine the position of the optical image (v') of v'' within the image space of the mirror optics. The projection of v' onto the image plane results in v_a.

In an ideal case, v_a is located at the position that also maps to the texture coordinates u,v within the untransformed patch. We can identify the location which does this as the desired image position v_d. However, if $v_a \neq v_d$, the image space error δ_{is} for this pixel is non-zero.

The image space errors for the four points on the transformed patch (Figure 6.27(a)) that should map to the patch's T-vertices on the untransformed patch (Figure 6.27(b)) as if the image space errors were zero for these positions can be computed. Obviously, this is not the case in the example, shown in Figure 6.27(b).

Since the untransformed patch is a *rectangular quad*, small image space errors suggest that the optical mapping between the transformed and untransformed patch is linear. Furthermore, it can then be concluded that a linear texture interpolation within the displayed transformed patch produces approximately correct results while being mapped (i.e., reflected and refracted) into image space. Consequently, it can be assumed that the resulting image space errors describe a patch's curvilinearity at the representative pixel locations.

To decide whether or not a patch has to be further refined, the largest of the four image space errors is determined. If it is above a predefined threshold value $\bar{\delta}_{is}$ the patch has to be further refined and `RefineFurther` returns *true*.

Computing object-image reflections. To compute v_a from a given viewpoint e, the object point v'' and the optic's geometry is equivalent to finding the *extremal Fermat path* from v'' to e via the optics. In general, this would be a difficult problem of *variational calculus*.

Beside ray- and beam-tracing approaches, several methods have been proposed that approximate reflection on curved mirrors to simulate global illumination phenomena within rendered three-dimensional scenes. All of these methods face the previously mentioned problem in one or the other way.

Mitchell and Hanrahan [111], for instance, solve a multidimensional non-linear optimization problem for explicitly defined mirror objects ($g(x) = 0$) with *interval techniques*. To compute reflected illumination from curved mirror surfaces, they seek the *osculation ellipsoid* that is tangent to the mirror surface, whereby its two focal points match the light source and the object point.

For a given viewpoint e, Ofek and Rappoport [122] spatially subdivide the screen space into *truncated tri-pyramid shaped cells*. In contrast to solving an optimization problem, they apply *accelerated search techniques* to find the corresponding cell that contains the object v'' at interactive rates.

While Mitchell's *multidimensional optimization* approach is far from being used at interactive rates, Ofek's search method offers a good approximation for rendering global illumination effects (such as reflections), but does not provide the precision required by an optical display.

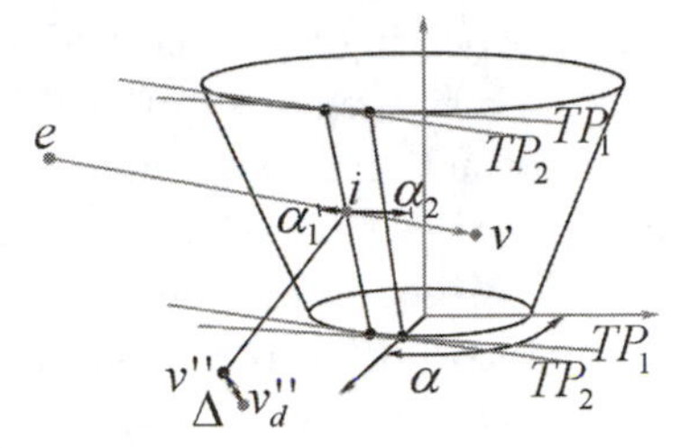

Figure 6.28. Object-image reflection via numerical minimization. (*Image reprinted from [16] © Elsevier.*)

In the following, a *numerical minimization* method to compute the object-image reflection for specific parametric mirror surfaces (such as cones and cylinders) is described. For such surfaces, the optimization problem is reduced to only one dimension. Consequently, the method provides an appropriate precision at interactive rates.

For the subsequent example, the same cone-shaped mirror surface is chosen as used for the examples in Sections 6.6.1–6.7.8.

Cones and similar *bodies of revolution* have the property that multiple surface points lie on a common plane. For example, all points on the straight line spanned by a cone's peak and an arbitrary point on its bottom circle lie on the same plane. Consequently, individual plane parameters can be determined for all angles around the cone's principle axis.

To determine an object's (v''_d) image for a given viewpoint e and mirror surface $g(x) = 0$, an arbitrary angle α around the cone's principle axis is assumed. Then the surface's tangent plane TP_1 is determined at α by computing a surface point and the surface normal at α. Since $g(x)$ is an explicit function, the surface normal can be computed by using its first-order derivatives $g/(\partial x)$. Next, v''_d is reflected over TP_1 to its corresponding position within the image space, and it is projected onto the image plane. In Figure 6.28, the projected image point is denoted by v.

To verify the quality of this assumption, v is reflected back into the screen space and projected onto the display surface. For a given v, $g(x)$ and e, a simple analytical solution exists to determine this image-object transformation. The ray spanned by e and v is intersected with $g(x)$ by solving a simple quadratic equation. In the case of a conical beam combiner, the quadratic equation is given by inserting the linear ray equation into the quadratic equation of a cone and solving for x. The surface intersection i, together with the normal vector at i (again determined using the surface's first-order derivatives), gives the tangent plane TP_2 at i. Reflecting v and e over TP_2 and projecting the reflection of v onto the display surface using the reflection of e as the center of projection, results in point v''. The image-object transformation is illustrated in Figure 6.24(c). Note that for simplicity, the image-object transformation and the *object-image transformation* have been described as simple reflection/projection transformations. Normally, they incorporate refraction as well, as described in Section 6.7.3.

If the tangent plane at α produces the extremal Fermat path between v''_d and e, the geometric distance Δ between v''_d and v'' is zero, $TP_1 = TP_2$, and $v_a = v$. Otherwise Δ is non-zero.

To approximate the correct α, the function is minimized for Δ. Since this function depends only on α, fast numerical optimizers for one dimension can be used. However, because its first-order derivatives cannot easily be derived but it appears to be nicely parabolic near its minimum, Brent's inverse parabolic interpolation [23] with bracketing is used.

To speed up the minimization process (i.e., to reduce the number of function iterations), the function range for a particular v''_d can be constrained. As illustrated in Figure 6.28, the minimization is restricted to find α between α_1 and α_2. These boundaries can be determined as follows: Given e and the untransformed patch that belongs to v''_d, the projections on the mirror surface of the untransformed patch at the next lower LOD are evaluated. The two angles α_1 and α_2 at the horizontal extrema of these projections are the corresponding *boundary angles*. Note that these projections are determined while transforming the patches (i.e., within `TransformPatchVertices`). Thus, for efficiency reasons, they are stored and looked up in the memory, rather than recomputed again.

Experiments showed that sufficiently precise solutions can be found after a small number of iterations. Typically, average errors of $\Delta = 0.002$ mm are achieved with an average of three iterations.

Error direction propagation. Although the number of iterations is relatively small, the computational expenses of four minimization steps per patch result in a noticeable loss of performance. Especially while evaluating the large number of higher LOD patches, such an approach might not produce a speed-up.

It can be noted that the direction in which the largest image space error (i.e., the one of the four patch sides where δ_{is} is maximal) propagates is the same for larger LOD patches that are derived from the same parent patch.

However, from which level-of-detail this occurs depends on the display's properties. If, for instance, the optics produce well-behaved image deformations, the propagation direction of the largest error is consistent for relatively high LODs. If, on the other hand, the optics produces noisy images, the error direction alternates randomly.

To benefit from this situation, we specify a LOD Λ depending on the optic's and the display surface's properties.

For patches that are below Λ, the image space error is determined as described above. The value of δ_{is} is computed at all four edges and the largest value is found. In addition, the error direction (i.e., the edge where δ_{is} is maximal) is recorded for each patch.

For patches that are above Λ, the error direction of the corresponding parent patch can be reused and δ_{is} can be computed only for the edge in this direction, rather than at all four edges. By doing this, it is assumed that the largest error will occur in the same direction as for the parent patch. Consequently, the number of minimization steps is reduced from four to one for all patches that are above Λ (i.e., for the majority of all relevant grid patches). Experiments have shown that this heuristic leads to a significant speed-up while producing the same final visual output.

Note that although the examples deal with a simple mirror shape (i.e., a cone), an algorithm that uses a precomputed look-up table (e.g., such as the one described by Ofek [122]) instead of dynamic numerical minimizations would either result in fairly large data structures that cannot be efficiently searched, or in a precision that is not adequate for an optical display.

Display specific components. While the described algorithm is valid for other displays that require non-linear predistortion, its display-specific components have been explained based on a particular mirror display, a cone-shaped beam combiner. The nature of the additional mirror optics makes the transformation of the grid patches and the computation of the resulting displacement error fairly complex. In fact, the implementation of the `TransformPatchVertices` and the `RefineFurther` functions have been explained with an emphasis on cone-shaped mirror surfaces. These two functions represent the display-specific components of our algorithm. For a mirror display, `TransformPatchVertices` and `RefineFurther` have to consider laws of geometric optics (such as reflection and refraction transformations) to map grid vertices from the image plane onto the display surface and vice versa. They can be generalized to do the same for other displays without modifying the general algorithm.

If, for instance, the algorithm is used to project images onto curved screens (e.g., a cylindrical or spherical projection device), `TransformPatchVertices` would incorporate only projection transformations (i.e., it would only determine intersections of vertex-individual geometric lines of sight with the display surface). The resulting displacement error that is computed by `RefineFurther` can then be determined within the screen space, rather than within an image space. Compared to our numerical approach for convex mirror displays, this would be less complex since it involves only simple intersection computations for which analytical solutions exist. If view dependence is not required, `TransformPatchVertices` and `RefineFurther` could also retrieve precomputed values from a look-up table.

For curved mirror beam combiners that require a correction of a non-linear distortion by applying multi-pass rendering, generating appropriate local levels of detail instead of applying uniform grids or constant image geometries allows us to consider the error that is caused from a piecewise linear texture interpolation and to minimize it by adapting the underlying geometry.

The method described previously prevents oversampling and texture artifacts that result from undersampling. If executed on CPUs, a significant speed-up can be achieved by rendering a selectively refined image geometry instead of a high-resolution uniform image grid. However, the capability of per-vertex operations of today's GPUs beats the computational overhead that is required for recursion and minimization with a faster transformation of high-resolution uniform image geometries, as described in Sections 6.6.2 and 6.7.3.

6.8 Summary and Discussion

The basic principles of geometric optics represent the mathematical, physical, and physiological foundations for this chapter. Planar mirror beam combiners that produce true stigmatic reflections between all object-image pairs, and convexly curved mirror beam combiners that represent non-absolute optical systems represent the main components for the creation of optical overlays in combination with spatial secondary screens.

If such beam combiners are integrated into the optical path between observer and secondary screen, the perceived graphical images are transformed and distorted by the optical elements. We have discussed real-time rendering techniques that are capable of neutralizing these effects for different display configurations in such a way that orthoscopic, stereoscopically, and perspectively correct and undistorted images can be seen from any viewing position. Transparent primary screens, such as active or passive screens, which can be utilized for spatial optical combination have been outlined briefly, but details on specific rendering techniques that support them are explained in Chapters 3 and 5.

Mirror beam combiners that consist of a half-silvered mirror layer coated or impregnated onto or into a transparent base carrier represent a combination of lens and mirror. Consequently, reflection transformation and refraction distortion have to be considered.

Since planar mirrors are absolute optical systems, affine model and viewpoint transformations can be applied for such elements. These transformations can be integrated into traditional fixed function rendering pipelines and, consequently, can be fully supported by hardware acceleration without causing additional computational costs.

In contrast to this, elements that cause significant refraction and warped reflection distortion require curvilinear image transformations. A multi-pass image-warping method avoids a direct access to the scene geometry, and prevents the time-consuming transformations of many scene vertices. In addition, it is not restricted to a geometry-based first rendering pass, but rather supports any perspective image generating rendering method (such as point-based rendering, image-based rendering, interactive ray tracing and radiosity, or non-photorealistic rendering). It can be easily configured and extended to support new optics and display technology, and it is flexible enough to be smoothly integrated into existing software frameworks. Additionally, it takes as much advantage of hardware-implemented rendering concepts (such as vertex shaders and render-to-texture capability) as currently possible. Such optics also cause aberrations of non-stigmatic mappings that cannot be corrected in software. However, due to the limited retinal resolution of the eyes, small aberrations are not detected.

To efficiently support simultaneous rendering for multiple viewers on cost-effective rendering platforms (e.g., PCs), a networked cluster of rendering nodes could be used. Each node would carry out the image generation and deformation for a single observer. However, in terms of displaying the final images on arbitrary locations of a single projection screen, they need to be composed within one image. One possibility for doing this is to utilize coding and streaming technology to transmit the images to a central node that is responsible for their composition and the presentation of the resulting image. However, the transmission of the images over an all-purpose network (e.g., an Ethernet) as well as their explicit composition might represent bottlenecks that slow down the overall rendering performance significantly. Recent advances in hardware image composition technology (such as the Lightning-2 device[175]) solve this problem. Instead of connecting the rendering nodes via an all-purpose network, they can be connected to a specific display subsystem. These subsystems allow the image data generated by the connected nodes to be mapped to any location of the connected displays without losing much time for transmission and composition. Some devices (such as the Lightning-2 device [175]) are able to scale in both dimensions—the number of rendering nodes and the number of displays. Such technology also supports efficiently driving high-resolution displays that are built from a collection of multiple display units.

For non-mobile applications, several advantages of spatial optical combination over current head-attached optical see-through approaches exist.

- *Easier eye accommodation and vergence.* Using mirror beam combiners, for instance, the reflected image plane can be brought very close to the physical location that has to be overlaid. Consequently, multiple focal planes do not differ greatly. The reflected image remains at a static position in space if neither mirror combiner nor secondary screen is not moved. This position is invariant of the user's location (Note, the presented content is not). Moving closer to or further away from the physical location and the reflected image yields the same natural accommodation and vergence effects. This is not the case for head-attached displays that present the image at a constant distance in front of the observer. In this case, the image's position and its focal plane do not change for a moving user. Consequently multiple focal planes (i.e., in the physical world and on the image plane) make it impossible to focus on real and virtual content at the same time. Either the real or virtual environment has to be perceived unfocused. This requires that the observer continuously switches focus.

- *High and scalable resolution.* Spatial screens, such as projection display or monitors, provide a high resolution. If, however, a single display does not offer the resolution that is required for a particular application, multiple image sources (such as a video beamer) can be combined. Packing issues that are related to the miniature displays which have to be integrated into the limited space for head-attached devices are not a problem for spatial configurations. In addition, the compression characteristics of a convex mirror that map a larger screen portion into a smaller reflection area causes a high density of pixels. The spatial resolution of such a display is higher than the resolution of the secondary screen.

- *Large and scalable field of view.* The same argumentation for the resolution applies to the field of view. Screens and optics are theoretically not constrained in their dimensions. Spatial surround screen display, for instance, can fill out the entire field of view, a characteristic that cannot be achieved with today's head-attached devices.

- *Improved ergonomic factors.* Today's head-attached displays suffer from an unbalanced ratio between cumbersome and high-quality optics. Spatial displays do not require the user to wear much technology. While stereoscopic displays imply the application of light-weight passive or active glasses, autostereoscopic displays detach any kind of technology from the user.

- *Easier and more stable calibration.* While some head-attached displays can have up to 12 degrees of freedom, spatial optical see-through configurations generally have significantly less. This implies an easier and user- or session-independent calibration, and, consequently, yields more precise overlays.

- *Better controllable environment.* A limited indoor space can be more easily controlled than large-scale outdoor environments. This applies to tracking and illumination conditions and leads to higher precision and consistency, as well as to better quality.

Beside these advantages, several disadvantages can be found with respect to head-attached approaches:

- *Mutual occlusion.* Optical beam combiners cannot present mutual occlusion of real and virtual environments. Consequently, bright real surfaces overlaid with dark virtual objects optically emphasize the real environment and let the virtual objects disappear or appear semi-transparent. A general solution to this problem for spatial optical see-through displays is presented in Chapter 7;

- *Window violation.* The window violation or clipping problem is linked to fish-tank-sized and semi-immersive screens. Graphics that, due to a disadvantageous user position, are projected outside the display area are unnaturally cropped. This effect also appears with the mirror beam combiners; displayed images that, due to a disadvantageous user position, cannot be reflected are cropped at the mirror edges as well.

- *Multi-user viewing.* The number of observers that can be supported simultaneously is restricted by the applied optics and screen configuration.

- *Support of mobile applications.* Spatial displays do not support mobile applications because of the spatially aligned optics and display technology.

- *Direct interaction.* In most cases, the applied optics prevents a direct manipulative interaction. Real objects might not be touchable because they are out of arm's reach or because the additional optics represents a physical barrier. Indirect interaction methods have to be used in these cases.

In general, it should be noted that new technology will not only open new possibilities for the spatial optical see-through concept, but also for other display concepts, such as head-attached displays. Organic light emitting diodes (OLEDs), for instance, may replace the crystalline LEDs that are currently being used to build the miniature displays for HMDs.

OLEDs and other light emitting polymers (LEPs) provide the opportunity for cheap and very high-resolution full-color matrix displays that can give head-attached displays a technological advantage. A variant of OLEDs are light emitting polymers; they provide the opportunity for the fabrication of large, flexible, full-color emissive displays with a high resolution, a wide viewing angle, and a high durability. Large-scale transparent LEPs may present a future variation of an optical combination with an active screen. This is also true for other polymers with transparent, light-emitting semi-conductors. However, LEPs have not yet left the basic research stages and will not be applicable to build stereoscopic AR displays in the near future. But, in combination with autostereoscopic or holographic techniques, this problem may be solved.

Fog[135] and *air*[74] *displays* offer an interesting new means for optical augmentation by projecting an image into screens composed of fog or "modified" air (see Figure 6.29).

Figure 6.29. First generation "Heliodisplay," an interactive free space display. *(Images courtesy of IO2 Technology [74] © 2004.)*

In the short run, especially high-resolution and bright display devices, high-performance and cost-efficient rendering hardware, reliable and precise tracking technology, and advanced interaction techniques and devices will pave the way for future spatial optical see-through configurations. However, the underlying technology must be robust and flexible, and the technology that directly interfaces to users should adapt to users, rather than forcing users to adapt to the technology. Therefore, human-centered and seamless technologies, devices, and techniques will play a major role for augmented reality in the future.

7

Projector-Based Illumination and Augmentation

Because of their increasing capabilities and declining cost, video projectors are becoming widespread and established presentation tools. The ability to generate images that are larger than the actual display device virtually anywhere is an interesting feature for many applications that cannot be provided by desktop screens. Several research groups exploit this potential by using projectors in unconventional ways to develop new and innovative information displays that go beyond simple screen presentations.

The *Luminous Room* [192], for instance, describes an early concept for providing graphical display and interaction at each of an interior architecture space's surface. Two-way optical transducers, called *I/O bulbs*, that consist of pairs of projector-cameras capture the user interactions and display the corresponding output. With the *Everywhere Displays projector* [138], this concept has been extended technically by allowing a steerable projection using a pan/tilt mirror. Recently, it was demonstrated how context-aware hand-held projectors, so-called *iLamps*, can be used as mobile information displays and interaction devices [158].

Another concept called *Shader Lamps* [155] attempts to lift the visual properties of neutral diffuse objects that serve as a projection screen. The computed radiance at a point of a non-trivial physical surface is mimicked by changing the BRDF and illuminating the point appropriately with projector pixels. Animating the projected images allows the creation of the perception of motion without physical displacement of the real object [157]. This type of spatial augmentation is also possible for large, human-sized environments, as demonstrated by Low [93].

Projector-based illumination has become an effective technique in augmented reality to achieve *consistent occlusion* [121, 12] and *illumination* effects [15] between real artifacts and optically overlaid graphics. Video projectors, instead of simple light bulbs, are used to illuminate physical objects with arbitrary diffuse reflectance. The *per-pixel illumination* is controllable and can be synchronized with the rendering of the graphical overlays. It also makes the combination of high-quality optical holograms with interactive graphical elements possible [18]. Using a video projector to produce a controlled reference wave allows a partial reconstruction of the hologram's object wave, except at those portions that are overlaid by integrated graphical elements.

New optical combiners, together with real-time color correction methods, allow us to effectively *superimpose arbitrarily textured*, but flat, surfaces, such as paintings [19]. This is enabled by thin, transparent film materials that diffuse a fraction of the light projected onto them. Such a film can be seamlessly integrated into the frame that holds the artwork. Any kind of visual information, such as animations or interactive graphical elements, can be overlaid in correct colors.

A virtual retouching technique that uses video projectors for reconstructing and enhancing the faded colors of paintings by projecting light directly onto the canvas is described in Yoshida [209]. An affine correlation between projection and the result captured by a camera is established manually for each pixel. Users are then able to retouch the original painting interactively via a desktop GUI.

Other methods allow augmenting *arbitrary, geometrically non-trivial, textured surfaces* without the application of special optical combiners, such as transparent screen materials [20, 59, 116]. They scan the surfaces' geometric and radiometric properties and carry out a real-time per-pixel displacement and color correction before the images are projected.

In addition to these examples, a variety of other techniques and applications have been described that utilize projectors and a structured illumination to achieve special effects. To describe them all is beyond the scope of this book. Instead, we want to discuss the basic concepts in this chapter. We will discuss some of their applications in Chapter 8.

7.1 Image-Based Illumination: Changing Surface Appearance

When we project images of 3D objects onto physical display surfaces, we in effect introduce those virtual 3D objects into the real world around

Figure 7.1. The surface appearance of a neutral colored object is changed by 'painting' it with projected light. (*Images reprinted from [155] © Springer-Verlag; see Plate III.*)

us. In traditional projector-based displays, the virtual objects appear on a flat screen. However, once we understand the relationship between the projector and the geometry of the surface it illuminates, we can project correct images of virtual objects on physical surface of various geometric shapes. We can essentially treat those physical surfaces as 3D screens.

Normally in the physical world, the color, texture, and lighting associated with the surface of a physical object are an integral part of the object. In computer graphics, this is typically modelled with a bidirectional reflection distribution function (BRDF) for the surface. When we illuminate the object with a white light, the surface reflects particular wavelengths of light, and we perceive the respective surface attributes. Because our perception of the surface attributes is dependent only on the spectrum of light that eventually reaches our eyes, we can shift or rearrange items in the optical path, as long as the spectrum of light that eventually reaches our eyes is the same.

Many physical attributes can be incorporated into the light source to achieve a perceptually equivalent effect on a neutral object. Even non-realistic appearances can be realized. We can use digital light projectors and computer graphics to form *Shader Lamps* that effectively reproduce or synthesize various surface attributes, either statically, dynamically, or interactively. While the results are theoretically equivalent for only a limited class of surfaces and attributes, the results, described below, are quite realistic and compelling for a broad range of applications.

The need for an underlying physical model is arguably unusual for computer graphics; however it is not so for architects [69], artists, and computer animators. In addition, various approaches to automatic three-dimensional fabrication are steadily becoming available. Methods include laminate object manufacturing, stereolithography, and fused deposition. It is not unreasonable to argue that three-dimensional printing and faxing is coming.

In the mean time, if necessary, one can use a 3D probe device. We used such a device (a Faro touch probe arm) for the Taj Mahal model shown in Figure 8.2.

7.1.1 The Rendering Process

We introduce the idea of rearranging the terms in the relationship between illumination and reflectance to reproduce equivalent radiance at a surface. As shown in flatland in Figure 7.2, the radiance in a certain direction at point x , which has a given BRDF in the physical world (left), can be mimicked by changing the BRDF and illuminating the point with an appropriately chosen light source, e.g., a projector pixel (right). Below we identify a radiance adjustment equation for determining the necessary intensity of a projector pixel, given the position and orientation of the viewer and the virtual scene. For a more systematic rendering scheme, we describe the notion of separating the rendering view—the traditional virtual camera view—from the shading view—the position of the viewer for lighting calculations.

First, let us consider the rendering equation, which is essentially a geometrical optics approximation as explained in [76]. The radiance at a visible surface point x in the direction (θ, ϕ) that would reach the observer of a physical realization of the scene is

$$L(x, \theta, \phi) = g(x, \theta, \phi)(L_e(x, \theta, \phi) + h(x, \theta, \phi)),$$

where

$$h(x, \theta, \phi) = \int_i F_r(x, \theta, \phi, \theta_i, \phi_i) L_i(x, \theta_i, \phi_i) \cos(\theta_i) d\omega_i$$

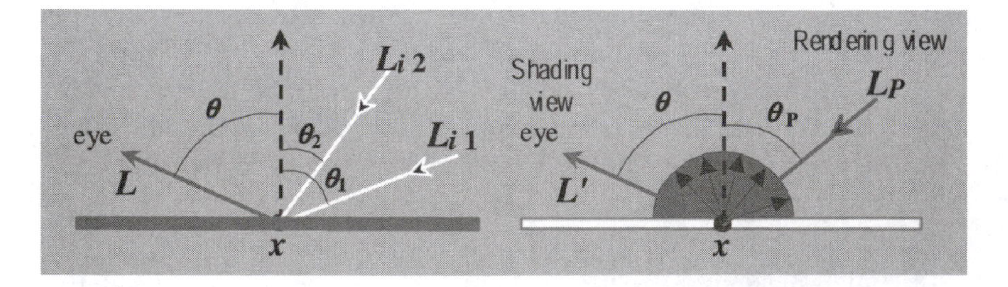

Figure 7.2. The radiance at a point in the direction (right) The radiance as a result of illumination from a projector lamp. By rearranging the parameters in the optical path, the two can be made equal. (*Image reprinted from [155] © Springer-Verlag.*)

and $g(x, \theta, \phi)$ is the geometry term (visibility and distance), $L_e(x, \theta, \phi)$ is the emitted radiance of the point (non-zero only for light sources), and $F_r(x, \theta, \phi, \theta_i, \phi_i)$ is the bidirectional reflection distribution function (BRDF) for the point. The integral in $h(x, \theta, \phi)$ accounts for all reflection of incident radiance $L_i(x, \theta_i, \phi_i)$ from solid angles $d\omega_i$. Radiance has dimensions of energy per unit time, area and solid angle.

Treating the projector lamp as a point emitter, the radiance due to direct projector illumination at the same surface point at distance $d(x)$ but with diffuse reflectance $k_u(x)$ is given by

$$L'(x, \theta, \phi) = g(x, \theta, \phi)k_u(x)\frac{I_P(x, \theta_P, \phi_P)cos(\theta_P)}{d(x)^2}, \tag{7.1}$$

where $I_P(x, \theta_P, \phi_P)$ is the radiant intensity of the projector in the direction (θ_P, ϕ_P) and is related to a discretized pixel value via filtering and tone representation.

We can reproduce radiance $L'(x, \theta, \phi)$ equivalent to $L(x, \theta, \phi)$ for a given viewer location, by solving Equation (7.1) for I_P:

$$I_P(x, \theta_P, \phi_P) = \frac{L(x, \theta, \phi)d(x)^2}{k_u(x)cos(\theta_P)}, \qquad \text{for, } k_u(x) > 0. \tag{7.2}$$

Thus, as long as the diffuse reflectance $k_u(x)$ is non-zero for all the wavelengths represented in $L(x, \theta, \phi)$, we can effectively represent the surface attribute with appropriate pixel intensities. In practice, however, the range of values we can display is limited by the brightness, dynamic range, and pixel resolution of the projector. Thus, for example, we can not make a red surface appear green, because the wavelength of light corresponding to green color, the diffuse reflectance of a red object, $k_u^{green} = 0$. We also cannot project onto objects with mirror-like BRDF because the diffuse reflectance is zero.

The rendering process here involves two viewpoints: the user's and the projector's. A simple approach would be to first render the image as seen by the user, which is represented by $L(x, \theta, \phi)$, and then use traditional image-based rendering techniques to warp this image to generate the intensity-corrected projected image, represented by $I_P(x, \theta_P, \phi_P)$ [29, 108]. Thus, in one way, this is equivalent to rendering and warping the image followed by intensity correction. This is the 2-pass method of the general rendering framework. For a changing viewer location, view-dependent shading under static lighting conditions can also be implemented [37, 88, 58]. However, the warping can be avoided in the case where the display medium is the same as the virtual object. A special case of the general rendering method is required: for single-pass rendering, we treat the moving user's viewpoint as the shading view. Then, the image synthesis process involves rendering the scene from the projector's view by using a perspective projection matrix that matches the projector's intrinsic and extrinsic parameters, followed by radiance adjustment. The separation of the two views offers an interesting topic of study. For example, for a static projector, the visibility and view-independent shading calculations can be performed just once even when the user's viewpoint is changing.

To realize a real-time interactive implementation, we can use conventional 3D rendering APIs, which only approximate the general rendering equation. The BRDF computation is divided into view-dependent specular and view-independent diffuse and ambient components. View-independent shading calculations can be performed by assuming the rendering and shading view are the same. (The virtual shadows, also view-independent, are computed using the traditional two-pass shadow-buffer technique.) For view-dependent shading, such as specular highlights (Figure 7.1), however, there is no existing support to separate the two views.

View-dependent shading. While the rendering view defined by the projector parameters remains fixed, the shading view is specified by the head-tracked moving viewer. For view-independent shading (e.g., diffuse shading), calculations of the two views can be assumed to be the same. For view-dependent shading calculations, we show a minor modification to the traditional view setup to achieve the separation of the two views in a single pass rendering. There is no additional rendering cost. The pseudocode below shows the idea using, as an example, OpenGL API.

```
glMatrixMode( GL_PROJECTION );
glLoadMatrix( intrinsic matrix of projector );
glMultMatrix( xform for rendering view );
glMultMatrix( inverse(xform for shading view) );
glMatrixMode( GL_MODELVIEW  );
glLoadMatrix( xform for shading view );
// set virtual light position(s)
// render graphics model
```

Secondary scattering. Shader Lamps are limited in the type of surface attributes that can be reproduced. In addition, since we are using neutral surfaces with (presumed) diffuse characteristics, secondary scattering is unavoidable and can potentially affect the quality of the results. When the

(a) (b)

Figure 7.3. (a) A green paper illuminated with white light; (b) the white diffuse surface on the right is illuminated with green light. In this special case, the secondary scattering off the white surface below is similar for both parts. (*Images reprinted from [155] © Springer-Verlag; see Plate VII.*)

underlying virtual object is purely diffuse, sometimes the secondary scattering can be used to our advantage. The geometric relationships, also known as form factors, among parts of the physical objects, are naturally the same as those among parts of the virtual object. Consider the radiosity solution for a patch i in a virtual scene with m light sources and n patches:

$$\begin{aligned} B_{i-intended} &= k_{d_i} \sum_j B_j F_{i,j} \\ &= k_{d_i} \Big(\sum_{1 \le j \le m} B_j F_{i,j} + \sum_{m+1 \le j \le m+n} B_j F_{i,j} \Big). \end{aligned}$$

Here k_d is the diffuse reflectance, B_j is the radiance of patch j, and $F_{i,j}$ is the form factor between patches. Using Shader Lamps to reproduce simply the effect of direct illumination (after radiance adjustment), we are able to generate the effect of m light sources:

$$B_{i-direct} = k_{d_i} \sum_{1 \le j \le m} B_j F_{i,j}.$$

However, due to secondary scattering, if the neutral surfaces have diffuse reflectance k_u, the perceived radiance also includes the secondary scattering due to the n patches, and that gives us

$$\begin{aligned} B_{i-actual} &= B_{i-direct} + B_{i-secondary} \\ &= k_{d_i} \sum_{1 \le j \le m} B_j F_{i,j} + k_u \sum_{m+1 \le j \le m+n} B_j F_{i,j}. \end{aligned}$$

The difference between the desired and perceived radiance is

$$|(k_{d_i} - k_u)(\sum_{m+1 \leq j \leq m+n} B_j F_{i,j})|. \tag{7.3}$$

Thus, in scenarios where k_d and k_u are similar, we get approximate radiosity for "free"—projection of even a simple direct illumination rendering produces believable "spilling" of colors on neighboring parts of the physical objects. From Equation (7.3), the secondary contribution from the neutral surfaces is certainly not accurate, even if we reproduce the first bounce exactly. The difference is even larger when the virtual object has non-lambertian reflectance properties. In some cases it may be possible to use inverse global illumination methods so that the projected image can more accurately deliver the desired global illumination effect. Figure 7.3 shows a green and a white paper with spill over from natural white and projected green illumination. In this special case, the secondary scattering off the horizontal white surface below is similar for both parts.

Illumination of all visible surfaces. One may wonder, given a physical object, what is a good set of viewpoints for the lamps, so that every visible surface is illuminated by at least one lamp. This problem is addressed by Stürzlinger [177], where he finds, using a hierarchical visibility algorithm, a set of camera viewpoints such that every visible part of every surface is imaged at least once. The problem of determining an optimal set of viewpoints is NP-hard and is related to the art gallery problem [127] known in the field of computational geometry.

7.2 Creating Consistent Occlusion

Spatial augmented reality systems share many positive properties of spatial virtual reality systems. These displays provide high-resolution, improved consistency of eye accommodation and convergence, little motion sickness potential, and the possibility of an integration into common working and living environments. One of the main challenges for spatial AR systems, as well as for head-mounted optical see-through displays, is the generation of *correct occlusion effects* between virtual and real objects.

The basic problem is that in most cases, the existing environment illumination is applied to lighten the physical scenery. However, light that is reflected off the real objects' surface interferes with the optically overlaid graphics that appear as semi-transparent ghosts which are unrealistically

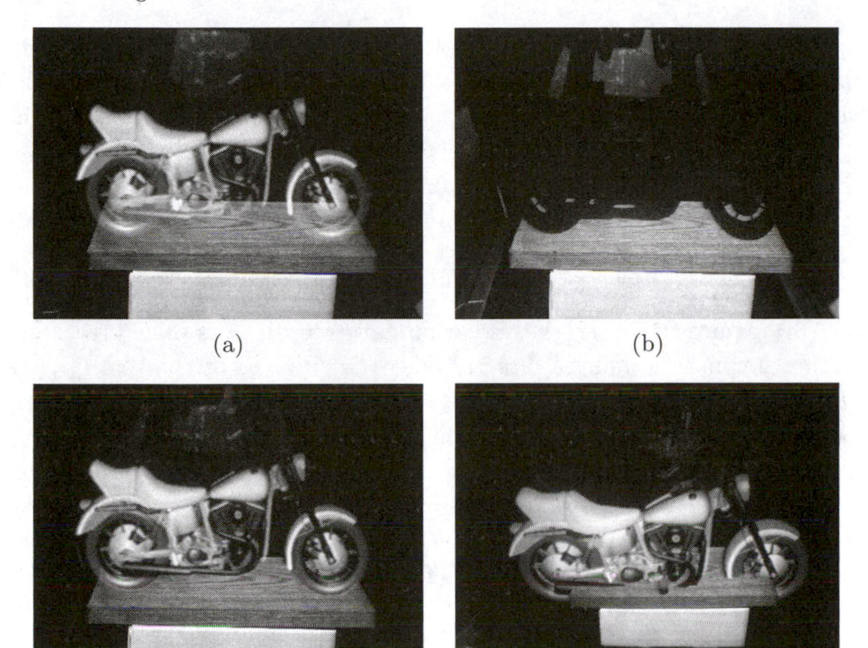

Figure 7.4. (a) Wrong occlusion effects with normal illumination; (b) occlusion shadow generated with projector-based illumination; (c) realistic occlusion of the real object by the virtual one; (d) knowing depth information of real objects allows the occlusion of virtual objects by real ones. (*Images reprinted from [12] © IEEE.*)

floating in midair (Figure 7.4). For indoor situations, the environment illumination can be controlled and synchronized with the rendering process if the simple light bulbs are replaced by video projectors (sometimes referred to as *light projectors*). This concept is called *projector-based illumination.* It requires a physical environment that is initially not illuminated. While this condition can be man-made for many stationary display situations, some setups already provide a dark surrounding by default, like the Virtual Showcase [9].

For optical see-through head-mounted displays, a special optics has been developed by Kiyokawa et al. [79] called ELMO that supports *mutual occlusion*. ELMO uses half-silvered mirrors as optical combiners and an additional semi-transparent LCD panel in front of the conventional optics. The LCD panel is used to selectively block the incoming light on a per-pixel basis. This enables virtual objects to occlude real ones. A head-attached depth sensor allows them to acquire depth maps of the real environment in real time. This makes the occlusion of virtual objects by real ones possible. ELMO faces a number of problems that are linked to the LCD panel: light attenuation caused by the LCD panel and low response time and resolution of the LCD panel. However, as the first functioning system of its kind, it effectively addresses the occlusion problem of optical see-through head-mounted displays.

Head-mounted projective displays (such as those described by Hua et al. [70]) require the observer to wear miniature projectors. The projectors beam the synthetic images directly onto the surfaces of the real objects that are within the user's field of view. Since the observer's viewing frustum can be optically matched with the projection frustum, view-dependent rendering is possible while benefiting from a view-independent projection (i.e., depth information for real objects is not required). However, the real objects' surfaces have to be coated with a retro-reflective material; they allow stereoscopic rendering, multi-user applications, and the usage of such displays within uncontrolled illuminated environments. The occlusion problem of optical see-through displays is not an issue for HMPDs, since the retro-reflective material avoids the problem of environment light interfering with the graphical overlays.

Video-projectors have been used to address the occlusion problem for spatial AR configurations. Noda et al. [121], for instance, have presented a stationary optical see-through display that uses a video projector to illuminate real objects selectively, not lighting those areas that are overlaid by graphics. However, view-dependent rendering is not possible in this case. The observer's viewpoint has to match with the center of projection of the video projector since the illumination pattern is rendered from this point using a normal on-axis projection. In this special case, no depth information of the real environment is required for a correct rendering. In addition, stereoscopic rendering is not provided. Later, Naemura et al. [114] proposed an approach that is technically similar to Noda's. The conceptual difference, however, is that they apply a hand-held video projector as a real flashlight to interactively generate shadow effects of virtual objects on real surfaces. They do not address the occlusion problem of optical see-through displays, but focus on enhancing such interactive mixed reality applications by providing additional visual cues through shadows. As in Noda's case, no depth information of the real objects is needed.

In the following sections, general projector-based illumination techniques are described that can be applied in combination with spatial optical see-through AR displays. Off-axis and on-axis situations are treated in exactly the same way. By using computer-controlled video projectors as replacements for simple light bulbs, the lighting situation can be fully controlled on a per-pixel basis. This allows us to produce correct occlusion effects between virtual and real objects by projecting view-dependent shadows, called *occlusion shadows* [12], directly onto real objects located behind virtual ones using projector-based illumination (Figure 7.4).

7.2.1 Single Users

For rendering occlusion shadows, the viewpoints of each user, the intrinsic and extrinsic parameters of each light projector, as well as the virtual and the real scene must be known.

The viewpoints are continuously measured with head-tracking technology, while the projectors' parameters are determined only once during a calibration phase (see Chapter 3 for details on projector calibration). Virtual objects can be interactively manipulated during run-time.

Knowing the scene and the view transformation lets us compute the perspective projection matrix (V) of the corresponding viewpoint that incorporates the model-view transformation with respect to the scene's origin.

The projectors can be calibrated to a geometric representation of the real scene that is registered to its physical counterpart. To do this a semi-manual calibration routine can be used. The two-dimensional projections of known three-dimensional fiducials can be marked on the projector's image plane. Using these mappings, a numerical minimization method (such as Powell's direction set method [144]) is used to solve a *perspective N-point problem*. This results in the perspective 4×4 projection matrices, P, of the projector that incorporates the correct model-view transformation with respect to the scene origin.

If multiple projectors are used, the calibration process has to be repeated for each projector separately. More details on geometric calibration and correct blending of projectors are presented in Chapters 3 and 5.

The basic algorithm given in this section illustrates how to generate occlusion shadows for a single point of view with the aid of multi-pass rendering. The depth information of both the real and the virtual content have to be known. A *shadow mask* that contains the silhouette of the virtual content is generated in the first pass (Figure 7.5(a–b)) which is then mapped onto the known geometry of the real content in the second pass

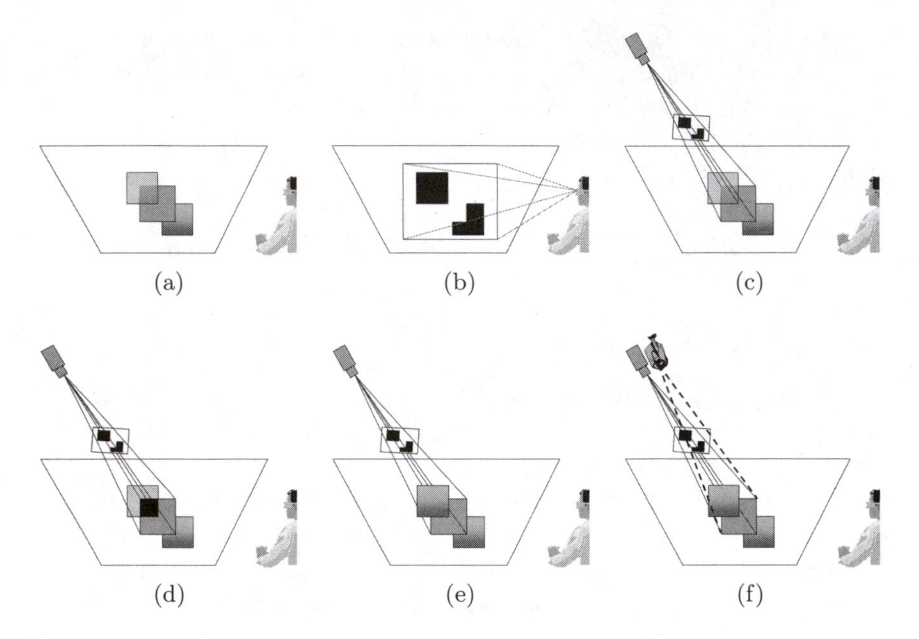

Figure 7.5. Multi-pass rendering and perspective texture mapping for creating occlusion shadows.

via *perspective texture mapping* (Figure 7.5(c–d)). Then, the illumination for the real content is rendered into the frame buffer.

For now we assume that we project only uniformly colored light onto the real surfaces from the light projector's point of view while virtual objects are illuminated from the positions of the virtual light sources. This illumination, however, could be computed with a more advanced BRDF model, producing a correct and matching radiance on real and virtual surfaces with respect to virtual light sources. We will describe this in Section 7.3.

```
generate shadow mask (first pass):
  set projection matrix to V
  render real content into depth buffer
  render virtual content into stencil buffer
  render illumination for real content into...
    ...frame buffer (previously cleared to black)

read-back:
  transfer frame buffer into texture memory T

render shadow mask (second pass):
  set projection matrix to P
```

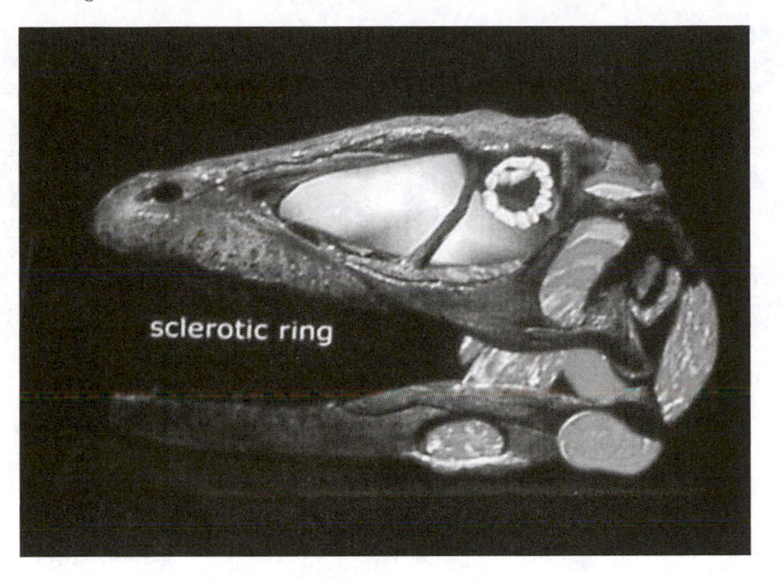

Figure 7.6. Correct occlusion effects for a complex scene: a real dinosaur with integrated soft-tissue. (*Image reprinted from [13] © IEEE.*)

```
set texture matrix to V+ normalization space correction
clear frame buffer to black
render real content into frame buffer using...
  ...projective texture T
```

Rendering the real content into the depth buffer ensures a correct occlusion of virtual objects by real ones. The normalization space correction consists of a scaling by [0.5, 0.5, 1.0], followed by a translation of [0.5, 0.5, 0.5] to map from normalized screen space to normalized texture space[1].

Note, that if the graphics card provides a render-to-texture option, the read-back operation from the frame buffer into texture memory can be bypassed.

While Figure 7.4 illustrates the entire process on a trivial real scene, Figure 7.6 shows that the same process is also efficient for complex objects.

7.2.2 Multiple Users

A clear limitation of the basic method described in Section 7.2.1 is the following fact: If the same real surfaces are simultaneously visible from

[1]This applies to OpenGL-like definitions of the texture and normalized device coordinates.

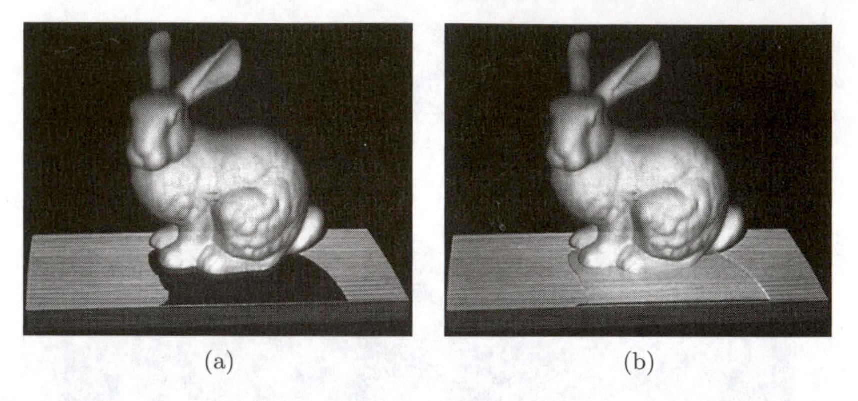

(a) (b)

Figure 7.7. (a) Occlusion shadow of second observer is clearly visible; (b) wrongly visible occlusion shadow is covered by optically overlaying the corresponding part of the reflectance map. (*Images reprinted from [15] © IEEE.*)

multiple points of view (e.g., for different observers), individual occlusion shadows that project onto these surfaces are also visible from different viewpoints at the same time.

Consider two observers: for instance, Observer A might be able to see the occlusion shadow that is generated for Observer B and vice versa. In addition, the shadows move if the viewers are moving, which might be confusing (Figure 7.7(a)).

One can think of several methods to reduce or avoid these effects.

Method I: [12] Occlusion shadows generated for other viewpoints are the *umbral hard shadows* that are cast by the virtual scene with a light source positioned at the other viewpoints' locations. This fact can be utilized by attaching a point light to each viewpoint, thereby generating correct lighting effects on the virtual scene's surfaces in addition to matching hard shadows on the real scene's surfaces.

Method II: [12] The interference between individual occlusion shadows can be minimized by ensuring that they are generated only on those real surfaces that are visible from the corresponding viewpoint. However, since the occlusion shadows are finally rendered from the viewpoint of the projector(s), all view-dependent computations (e.g., back-face culling and depth buffering) are done for this perspective, not for the perspectives of the actual viewpoints.

Method III: [15] Each projector is complemented by a video camera that is used to dynamically scan *reflectance* information from the surfaces of the real objects (Figure 7.5(f)). Knowing this reflectance information, however, leads to an effective and general solution. In addition to the virtual scene, the portions of the real scene (i.e., its registered reflectance map) that are covered by the occlusion shadows of all other observers are rendered. If this is done well, we can create a seamless transition between the real and the virtual portions (Figure 7.7(b)). For each observer, the occlusion shadows of all other observers are rendered into the stencil buffer first. This is done by rendering the real scene's geometry from each observer's perspective and adding the corresponding occlusion shadows via projective texture mapping. The stencil buffer has to be filled in such a way that the area surrounding the occlusion shadows will be blanked out in the final image. Then, the real scene's reflectance map is rendered into the frame buffer (also from the perspective of the observer) and is shaded under the virtual lighting situation. After stenciling has been disabled, the virtual objects can be added to the observer's view. A possibility of obtaining the reflectance map to diffuse real scenes with projector-camera combinations is described in Section 7.3.

Due to *self-occlusion*, not all portions of the real content can be lit by a single light projector. A solution to this problem is to increase the number of projectors and place them in such a way that the projected light is distributed over the real content. To guarantee a *uniform illumination*, however, surfaces should not be lit by more than one projector at the same time or with the same intensity. Otherwise, the projected light accumulates on these surfaces and they appear brighter than others. In Chapter 5, a *crossfeathering* method is discussed that balances the contribution of each projector equally.

7.3 Creating Consistent Illumination

Achieving a consistent lighting situation between real and virtual environments is important for convincing augmented reality applications. A rich pallet of algorithms and techniques have been developed that match illumination for video- or image-based augmented reality. However, very little work has been done in this area for optical see-through AR.

Inspired by the pioneering work of Nakamae et al. [115] and, later, Fournier et al. [50], many researchers have attempted to create consistent illumination effects while integrating synthetic objects into a real environment. Most of these approaches represent the real environment in the form of images or videos. Consequently, mainly image processing, inverse rendering, inverse global illumination, image-based, and photorealistic rendering techniques are used to solve this problem. Due to the lack of real-time processing, these approaches are only applicable in combination with desktop screens and an unresponsive[2] user interaction. Devices that require interactive frame rates, such as head-tracked personal or spatial displays, cannot be supported. Representative for the large body of literature that exists in this area, several more recent achievements can be highlighted.

Boivin et al. [22] present an interactive and hierarchical algorithm for reflectance recovery from a single image. They assume that the geometric model of the scenery and the lighting conditions within the image are known. Making assumptions about the scene's photometric model, a virtual image is generated with global illumination techniques (i.e., ray tracing and radiosity). This synthetic image is then compared to the photograph and a photometric error is estimated. If this error is too large, the algorithm will use a more complex BRDF model (step-by-step using diffuse, specular, isotropic, and finally, anisotropic terms) in the following iterations, until the deviation between synthetic image and photograph is satisfactory. Once the reflectance of the real scenery is recovered, virtual objects can be integrated and the scene must be re-rendered. They report that the analysis and re-rendering of the sample images takes between 30 minutes and several hours depending on the quality required and the scene's complexity.

Yu et al. [210] present a robust iterative approach that uses global illumination and inverse global illumination techniques. They estimate diffuse and specular reflectance as well as radiance and irradiance from a sparse set of photographs and the given geometry model of the real scenery. Their method is applied to the insertion of virtual objects, the modification of illumination conditions, and to the re-rendering of the scenery from novel viewpoints. As for Boivin's approach, BRDF recovery and re-rendering are not supported at interactive frame-rates.

Loscos et al. [92] estimate only the diffuse reflectance from a set of photographs with different but controlled real world illumination conditions. They are able to insert and remove real and virtual objects and shadows and to modify the lighting conditions. To provide an interactive manipulation of the scenery, they separate the calculation of the direct and indirect illumination. While the direct illumination is computed on a per-pixel basis, indirect illumination is generated with a hierarchical radiosity system

[2]Not real-time.

that is optimized for dynamic updates [42]. While the reflectance analysis is done during an offline preprocessing step, interactive frame rates can be achieved during re-rendering. Depending on the performed task and the complexity of the scenery, they report re-rendering times for their examples between 1–3 seconds on a SGI R10000. Although these results are quite remarkable, the update rates are still too low to satisfy the high response requirements of stereoscopic displays that support head-tracking (and possibly multiple users).

Gibson and Murta [55] present another interactive image composition method to merge synthetic objects into a single background photograph of a real environment. A geometric model of the real scenery is also assumed to be known. In contrast to the techniques described previously, their approach does not consider global illumination effects to benefit from hardware-accelerated multi-pass rendering. Consequently, a reflectance analysis of the real surfaces is not required, indirect illumination effects are ignored, and a modification of the lighting conditions is not possible. The illumination of the real environment is first captured in the form of an omnidirectional image. Then, a series of high dynamic basis radiance maps are precomputed. They are used during runtime to simulate a matching direct illumination of the synthetic objects using sphere mapping. Shadow casting between real and virtual objects is approximated with standard shadow mapping. With their method, convincing images can be rendered at frame rates up to 10fps on an SGI Onyx 2. However, it is restricted to a static viewpoint.

The main problem for a consistent illumination for optical see-through approaches is that the real environment is illuminated by physical light sources while the virtual environment is illuminated based on synthetic light sources (Figure 7.8(a)). This results in inconsistent shading and shadow effects unless the virtual light sources approximate the properties (such as position, direction, intensities, color, etc.) of the physical ones. This, however, is very inflexible. In contrast to video see-through, the pixel appearance of the real environment in the video image cannot be modified to achieve a matching illumination. Consequently, the physical contribution of the real light sources has to be neutralized to illuminate the entire environment based on the virtual lighting situation (Figures 7.8(a) and 7.8(c)). This is only possible with controllable physical light sources, such as video projectors.

This section describes methods which create a consistent illumination between real and virtual components within an optical see-through envi-

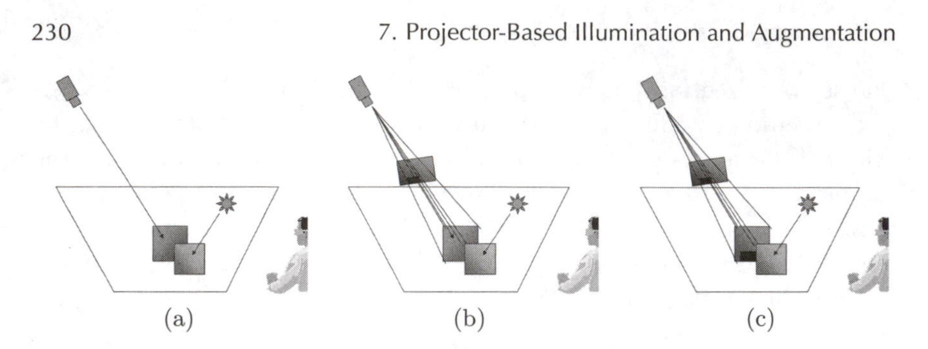

Figure 7.8. Multi-pass rendering and perspective texture mapping for creating occlusion shadows.

ronment [15] for example realized by displays such as the ones presented in Chapters 4, 6, and 8. Combinations of video projectors and cameras can be used to capture *reflectance information* from diffuse real objects and to illuminate them under new *synthetic lighting* conditions (Figure 7.9). For diffuse objects, the capturing process can also benefit from hardware

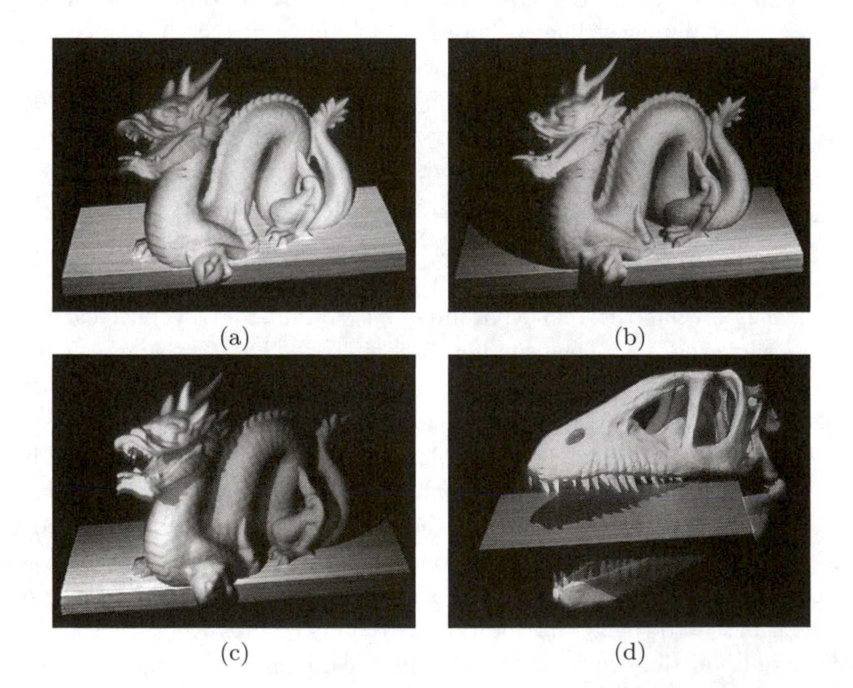

Figure 7.9. (a) Unrealistic illumination with correct occlusion; (b)–(d) realistic illumination under varying virtual lighting conditions with matching shading and shadows. (*Images reprinted from [15] © IEEE.*)

acceleration supporting dynamic update rates. To handle *indirect lighting* effects (like *color bleeding*) an offline radiosity procedure is outlined that consists of multiple rendering passes. For *direct lighting* effects (such as simple shading, shadows, and reflections) hardware-accelerated techniques are described which allow interactive frame rates. The reflectance information can be used in addition to solve the multi-user occlusion problem discussed in Section 7.2.

7.3.1 Diffuse Reflectance Analysis

As in Section 7.2, we want to assume that the geometry of both object types, real and virtual, has been modelled or scanned in advance. While the material properties of virtual objects are also defined during their modelling process, the diffuse reflectance of physical objects is captured on- or offline with a set of video projectors and cameras and a structured light analysis. This sort of analysis is standard practice for many range scanner setups. But since only diffuse real objects can be considered (a projector-based illumination will generally fail for any specular surface), a simple diffuse reflectance analysis can benefit from hardware-accelerated rendering techniques. In contrast to conventional scanning approaches, this leads to dynamic update rates. Figure 7.10 illustrates an example.

Again, the intrinsic and extrinsic parameters of projectors and cameras within the world coordinate system have to be estimated first. Each device has to be calibrated separately. As described in Section 7.2, two-dimensional projections of known three-dimensional fiducials can be interactively marked on a projector's/camera's image plane. Using these mappings, a minimization method is used to solve a perspective n-point problem for each device. This results in the perspective projection matrices P, C of a projector and a camera. Both matrices incorporate the correct model-view transformations with respect to the origin of the world coordinate system. Once calibrated, a projector/camera combination can be used to perform the diffuse reflectance analysis.

Capturing a radiance map. The video projector is used to send structured light samples to the diffuse physical object and illuminate it with a predefined color C_p and an estimated intensity η. Synchronized to the illumination, the video camera captures an input image. Since this image contains the diffuse reflectance of the object's surface under known lighting conditions it represents a *radiance map*. *White-balancing* and other *dynamic correction functions* have been disabled in advance. The parameters of

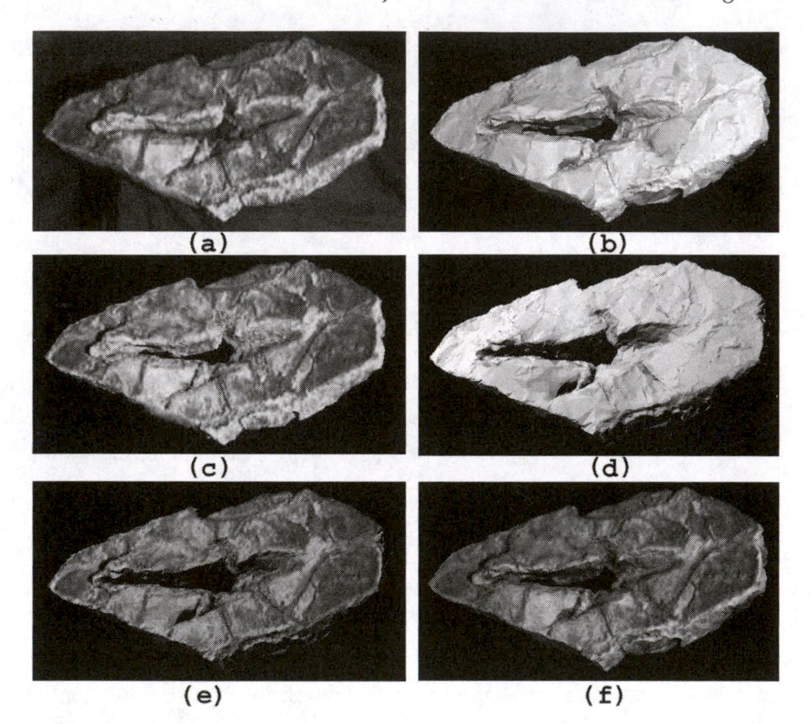

Figure 7.10. (a) Captured radiance map of a fossilized dinosaur footprint; (b) intensity image rendered for calibrated projector from (a); (c) computed reflectance map; (d) novel illumination situation; (e) reflectance map under novel illumination from (d); (f) reflectance map under virtual illumination from (b). (*Images reprinted from [15] © IEEE.*)

the camera's *response function* can be adjusted manually in such a way that the recorded images approximate the real world situation as close as possible.

Some types of video projectors (such as digital light projectors (DLP)) display a single image within sequential, time-multiplexed light intervals to achieve different intensity levels per color. If such projectors are used, a single snapshot of the illuminated scene would capture only a slice of the entire display period. Consequently, this image would contain incomplete color fragments instead of a full-color image. The width of this slice depends on the exposure time of the camera. To overcome this problem, and to be independent of the camera's exposure capabilities, we capture a sequence of images over a predefined period of time. These images are then combined to create the final diffuse radiance map I_{rad} (Figure 7.10(a)).

Creating an intensity image. To extract the diffuse material reflectance out of I_{rad}, the lighting conditions that have been created by the projector have to be neutralized. OpenGL's diffuse lighting component, for instance, is given by Neider [118]:

$$I_i = \frac{1}{r_i^2} \cos(\theta_i) \, (D_l D_m)_i \, ,$$

where I_i is the final intensity (color) of a vertex i, D_l is the diffuse color of the light, D_m is the diffuse material property, the angle θ_i is spanned by the vertex's normal and the direction vector to the light source, and the factor $1/r_j^2$ represents a square distance attenuation.

As described in Section 7.1, an intensity image I_{int} that contains only the diffuse illumination can be created by rendering the object's geometry (with $D_m = 1$) from the perspective of the video camera, illuminated by a point light source (with $D_l = C_p\eta$) that is virtually located at the position of the projector (Figure 7.10(b)).

In addition, hard shadows can be added to the intensity image by applying standard shadow mapping techniques. Consequently, the background pixels of I_{int}, as well as pixels of regions that are occluded from the perspective of the light source, are blanked out ($I_{int}(x, y) = 0$), while all other pixels are shaded under consideration of the previous diffuse lighting equation.

Extracting and re-rendering diffuse reflectance. Given the captured radiance map I_{rad} and the rendered intensity image I_{int}, the diffuse reflectance for each surface that is visible to the camera can be computed by

$$\begin{aligned} I_{ref}(x, y) &= \frac{I_{rad}(x, y)}{I_{int}(x, y)} && \text{for } I_{int}(x, y) > 0, \\ I_{ref}(x, y) &= 0 && \text{for } I_{int}(x, y) = 0. \end{aligned}$$

Note that the division of the two images can be implemented by a pixel shader and can be executed in real time on graphics hardware.

The reflectance image I_{ref} is stored, together with the matrix C and the real object's world-transformation O_c that is active during the capturing process, within the same data structure. This data structure is referred to as the *reflectance map* (Figure 7.10(c)).

The captured reflectance map can be re-rendered together with the real object's geometric representation from any perspective with an arbitrary world-transformation O_a. Thereby, I_{ref} is applied as a projective texture map with the texture matrix[3] set to $O_a^{-1}O_cC$. Enabling texture modulation, it can then be relit virtually under novel illumination conditions (Figures 7.10(d), 7.10(e), and 7.10(f)).

The basic diffuse reflectance analysis method has the following problems:

1. Due to under-sampling, surfaces which span a large angle ϕ_i between their normal vectors and the direction vectors to the camera can cause texture artifacts if I_{ref} is remapped from a different perspective.

2. A single reflectance map covers only the surface portion that is visible from the perspective of the camera.

3. The radiance map can contain indirect illumination effects caused by light fractions that are diffused off other surfaces (so-called *secondary scattering*). The intensity image I_{int}, however, does not contain secondary scattering effects since a global illumination solution is not computed. The extracted reflectance is incorrect in those areas that are indirectly illuminated by secondary scattering.

4. The projector intensity η has to be estimated correctly.

To overcome the under-sampling problem, the constraint can be made that only surfaces with $\phi_i \leq \phi_{\max}$ are analyzed. All other surfaces will be blanked out in I_{ref} (i.e., $I_{ref}(x, y) = 0$). Experiments have shown that $\phi_{\max} = 60°$ is an appropriate value.

Multiple reflectance maps that cover different surface portions can be captured under varying transformations, O_c or C. They are merged and alpha blended during remapping via *multi-texturing* onto the object's geometric representation. This ensures that regions which are blanked out in one reflectance map can be covered by other reflectance maps. To generate seamless transitions between the different texture maps, bi- or tri-linear texture filtering can be enabled.

Illuminating the entire scene can cause an extreme secondary scattering of the light. To minimize the appearance of secondary scattering in I_{rad}, the scene can be divided into discrete pieces and their reflectance can be captured one after the other. For this, the same algorithm as described above can be used. The difference, however, is that only one piece at a time

[3]Including the corresponding mapping transformation from normalized device coordinates to normalized texture coordinates.

is illuminated and rendered, I_{rad} and I_{int}. By evaluating the blanked out background information provided in I_{int}, the selected piece we effectively segmented in I_{int} and its reflectance can be computed. This is repeated for each front-facing piece, until I_{ref} is complete.

The projector's intensity η can be estimated as follows: First, a reflectance map is generated with an initial guess of η. This reflectance map is then remapped onto the object's geometric representation, which is rendered from the perspective of the camera and illuminated by a virtual point light source with η located at the projector. The rendered radiance map I_{radv} is then compared to the captured radiance map I_{rad} by determining the average square distance error Δ among all corresponding pixels. Finally, an approximation for η can be found by minimizing the error function f_{Δ}. For this, Brent's inverse parabolic minimization method with bracketing [23] can be used, for example. By estimating η, the constant black-level of the projector can also be incorporated.

7.3.2 Augmenting Radiance

In computer graphics, the *radiosity method* [57] is used to approximate a *global illumination solution* by solving an *energy-flow* equation. Indirect illumination effects, such as secondary scattering, can be simulated with radiosity. The general radiosity equation for n surface patches is given by

$$B_i = E_i + \rho_i \sum_{j=1}^{n} B_j F_{ij},$$

where B_i is the radiance of surface i, E_i is the emitted energy per unit area of surface i, ρ_i is the reflectance of surface i, and F_{ij} represents the fraction of energy that is exchanged between surface i and surface j (the *form-factor*).

The simulation of radiosity effects within an optical see-through environment that consists of diffuse physical and virtual objects, faces the following challenges and problems:

1. Light energy has to flow between all surfaces—real ones and virtual ones;

2. Physical objects are illuminated with physical light sources (i.e., video projectors in our case) which do not share the geometric and radiometric properties of the virtual light sources;

3. No physical light energy flows from virtual objects to real ones (and vice versa). Consequently, the illuminated physical environment

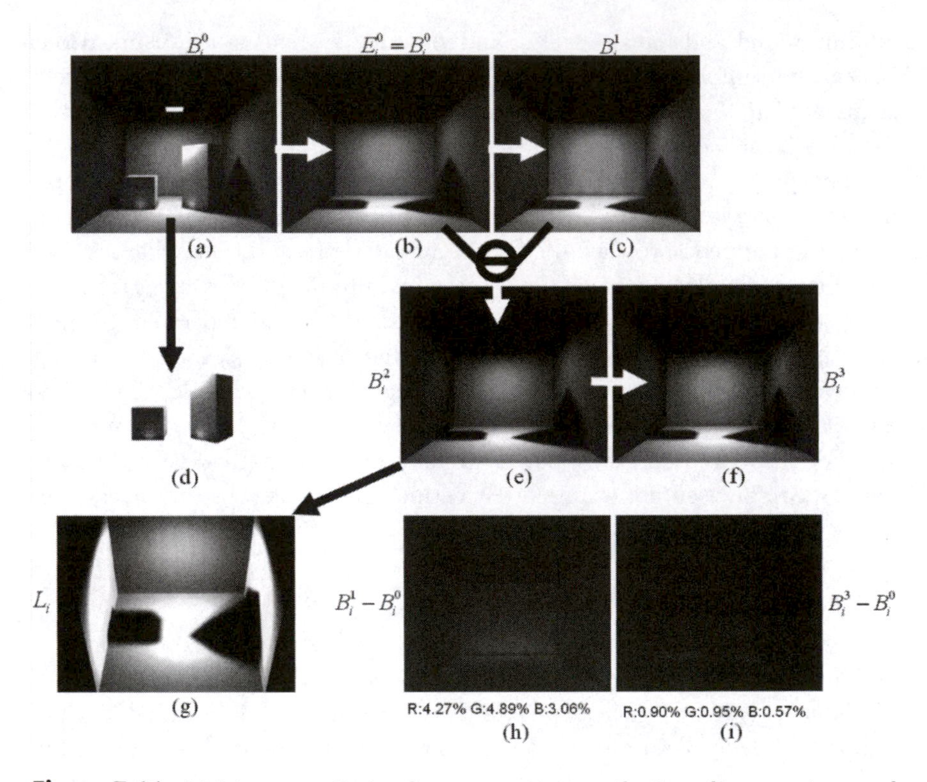

Figure 7.11. Multi-pass radiosity for augmenting synthetic radiance onto a real environment. (*Images reprinted from [15] © IEEE.*)

causes (due to the absence of the virtual objects) a different radiometric behavior than the entire environment (i.e., real and virtual objects together).

An example is illustrated in Figure 7.11(a)[4]. The entire environment consists of three walls, a floor, two boxes, and a surface light source on the ceiling. We want to assume that the walls and the floor are the geometric representations of the physical environment, and the boxes, as well as the light source, belong to the virtual environment. While the diffuse reflectance ρ_i of the physical environment can be automatically captured, it has to be defined for the virtual environment. After a radiosity simulation[5]

[4]A physical mock-up of the Cornell room has been chosen since it is used in many other examples as a reference to evaluate radiosity techniques.

[5]A hemi-cube–based radiosity implementation has been used with progressive refinement, adaptive subdivision, and interpolated rendering for our simulations.

of the entire environment, the radiance values B_i^0 for all surfaces have been computed[6]. Color-bleeding and shadow-casting are clearly visible.

For virtual objects, the computed radiance values are already correct (Figure 7.11(d)). The rendered image represents a radiance map that is generated from one specific perspective. Rendering the virtual objects from multiple perspectives results in multiple radiance maps that can be merged and alpha-blended during remapping via multi-texturing onto the virtual geometry (as described for reflectance maps in Section 7.3.1). In this case, our radiance maps are equivalent to light maps that are often used during prelighting steps to speed up the online rendering process.

The prelit virtual objects can simply be rendered together with their light maps and can be optically overlaid over the physical environment.

The physical surfaces, however, have to emit the energy that was computed in B_i^0 (Figure 7.11(b)). To approximate this, it can be assumed first that every physical surface patch directly emits an energy E_i^0 that is equivalent to B_i^0. If this is the case, fractions of this energy will radiate to other surfaces and illuminate them. This can be simulated by a second radiosity-pass (Figure 7.11(c)), which computes new reflectance values B_i^1 for all the physical surfaces, by assuming that $E_i^0 = B_i^0$, and not considering the direct influence of the virtual light source.

If we subtract the radiance values that have been computed in both passes, we receive the scattered light only; that is, the light energy radiated between the physical surfaces $B_i^1 - B_i^0$ (Figure 7.11(h)).

Consequently,

$$B_i^2 = B_i^0 - \left(B_i^1 - B_i^0\right)$$

approximates the energy that has to be created physically on every real surface patch. To prove this, a third radiosity pass can be applied to simulate the energy flow between the patches (Figure 7.11(f)). It can be seen that the remaining energy $B_i^1 - B_i^0$ will be nearly added, and we have

$$B_i^3 = B_i^2 + \left(B_i^1 - B_i^0\right) \approx B_i^0.$$

By removing the virtual objects from the environment and simulating the second radiosity pass, light energy will also be radiated onto surfaces which were originally blocked or covered by the virtual objects (either completely or partially). An example is the shadow areas that have been cast by the virtual objects. This can be observed in Figure 7.11(h) and Figure 7.11(i).

[6]Note, that the upper index represents the radiosity pass.

Negative radiance values are possible for such areas. To avoid this, the resulting values have to be clipped to a valid range.

The average deviations between B_i^0 and B_i^1, as well as between B_i^0 and B_i^3, within the three spectral samples red (R), green (G), and blue (B) are presented in Figure 7.11(h) and 7.11(i), respectively. Treating a video projector as a point light source B_i^2 can be expressed as follows:

$$B_i^2 = \rho_i L_i F_i,$$

where L_i is the irradiance that has to be projected onto surface i by the projector; F_i is the form factor for surface i, which is given by

$$F_i = \frac{\cos(\theta_i)}{r_i^2} h_i,$$

where θ_i is the angle between a surface patch's normal and the direction vector to the projector, r_i is the distance between a surface patch and the projector, and h_i is the visibility term of the surface patch, seen from the projector's perspective.

Extending and solving the previous equations for L_i, we obtain (Figure 7.11(g)):

$$L_i = \frac{B_i^2}{\rho_i F_i} \eta.$$

To cope with the individual brightness of a video projector, the intensity factor η can be added. How to estimate η for a specific projector is described in Section 7.3.1. To be consistent with our previously used terminology, we call L_i the *irradiance map*.

The computed radiance and irradiance values are view independent. Consequently, irradiance maps for the real objects and radiance maps for the virtual objects can be precomputed offline.

The real objects are illuminated with projected light during run time by rendering the generated irradiance map from the viewpoint of the projector (e.g., as illustrated in Figure 7.11(g)). Virtual objects are rendered with the computed light maps (e.g., as illustrated in Figure 7.11(d)) and are then optically laid over the real environment. Due to the view independence of the method, the augmented scene can be observed from any perspective (i.e., head-tracking and stereoscopic rendering are possible). However, the scene has to remain static, since any modification would require the recomputing of new radiance and irradiance maps throughout multiple radiosity passes. This is not yet possible at interactive rates.

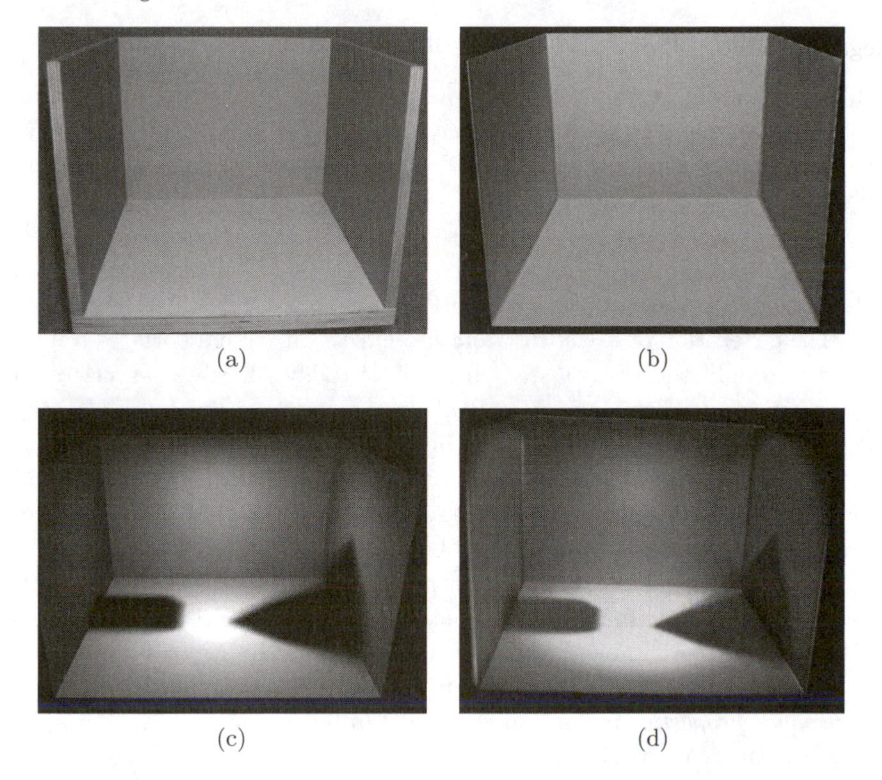

Figure 7.12. (a) Photograph of original object under room illumination; (b) screen shot of captured reflectance relit with virtual point light source and Phong shading; (c) screen shot of simulated radiosity solution with captured reflectance, virtual surface light source (shown in Figure 7.11), and two virtual objects (shown in Figure 7.11); (d) photograph of original object illuminated with the computed irradiance. (*Images reprinted from [15] © IEEE; see Plate IV.*)

Figure 7.12 shows a photograph of (a) the physical object under room illumination, (b) a screen-shot of captured reflectance maps that have been re-rendered under novel lighting conditions, (c) a screen-shot of the simulated radiance situation B_i^0, and (d) a photograph of a physical object that has been illuminated with L_i. Note, that small deviations between the images can be contributed to the response of the digital camera that was used to take the photograph, as well as to the high black-level of the projector that, for instance, makes it impossible to create completely black shadow areas.

7.3.3 Interactive Approximations: Shading and Shadows

In the following sections, several interactive rendering methods are described that make use of hardware acceleration. In particular, techniques to create matching shading and shadow and reflection effects on real and virtual objects are discussed. Indirect lighting effects such as color bleeding, however, cannot be created with these techniques. Yet, they create acceptable results at interactive frame rates for multiple head-tracked users and stereoscopic viewing on conventional PCs.

The generation of *direct illumination effects* on virtual surfaces caused by virtual light sources is a standard task of today's hardware-accelerated computer graphics technology. Real-time algorithms, such as *Gouraud shading* or *Phong shading* are often implemented on graphics boards.

Consistent and matching shading effects on real surfaces from virtual light sources can be achieved by using video projectors that project appropriate irradiance maps onto the real objects. In Section 7.1, it is shown how to compute an irradiance map to lift the radiance properties of neutral diffuse objects with uniform white surfaces into a precomputed radiance map of a virtual scene illuminated by virtual light sources. An irradiance map that creates virtual illumination effects on diffuse real objects with *arbitrary reflectance properties* (color and texture) can be computed as follows:

First, the real objects' captured reflectance map (I_{ref}) is rendered from the viewpoint of the projector and is shaded with all virtual light sources in the scene. This results in the radiance map I_{rad_1}. Then I_{ref} is rendered again from the viewpoint of the projector. This time, however, it is illuminated by a single point light source ($D_l = 1 \cdot \eta$) which is located at the position of the projector. This results in the radiance map I_{rad_2}. Finally, the correct irradiance map is computed by

$$L = \frac{I_{rad\_1}}{I_{rad\_2}}.$$

Note that, in general, this method correlates to the method described in Section 7.3.2. The difference is the applied illumination model. While in Section 7.3.2 an indirect global illumination model (radiosity) is used, here a hardware-accelerated direct model (such as Phong or Gouraud shading) is used. It is easy to see that I_{rad_1} corresponds to B_i^2 and that I_{rad_2} corresponds to $\rho_i F \eta$.

Note, also, that this method is actually completely independent of the real objects' reflectance. This can be shown by balancing I_{rad_1} with I_{rad_2}. In this case, the diffuse material property D_m (i.e., the reflectance) is cancelled out. Consequently, I_{rad_1} and I_{rad_2} can be rendered with a constant (but equal) reflectance (D_m). If $D_m = 1$ is chosen, then the irradiance map is simply the quotient between the two intensity images I_{int_1} and I_{int_2} that result from the two different lighting conditions, the virtual one and the real one.

The irradiance map L should also contain consistent shadow information. Figure 7.9 illustrates examples with matching shading effects[7].

Six types of shadows can be identified within an optical see-through environment:

1. Shadows on real objects created by real objects and real light sources;
2. Shadows on virtual objects created by virtual objects and virtual light source;
3. Shadows on virtual objects created by real objects and virtual light sources;
4. Shadows on real objects created by real objects and virtual light sources;
5. Shadows on real objects created by virtual objects and virtual light sources; and
6. Occlusion shadows.

The first type of shadow is the result of occlusions and self-occlusions of the physical environment that is illuminated by a physical light source (e.g., a video projector). Since it is focused on controlling the illumination conditions within the entire environment via virtual light sources, these shadows have to be removed. This can be achieved by using multiple synchronized projectors that are able to illuminate all visible real surfaces. Chapter 5 discusses techniques that compute a correct color and intensity blending for multi-projector displays.

The second and third shadow types can be created with standard *shadow mapping* or shadow buffering techniques. To cast shadows from real objects onto virtual ones, the registered geometric representations of the real objects have to be rendered together with the virtual objects when the

[7]Note that a simple wooden plate has been chosen to demonstrate and to compare the different effects. However, all techniques that are explained in these sections can be applied to arbitrary object shapes.

shadow map is created (i.e., during the first shadow pass). Such geometric real world representations (sometimes called *phantoms* [24]) are often rendered continuously to generate a realistic occlusion of virtual objects by real ones. Note that these hardware-accelerated techniques create hard shadows while global illumination methods (such as radiosity) can create soft shadows. Texture blending, however, allows ambient light to be added to the shadow regions. This results in dark shadow regions that are blended with the underlying surface texture, instead of creating unrealistic black shadows.

Shadow types number (4) and (5) can also be created via shadow mapping. However, they are projected on the surface of the real object together with the irradiance map L, as discussed earlier. Therefore, I_{rad_1} has to contain the black (non-blended) shadows of the virtual and the real objects. This is achieved by rendering all virtual objects and all phantoms during the first shadow pass to create a shadow map. During the second pass, the shaded reflectance texture and the generated shadow texture are blended and mapped onto the objects' phantoms. A division of the black shadow regions by I_{rad_2} preserves these regions. Note that a blending of the projected shadows with the texture of the real objects occurs physically if the corresponding surface portions are illuminated (e.g., by a relatively small amount of projected ambient light).

Occlusion shadows have been described in Section 7.2. They are special view-dependent shadows created by the projectors on the real objects' surfaces to achieve a realistic occlusion of real objects by virtual ones within optical see-through augmented environments. They are normally not visible from the perspective of the observer, since they are displayed exactly underneath the graphical overlays. Occlusion shadow-maps, however, also have to be blended to the irradiance map L before it is projected.

The entire process can be summarized in the form of a three-pass rendering algorithm:

```
\\ first pass
create an intensity image I_rad_2 of the real object:
  render real object from perspective of light projector...
    ...having a white diffuse material...
    ...illuminated by a white virtual point light source...
    ...located at the projector

\\ second pass
create a shading image I_rad_1 of real and virtual objects:
  generate shadow map for real and virtual objects
```

```
  render real objects from perspective of light projector...
    ...having a white diffuse material...
    ...illuminated by the virtual light sources...
    ...with shadow mapping enabled

compute irradiance image L = I_rad_1 / I_rad_2

\\ third pass
render occlusion shadows from perspective of projector...
  ...and blend with L
```

Note that the implementation of the image division to compute the irradiance map L can be a pixel shader in order to achieve real-time performance.

7.3.4 Interactive Approximations: Reflections

Using hardware-accelerated *cube mapping* techniques, the virtual representation of the real environment (i.e., the objects' geometry together with the correctly illuminated reflectance map) can be reflected by virtual objects (Figure 7.13).

Therefore, only the registered virtual representation of the real environment has to be rendered during the generation step of the cube map. Virtual objects are then simply rendered with cube mapping enabled. Note, that for conventional cube mapping, reflection effects on a virtual object are physically correct for only a single point, the center of the cube map frusta. To create convincing approximations, this center has to be matched with

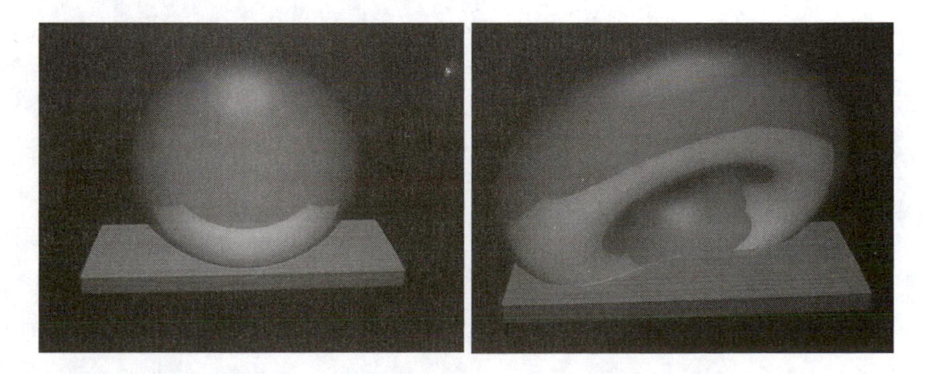

Figure 7.13. Virtual objects reflecting and occluding the real object (wooden plate). (*Images reprinted from [15] © IEEE.*)

the virtual object's center of gravity, and the cube map has to be updated every time the scene changes.

7.4 Augmenting Optical Holograms

A hologram is a *photometric emulsion* that records interference patterns of coherent light. The recording itself stores *amplitude*, *wavelength*, and *phase* information of light waves. In contrast to simple photographs (which can record only amplitude and wavelength information), holograms have the ability to reconstruct complete *optical wavefronts*. This results in a three-dimensional appearance of the captured scenery, which is observable from different perspectives.

Optical holograms are reconstructed by illuminating them with monochromatic (purity of color) light; the light has to hit the emulsion at the same angle as the reference laser beam that was used to record the hologram.

Figure 7.14 illustrates the basics of optical holographic recording and reconstruction. A laser beam is split into two identical beams. While one beam (called the *reference wave*) illuminates the holographic emulsion

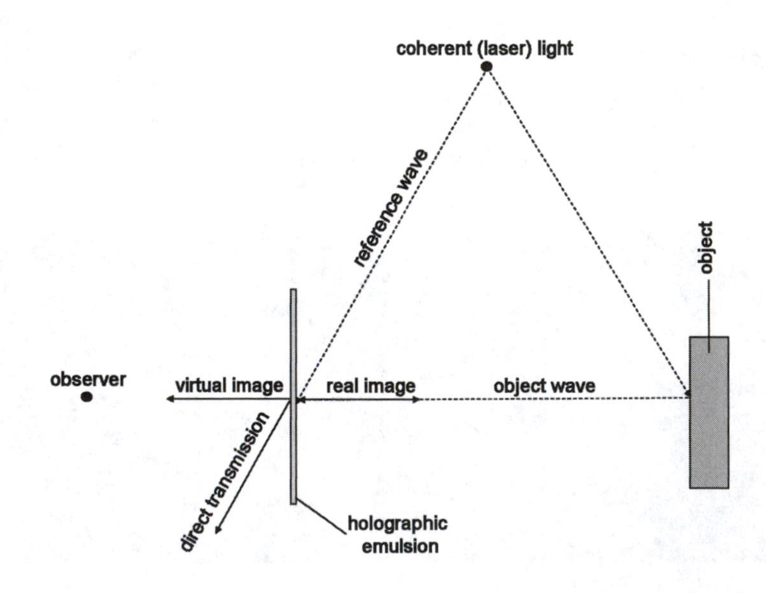

Figure 7.14. Recording and reconstruction of optical holograms. (*Image reprinted from [18] © IEEE.*)

directly, the other beam illuminates the object to be recorded. The light that is reflected from the object (called the *object wave*) shines on the emulsion and creates the fringe pattern together with the reference wave.

If the emulsion is illuminated with a copy of the reference wave, it interacts with the recorded interference fringes and reconstructs the object wave which is visible to the observer. The amplitude of the reconstructed object wave is proportional to the intensity of the reference wave. Its reconstruction can be controlled with a selective illumination.

There are two basic types of optical holograms: *transmission* and *reflection holograms*.

A transmission hologram is viewed with the light source and the observer on opposite sides of the holographic plate. The light is transmitted through the plate before it reaches the observer's eyes. Portions of the emulsion that are not recorded or not illuminated remain transparent.

Reflection holograms are viewed with the light source and the observer on the same side of the holographic plate. The light is reflected from the plate towards the eyes of the observer. As for transmission holograms, not recorded or not illuminated portions of the emulsion remain transparent (without an opaque backing layer).

Among these two basic types of holograms, a large pallet of different variations exists. While some holograms can only be reconstructed with laser light, others can be viewed under white light. They are called *white-light holograms*.

Some of the most popular white-light transmission holograms are *rainbow holograms*. With rainbow holograms, each wavelength of the light is diffracted through a different angle. This allows us to observe the recorded scene from different horizontal viewing positions, but also makes the scene appear in different colors when observed from different viewing positions.

In contrast to rainbow holograms, *white-light reflection holograms* can provide full parallax and display the recorded scene in a consistent but, in most cases, monochrome color for different viewing positions.

Color white-light holograms (both transmission and reflection types) can also be produced. Usually the same content is recorded on several emulsion layers. However, each layer is exposed to laser light with a different wavelength. When reconstructed, each object wave from each layer contributes with its individual wavelength. Together, they merge into a colored image.

Today, many applications for optical holograms exist. Examples include interferometry, copy protections, data storage, and holographic optical elements.

Due to their unique capability to present three-dimensional objects to many observers with almost no loss in visual quality, optical holograms are often used in museums. They can display artifacts that are not physically present without the need to build real replicas of the originals. In addition, medical, dental, archaeological, and other holographic records can be made, both for teaching and for documentation.

Optical holograms, however, are static and lack interactivity. Exceptions are *multiplex holograms* that are built from multiple narrow (vertical) strip holograms that contain recordings of the same scenery at different time intervals. While moving around the hologram (or spinning a cylindrically-shaped version around its principle axis), observers can perceive the recorded scene in motion; multiplex holograms are not interactive.

Three-dimensional computer graphics in combination with stereoscopic presentation techniques represents an alternative that allows interactivity. State of the art rendering methods and graphics hardware can produce realistic-looking images at interactive rates. However, they do not nearly reach the quality and realism of holographic recordings. Autostereoscopic displays allow for a glass-free observation of computer-generated scenes. Several autostereoscopic displays exist that can present multiple perspective views at one time thus supporting multiple users simultaneously. Resolution and rendering speed, however, decrease with the number of generated views. Holographic images, in contrast, can provide all depth cues (perspective, binocular disparity, motion parallax, convergence, and accommodation) and can be viewed from a theoretically unlimited number of perspectives at the same time.

Parallax displays are display screens (e.g., CRT or LCD displays) that are overlaid with an array of light-directing or light-blocking elements. Using these elements, the emitted light is directed to both eyes differently allowing them to see individual portions of the displayed image. The observer's visual system interprets corresponding light rays to be emitted by the same spatial point. Dividing the screen space into left-right image portions allows for a glass-free separation of stereo pairs into two or more viewing zones.

Some displays control the parallax array mechanically or electronically depending on the viewpoint of the observer to direct the viewing zones more precisely towards the eyes. Others generate many dense viewing zones, each of them showing a slightly different perspective of the rendered scene at the same time. Such displays support multiple users simultaneously, but do not yet allow high frame rates.

An example of a parallax display is a *parallax barrier display* that uses an array of light-blocking elements (e.g., a light-blocking film or liquid crystal barriers) in front of a screen. The light-blocking elements are used to cover portions of the screen for one eye that are visible to the other eye.

Another example is a *lenticular sheet display* that utilizes refraction of a lens array (e.g., small cylindrical, prism-like or spherical lenses) to direct the light into the different viewing zones. Images that are generated with lenticular sheet displays appear brighter than the ones displayed on barrier displays. While prisms and cylinders provide a horizontal parallax only, spherical lenses support a full parallax.

Several research groups are working on computer-generated holograms. There are two main types, *digital holography* and *electroholography* or *computer-generated holograms.*

In digital holography [207], holographic printers are used to sequentially expose small fractions of the photometric emulsion with a computer-generated image. Conventional holograms which display a computer-generated content are the result. This technique can also be used for the construction of large-scale, tiled holograms [80]. Although digital holograms can be multiplexed to display scenes in motion, they remain non-interactive.

Electroholography aims at the computer-based generation and display of holograms in real time [82, 96]. Holographic fringes can be computed by rendering multiple perspective images that are combined into a stereogram [95], or by simulating the optical interference and calculating the interference pattern [94]. Once computed, the fringes are dynamically visualized with a holographic display. Since a massive amount of data has to be processed, transmitted, and stored to create holograms, today's computer technology still sets the limits of electroholography. To overcome some of the performance issues, advanced reduction and compression methods have been developed. This results in electroholograms that are interactive, but small, low resolution, and pure in color. Recent advances on consumer graphics hardware may reveal potential acceleration possibilities [137].

Combining optical holograms with two-dimensional or three-dimensional graphical elements can be an acceptable trade-off between quality and interactivity (Figure 7.15). While the holographic content provides high quality but remains static, additional graphical information can be generated, inserted, modified, and animated at interactive rates.

Technically, optical combiners such as mirror beam combiners or semi-transparent screens (see Chapter 6) can be used to visually overlay the out-

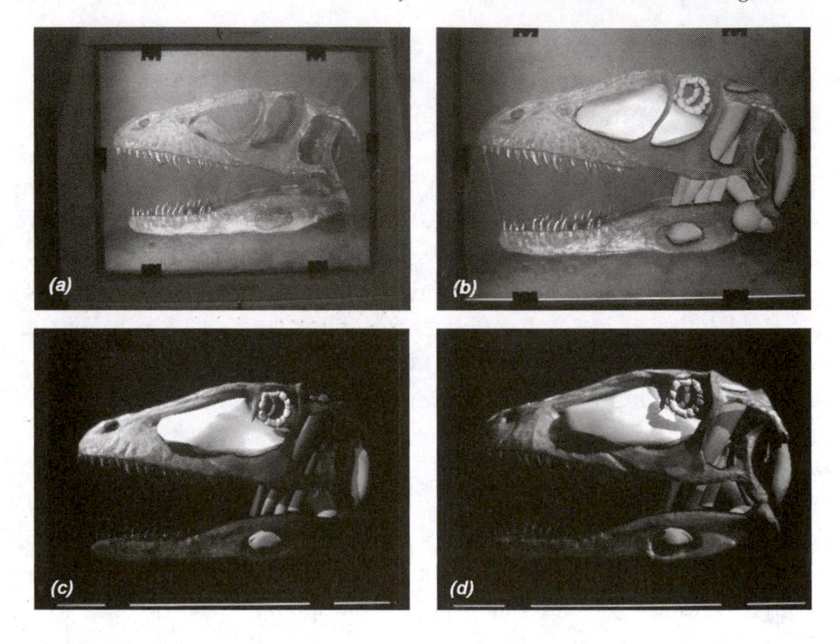

Figure 7.15. A rainbow hologram of a dinosaur skull combined with three-dimensional graphical elements and synthetic occlusion and shading effects: (a) the hologram only; (b) optically integrated graphical elements (muscles and other soft-tissue); (c) and (d) consistent illumination effects between holographic and graphical content. (*Images reprinted from [18] © IEEE.*)

put rendered on a screen over a holographic plate. In this case, however, the reconstructed light of the hologram will interfere with the overlaid light of the rendered graphics and an effective combination is impossible. However, the holographic emulsion can be used as an optical combiner itself, since it is transparent if not illuminated in the correct way. Consequently, it provides see-through capabilities.

This section describes how holograms can be optically combined with interactive, stereoscopic, or autostereoscopic computer graphics [18]. While Section 7.4.1 explains how correct occlusion effects between holographic and graphical content can be achieved, Section 7.4.2 outlines how a consistent illumination is created in such cases.

7.4.1 Partially Reconstructing Wavefronts

A key solution to this problem is to reconstruct the object wave only partially—not at those places where graphical elements have been inserted. This requires a point light source capable of selectively emitting light in

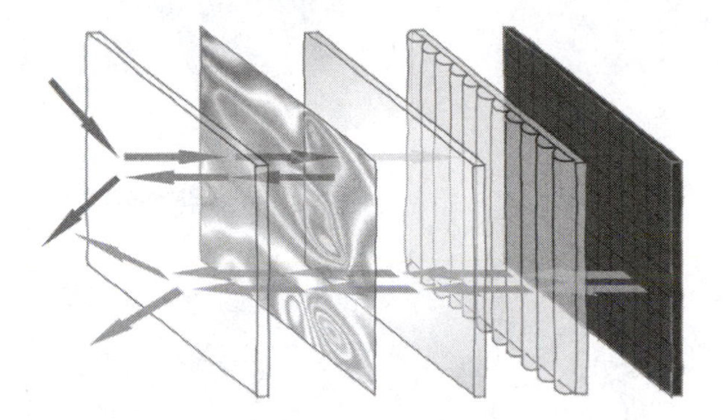

Figure 7.16. An explosion model of the stacked structure of optical layers. From left to right: glass protection, holographic emulsion, mirror beam combiner (only for transmission holograms), lenticular lens sheet, LCD array. The light-rays (darker arrows) are reflected and reconstruct the object wave on their way back through the emulsion. The stereoscopic images (gray arrows) pass through all layers until they are merged with the hologram. (*Image reprinted from [18] © IEEE.*)

different directions, creating an "incomplete" reference wave. Conventional video projectors are such light sources. They are also well suited for viewing white-light reflection or transmission holograms, since today's high-intensity discharge (HDI) lamps can produce a very bright light.

If autostereoscopic displays, such as parallax displays are used to render three-dimensional graphics registered to the hologram, then both holographic and graphical content appear three-dimensional within the same space. This is also the case if stereoscopic displays with special glasses that separate the stereo images are used.

Reflection holograms without an opaque backing layer and transmission holograms both remain transparent if not illuminated. Thus, they can serve as optical combiners themselves, leading to very compact displays. The illumination and rendering techniques that are described in these sections are the same for both hologram types.

Figure 7.16 illustrates an example of how a transmission hologram can be combined effectively with a flat-panel lenticular lens sheet display (a variation of a parallax display that utilizes refraction of a lens array to direct the light into the different viewing zones).

Placing a transmission hologram in front of a mirror beam combiner allows us to illuminate it from the front and to augment it with graphics from the back. For reflection holograms, this beam combiner is not necessary.

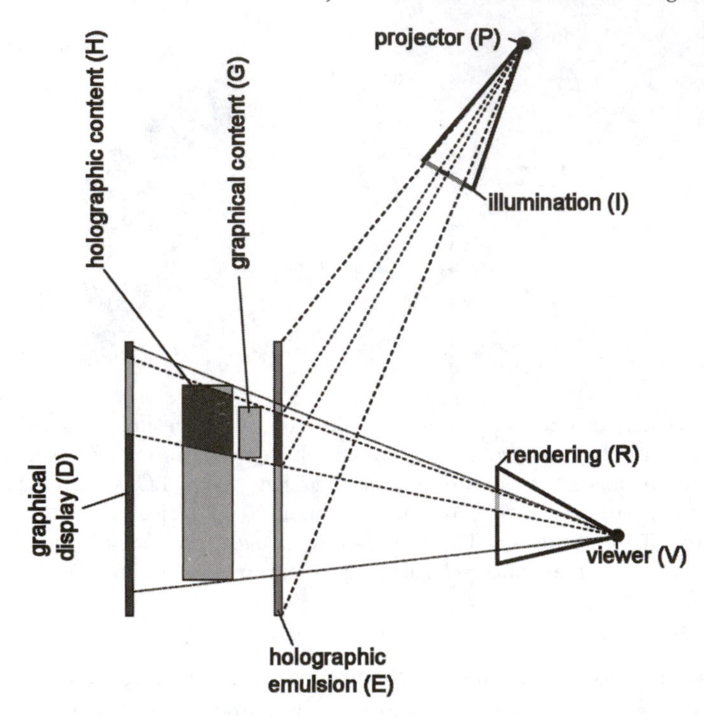

Figure 7.17. A conceptual sketch of the display constellation. The shaded areas on the graphical display and on the holographic emulsion illustrate which portion of the visible image is hologram (black) and which is graphics (gray). (*Image reprinted from [18] © IEEE.*)

A thin glass plate protects the emulsion from being damaged and keeps it flat to prevent optical distortion. The lenticular lens sheet directs the light emitted from the LCD array through all layers towards the eyes of the observer. The projected light is transmitted through the first two layers, and is partially reflected back (either by the beam combiner in combination with a transmission hologram or by a reflection hologram), reconstructing the recorded content. The remaining portion of light that is transmitted through all layers is mostly absorbed by the screen.

Figure 7.17 illustrates how the selective illumination on the holographic plate is computed to reconstruct the portion of the hologram that is not occluded by graphics.

In what follows, we outline how rendering and illumination can be realized with conventional graphics hardware. It is assumed that depth information of the holographic content (H), as well as a scene description of the graphical content (G) are available. Both contents are geometrically aligned during an offline registered step. If optical markers are recorded in the hologram together with the actual content, cameras can be used to perform this registration automatically. In addition, the extrinsic and intrinsic parameters of the video projector (P) with respect to the holographic emulsion (E) have to be known. They are also determined during an offline calibration step. If it is mechanically possible to mount the holographic emulsion close to the graphical display (i.e., the distance between E and D is small), then E and D can be approximated to be identical.

First, an intermediate texture image (T) is created from V over E by rendering H into the graphics card's depth buffer and filling the card's frame buffer entirely with predefined light color values. In addition, G is rendered into depth and stencil buffers. The stencilled areas in the frame buffer are cleared in black and the result is copied into the memory block allocated for T.

Note that if a render-to-texture option is provided by the graphics card, the read-back operation from the frame buffer into the texture memory is not necessary. The final illumination image (I) is rendered from P by drawing E into the frame buffer and texturing E's geometry with T. The illumination image (I) is beamed onto the holographic emulsion (E) with the projector (P).

Second, a rendering image (R) is generated from V over D (off-axis) by rendering H into the depth buffer and G into depth and frame buffers. The rendering image (R) is displayed on the graphical display (D).

This can be summarized with the following algorithm:

```
\\ first pass
create intermediate texture T from V over E:
  render H into depth buffer
  fill frame buffer with light color
  render G into depth and stencil buffer
  fill stenciled areas in frame buffer with black

\\ second pass
render E from P textured with T and display on projector

\\ optical inlay
create rendering image R from V over D:
  render H into depth buffer
  render G into depth buffer and frame buffer
  display G on projector
```

Figure 7.15(a) shows a photograph of the entire reconstructed hologram while it is illuminated with a projected uniform light.

By applying the presented techniques, illumination and stereoscopic images are generated in such a way that graphical and holographic content can be merged within the same space (Figure 7.15(b)).

7.4.2 Modifying Amplitude Information

The amplitude of the reconstructed object wave is proportional to the intensity of the reference wave. Beside using an incomplete reference wave for reconstructing a fraction of the hologram, intensity variations of the projected light allow us to locally modify the recorded object wave's amplitude.

In practice, this means that to create the illumination image (I), shading and shadowing techniques are used to render the holographic content, instead of rendering it with a uniform intensity.

To do this, the shading effects caused by the real light sources that were used for illumination while the hologram was recorded, as well as the physical lighting effects caused by the video projector on the holographic plate, have to be neutralized. Then the influence of a synthetic illumination has to be simulated. This can also be done with conventional graphics hardware (Figure 7.18). Three intensity images have to be rendered.

For the first image (I_1), H is rendered from V over E with a white diffuse material factor and graphical light sources that generate approximately the same shading and shadow effects on H as the real light sources that were used during the holographic recording process. This results in the intermediate texture T_1. I_1 is generated by rendering E from P and texturing it with T_1. It simulates the intensity of the recorded object wave. The same process is repeated to create the second image (I_2), but this time, graphical light sources are used to shade H under the new, virtual lighting situation. The ratio I_2/I_1 represents the required intensity of the reference wave at holographic plate E.

For the third image (I_3), E is rendered from P with a white diffuse material factor and a virtual point light source located at the projector's position. This intensity image represents the geometric relationship between the video projector as a physical point light source and the holographic plate:

$$F = \frac{\cos(\theta)}{r^2}.$$

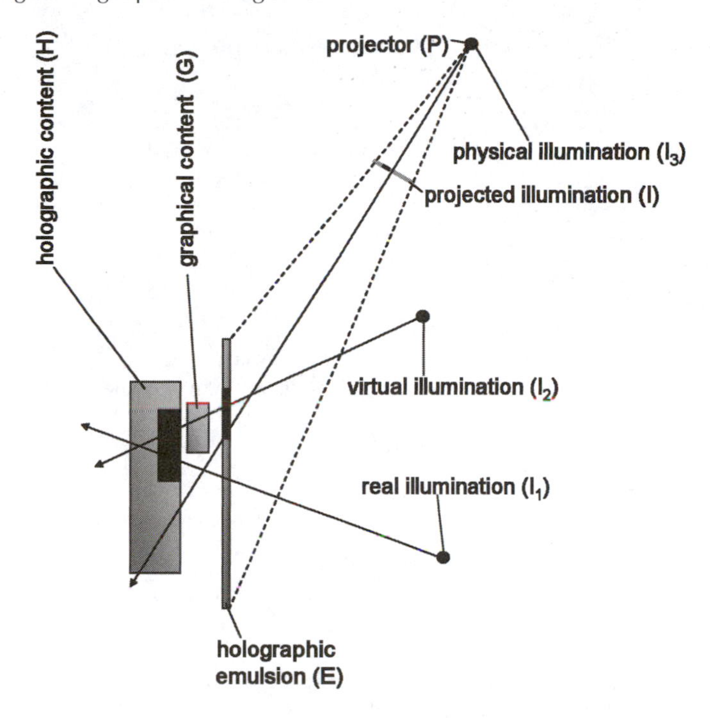

Figure 7.18. Light interaction between hologram and graphics: To simulate virtual shading and shadow effects on the holographic content, the recorded and the physical illumination effects have to be neutralized. (*Image reprinted from [18] © IEEE.*)

It contains form-factor components, such as the square distance attenuation (r^2) and the angular correlation ($\cos(\theta)$) of the projected light onto the holographic plate and can neutralize the physical effects of the projector itself.

The final illumination image (I) can be computed in real time with $I = I_2/I_1/I_3$ via pixel shaders. The projection of I onto E will neutralize the physical and the recorded illumination effects as well as possible and will create new shadings and shadows based on the virtual illumination. Note that as described previously, the graphical content has to be stencilled out in I before displaying it.

During all illumination and rendering steps, hardware-accelerated shadow mapping techniques are used to simulate real and virtual shadow effects on H and on G. Finally, synthetic shadows can be cast correctly from all elements (holographic and graphical) onto all other elements.

The following algorithm summarizes this approach:

```
\\ first pass
create intensity image I1:
  render H from V over E (white diffuse factor)...
    ...and graphical light sources that simulate real...
    ...shading on H -> T1
  render E from P textured with T1

\\ second pass
create intensity image I2:
  render H from V over E (white diffuse factor)...
    ...and graphical light sources that simulate virtual...
    ...shading on H -> T2
  render E from P textured with T1

\\ third pass
create intensity image I3:
  render E from P (white diffuse factor)...
    ...and graphical point light source attached to P

\\ fourth pass
create and display illumination image I from P:
  I = I2 / I1 / I3 (via pixel shader)
```

The capabilities of this technique are clearly limited. It produces acceptable results if the recorded scenery has been illuminated well while making the hologram. Recorded shadows and extreme shading differences cannot be neutralized. Furthermore, recorded color, reflections, and higher-order optical effects cannot be cancelled out either.

Projecting an intensity image that contains new shading and shadow effects instead of a uniform illumination allows us to neutralize most of the diffuse shading that is recorded in the hologram and produced by the projector. The holographic and graphical content can then be consistently illuminated (creating matching shading and shadow effects) under a novel lighting.

Figures 7.15(c) and 7.15(d) illustrate the synthetic shading effects that are caused by a virtual light source. In addition, virtual shadows are cast correctly between hologram and graphical elements. Although Figure 7.15 illustrates the results with a monochrome transmission hologram, the same effects are achieved with reflection holograms and color holograms.

Since all the discussed rendering techniques (like shadow mapping and shading) are supported by hardware-accelerated consumer graphics cards, interactive frame rates are easily achieved.

Using the concept described in this section, a pallet of different display variations can be developed. With only minor changes of the presented techniques, for example, arbitrarily curved shapes (such as cylindrical shapes being used for multiplex holograms) can be supported instead of simple planar plates. Even without graphical augmentations, the projector-based illumination alone has numerous potential uses. In combination with optical or digital holograms, it can be used to create visual effects. Certain portions of a hologram, for instance, can be made temporarily invisible while others can be highlighted. Emerging large-scale autostereoscopic displays and existing stereoscopic projection screens allow us to up-scale the proposed concept. Not only the display, but also the holograms can be composed from multiple smaller tiles to reach large dimensions and high resolutions.

7.5 Augmenting Flat and Textured Surfaces (Example: Pictorial Artwork)

A seamless and space efficient way to integrate visual information directly into *pictorial artwork* is to use the artwork itself as *information display* [19]. It can serve as a diffuse projection screen and conventional video projectors can be used to display computer graphics together with painted content (Figure 7.19).

The main difference of this approach is that an arbitrarily textured surface has to be augmented with colored information. To perceive the projected imagery in the correct colors and intensities, however, requires that the influence of the underlying physical *color pigments* is neutralized. In most situations, this is not possible if untreated images are simply projected directly onto arbitrary colored surfaces. The problem is that the projected light interacts with the color pigments on the canvas and is partially absorbed if the pigment's color isn't fully white. The process to correct these artifacts is called *radiometric compensation.* .

A solution to this problem is provided by a new film material which has two properties: (1) it is completely transparent, and (2) it diffuses a fraction of the light that is projected onto it. The film consists of an even deposition of fine particles on both sides of a polyester base with no visible artifacts. It was used for the creation of special effects in Hollywood movies, such as *Minority Report* (20th Century Fox, 2002) and *Paycheck* (Paramount Pictures, 2003), and sells for $350 per square foot. Initial

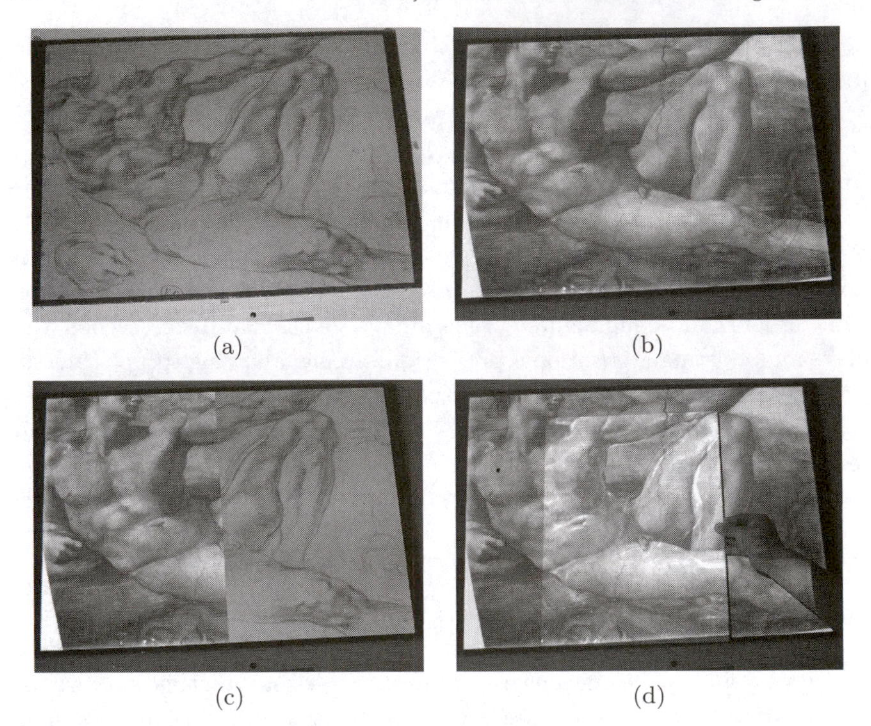

(a) (b)

(c) (d)

Figure 7.19. Results of a color correction process with a single projector on a real drawing: (a) real drawing (64 × 48 cm) under environment light; (b) output image emitted onto drawing; (c) partially augmented drawing; (d) output image on a white piece of paper. (*Displayed artwork courtesy of the British Museum, London; images reprinted from [19] © IEEE; see Plate V.*)

measurements have revealed that on average 20% (+/- 1%) of the light that strikes the film is diffused while the remaining fraction is transmitted towards the canvas (with or without direct contact). This 0.1 mm thin transparent film can be seamlessly overlaid over the canvas by integrating it into the frame that holds the artwork. Off-the-shelf 1100 ANSI lumen XGA digital light projectors have been used to display images on film and canvas.

7.5.1 Technical Approach and Mathematical Model

If a light beam with incident radiance L is projected onto the transparent film material that is located on top of the original artwork, a portion d of L is directly diffused from the film while the remaining portion t of L is transmitted through the film. The transmitted light tL interacts with

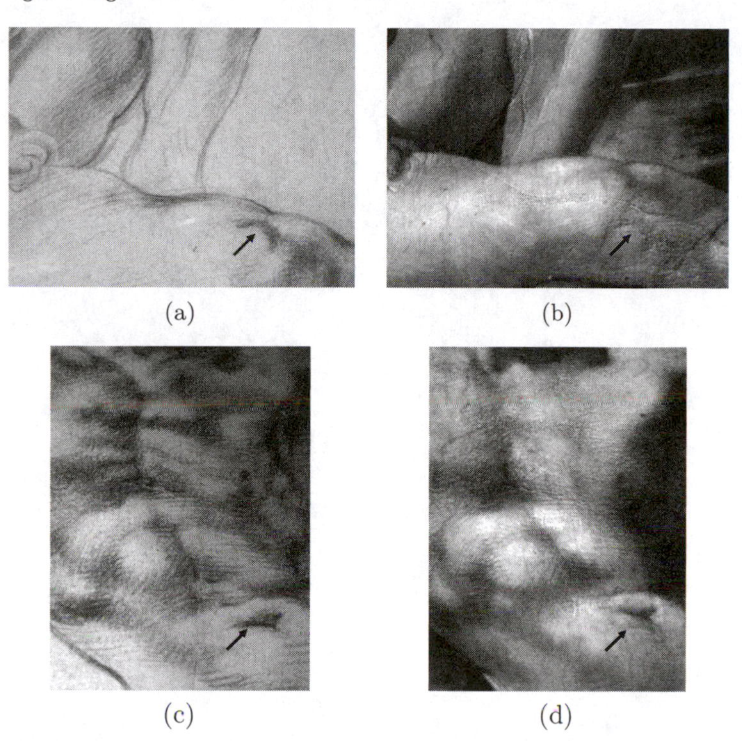

Figure 7.20. (a)–(d) Close-ups: while the upper body part coincides in drawing and painting, Michelangelo modified the lower body part. The arrows indicate the displaced knee and belly sections. They point at the same spot on the drawing. (*Displayed artwork courtesy of the British Museum, London; images reprinted from [19] © IEEE.*)

the underlying pigment's diffuse reflectance M on the canvas, and a color blended light fraction tLM is diffused. The portion $tLMt$ is then transmitted through the film, while the remaining part $tLMd$ is reflected back towards the canvas where it is color blended and diffused from the same pigment again. This ping-pong effect between film material and canvas is repeated infinitely while, for every pass, a continuously decreasing amount of light is transmitted through the film that contributes to the resulting radiance R. Mathematically, this can be expressed as an infinite geometric series that converges towards a finite value. The same is true for the environment light with incident radiance E that is emitted from uncontrollable light sources. Since these light sources also illuminate the canvas and the film material, the environment light's contribution to R has to be considered as well.

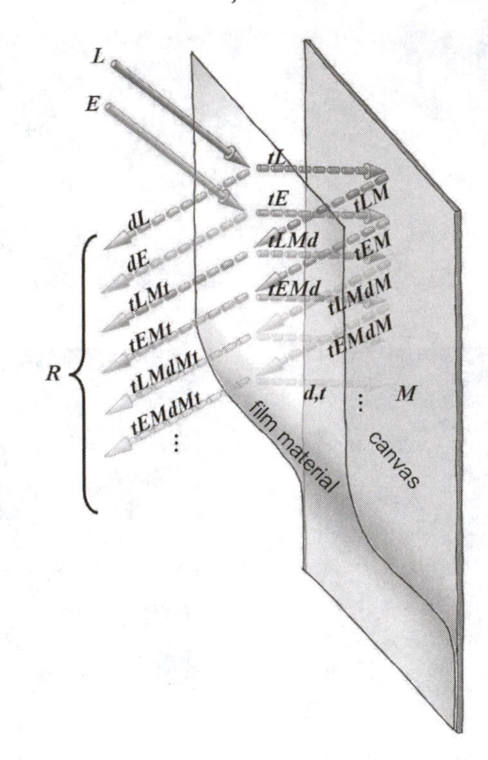

Figure 7.21. Interaction of projected light and environment light with the canvas and the film material (sequence diagram). (*Image reprinted from [19] © IEEE.*)

Figure 7.21 describes this process in the form of a sequence diagram. Note that in contrast to this conceptual illustration, there is no physical gap between film material and canvas, and that the light interaction occurs at the same spot.

If all parameters (L, E, M, t, and d) are known, we can compute the resulting radiance R that is visible to an observer in front of the canvas:

$$\begin{aligned} R &= (L+E)\,d + (L+E)\,t^2 M \sum\nolimits_{i=0}^{\infty} (Md)^i \\ &= (L+E)\left(d + \frac{t^2 M}{1 - Md}\right). \end{aligned} \tag{7.4}$$

Since R forms the image we expect to see is known, we need to solve Equation (7.4) for L:

$$L = \frac{R}{\left(d + \frac{t^2 M}{1-Md}\right)} - E.$$

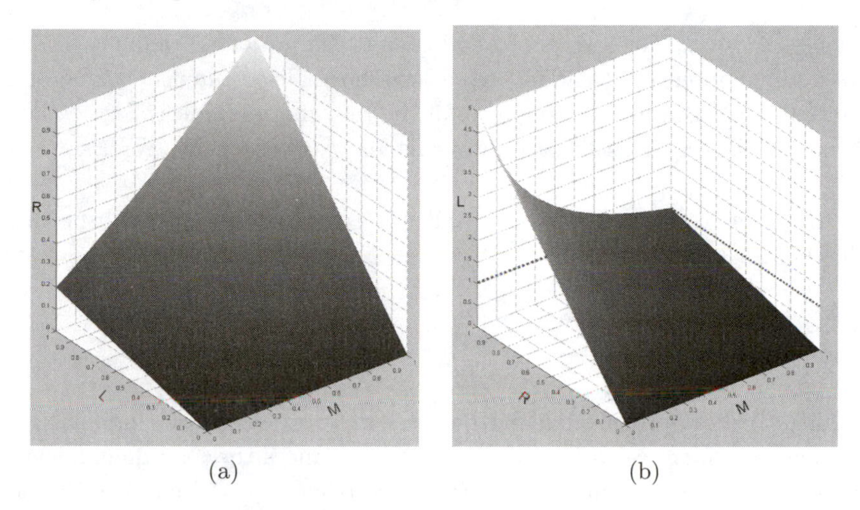

Figure 7.22. Visualization of: (a) Equation 7.9; (b) Equation 7.10 over R, L, and M (without E, t=80% and d=20%). (*Images reprinted from [19] © IEEE.*)

This allows us to compute the incident radiance L that needs to be projected onto the film and the canvas to create the known result R. The radiant intensity I of the projector to create L is related to a discretized pixel value and is given by

$$I = L\frac{r^2}{\cos\alpha}s,$$

where $r^2 \cos\alpha$ is the form-factor component: square distance attenuation and angular correlation of the projected light onto the canvas. The additional factor s allows us to scale the intensity to avoid clipping and to consider the simultaneous contributions of multiple projectors.

This approach has clear limitations, which are illustrated in Figures 7.22(a) and 7.22(b). Not all radiances R can be produced under every condition. If M is dark, most of L and E are absorbed. In an extreme case, the corresponding pigment is black (M=0). In this case, the right term of the equation that computes R is cancelled out. The remaining left term, which depends on the diffusion factor d of the film material, sets the boundaries of the final result that can be produced. The intersection of the surface with the RL-plane in Figure 7.22(a) illustrates these limitations. In the worst case of this example, only 20% of R can be generated. This situation is also reflected in Figure 7.22(b) as the intersection of the surface with the LR-plane. Here, we want to assume that $sr^2\cos\alpha = x$, which results in $L = I$. For a single video beamer, the projected radiance L and the radiant intensity I cannot exceed the normalized intensity value of 1 (dotted line). But for creating most of the resulting radiance values, L and I must be larger. This situation worsens for $r^2 \cos\alpha > x$ and for $E \to 1$ or $M \to 0$.

However, the contributions of multiple (n) projectors allow displacing this boundary with

$$L = \sum_{i=1}^{n} L_i \quad \text{and} \quad I_i = L_i \frac{r_i^2}{\cos\alpha_i} s_i.$$

If all projectors provide linear transfer functions (e.g., after gamma correction) and identical brightness, $s_i = 1/n$ balances the load among them equally. However, s_i might be decreased further to avoid clipping and to adapt for differently aged bulbs.

For both illustrations in Figure 7.22, the environment light E is not considered and is set to zero. Additional environment light would simply shift the surface in Figure 7.22(a) up on the R-axis, and the surface in Figure 7.22(b) down on the L-axis. Note that the mathematical model discussed previously has to be applied to all color channels (e.g., red, green, and blue for projected graphical images) separately.

7.5.2 Real-Time Color Correction

The equations described in Section 7.5.1 can be implemented as a pixel shader to support a color correction in real time. Figure 7.23 illustrates the rendering process based on an example of Michelangelo's masterpiece *Creation of Adam*. Although we will use this example to explain the process, it is universal and can be applied with arbitrary background images.

In this example, a copy of an early sanguine sketch of the Adam scene (now being displayed in the British Museum, London) serves as a real background image M. Our goal is to overlay it entirely with a registered photograph R of the actual ceiling fresco painted in the Sistine Chapel.

The first step of the rendering process is to create an input image I_i. This image can be either dynamically rendered (e.g., as part of a real-time animation or an interactive experience), it can be played back (e.g., frames of a prerecorded movie), or it can be static (e.g., a photograph of the corresponding ceiling portion, as is the case in our example).

The input image has to be registered to the physical drawing. Registration is achieved by texture mapping I_i onto a pre-distorted image geometry

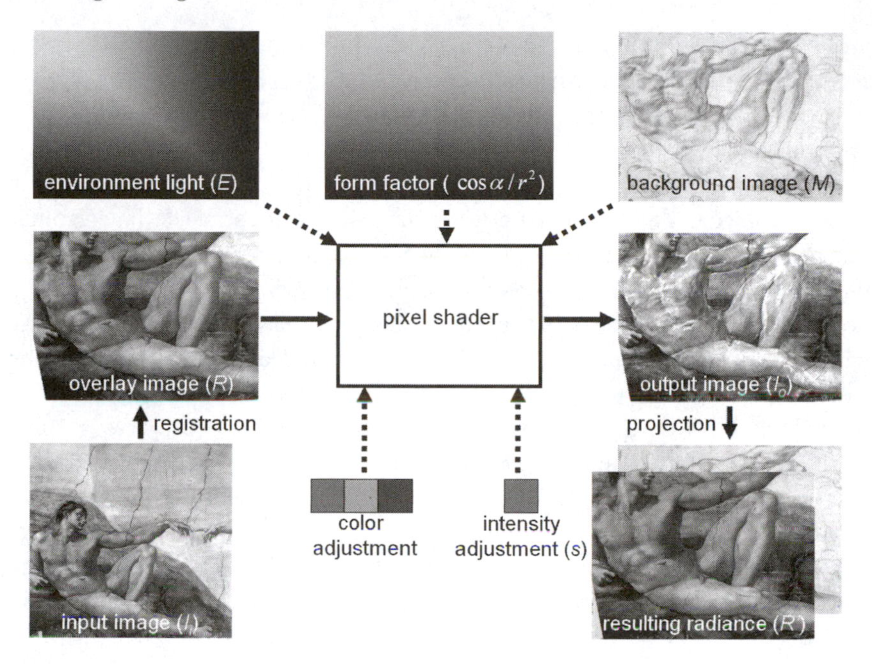

Figure 7.23. Real-time color-correction process with pixel shader. (*Displayed artwork courtesy of the Vatican Museum, Rome; image reprinted from [19] © IEEE.*)

that is precisely aligned with the physical drawing. The amount of distortion depends on the geometric relation between the video projector and the canvas. It extends from simple *keystone deformations* to more complex curvilinear warping effects (e.g., if lens distortion of the projector has to be neutralized). Several automatic approaches have been described that use video cameras and image analysis to align multiple images of tiled projection screens [31, 208]. A structured light registration benefits from controllable features, such as projected grid edges or Gaussian matchpoints that can easily be detected in the camera views with a sub-pixel precision. In the case described previously, however, a digital representation of the artistic content has to be registered against its physical representation on the canvas, rather than registering one projected structured light image against another one. To detect non-structured artistic features, such as fine lines, in the artwork and register them automatically against the corresponding features in the digital content represents a non-trivial task of computer vision, especially if projected pixels and physical pigments have to be aligned very precisely on the canvas. One can be critical about the feasibility and precision of an automatic method for this problem. It can be solved with a manual registration process that benefits from the resolution of the human visual system. Since the following steps have to be performed only once, they represent an acceptable solution.

An offline registration process allows us to interactively identify two-dimensional correspondences between artistic features in the background image M within the image space—that is, an image of the drawing displayed on a control screen—and the physical drawing on the wall. This is done using a two-dimensional input device, such as a conventional mouse whose pointer is visible on the control screen and as a projection on the canvas. A dual output graphics card and an additional video signal splitter are used to drive the control screen and one or two projectors. The result is a set of two-dimensional vertex fiducials with their corresponding texture coordinates within the image space. The fiducials are *Delauny triangulated* and the texture coordinates are used to map the correct image portions of I onto the image geometry. This results in the overlay image R. It has to be stressed that a precise correspondence between R and M is important to achieve qualitatively good results. The measurement of 50 to 70 fiducials proved to be sufficient for medium-sized canvases. In contrast to uniform grid methods normally applied for projector alignment, this general geometric registration allows the correlation of arbitrary features in the physical drawing with the corresponding pixels of M in the image space. Thus, it provides an accurate matching which can be regionally improved further if linear interpolation within single grid triangles fails to be precise enough. The registration results do not change if projector and background image are fixed. Before R is rendered, the color-correction pixel-shader has to be enabled. Five parameters are passed to it to ensure that the equations discussed in Section 7.5.1 can be computed.

The first parameter is the *environment light* E in the form of an intensity texture that has the same size as R. It contains intensity values that represent the uncontrollable lighting situation on the canvas. The intensity values can be determined by measuring the irradiance of the environment light with a light meter for a discrete number of sample spots on the canvas' surface, resulting in the lux values E'. To be processed by the shader, these values have to be normalized to an intensity space that ranges from 0 to 1. To do this, the same spots are measured again, but this time the highest intensity possible (i.e., a white image) is projected onto the light meter which is measured in addition to the environment light. These measurements are equivalent to the total irradiance $T' = L' + E'$, and also carry the unit lux. Since we know that $L' = T' - E'$ is equivalent to the scaled intensity value $\cos\alpha/r^2$, we can convert the measured radiance of the environment light from lux into the normalized intensity space with $E = E'/(T' - E')\cos\alpha/r^2$. To approximate the intensity values for the entire image area in E, all the measured spots are mapped into the image space, are Delauny triangulated, and the values for the remaining pixels are linearly interpolated by the graphics pipeline. The values for the remaining pixels are linearly interpolated by the graphics card while rendering the Delauny mesh. Note that E is constant if the environment light does not change. For the reasons that are described next, we can assume that $\cos\alpha/r^2$ is constant and equals 1.

The second parameter is the form-factor that represents the geometric relation between the video projector as a point light source and the canvas. Since it does not change for a fixed relation between projector and canvas, it can be precomputed and passed to the pixel shader in the form of an intensity texture with the same dimensions as E and R. Similar to the techniques described in the previous sections, this texture can be produced by the graphics pipeline. A geometric model of the canvas is rendered with a white diffuse reflectance from the viewpoint of the projector. Attaching a virtual point light source (also with a white diffuse light component) to the position of the projector and enabling square distance attenuation produces intensities that are proportional to $\cos\alpha/r^2$. The required reciprocal can be computed by the pixel shader. Practically (i.e., for normal-sized canvases and non-extreme projector configurations), the form factor can be assumed to be constant over all pixels. It is then contained by the intensity adjustment parameter s.

The third parameter is the *background image* M. It also has the same dimensions as E and R. This image can be generated by, for example, scanning the color values or taking a photograph of the original drawing under uniform illumination.

The fourth and fifth parameters contain *color and intensity adjustment values* that allow fine-tuning of the video projector's individual color response and prevent intensity clipping. They also allow adapting for color drifts that can be introduced during the capture of the background image and consideration of the contributions of multiple projectors. Note that *gamma correction* has to be applied in advance. This is true for projectors with non-linear transfer functions as well as for projectors with linear transfer functions that apply a *de-gamma mapping* on the video input. Gamma correction is usually supported by the graphics hardware and the video driver but can also be carried out by the pixel shader. These values are adjusted manually, but the support of automated methods for color-matching multiple projectors [100] is imaginable.

The output image I_o is the final result of this rendering process and will be displayed by the video projector. If projected geometrically onto the drawing correctly, the result R' will be visible. Both images, R and R' are mostly identical, except for slight artifacts that are due to the limitations discussed above. Figure 7.19 shows the result of our example projected onto the real drawing with a single video projector.

The underlying drawing can be made partially or completely invisible to display the graphical overlay in the correct colors on top of it. Close-ups are illustrated in Figures 7.19(e)–(h), in which diverging body parts (such as belly and knee) are overdrawn and displaced by the projection.

Some intensities and colors that are required to neutralize the underlying color pigments cannot be achieved by a single video projector. The worst case is to turn a black pigment on the canvas into a white color spot. Figures 7.22(a) and 7.22(b) illustrate that in such a case the required intensity can easily exceed the boundary of 1 in our normalized intensity space. The pixel shader clips these values to 1 which results in visible artifacts.

The simultaneous contributions of multiple projectors can reduce or even eliminate these effects. Figure 7.24 shows the extreme case of an input image that has no geometric correspondences to the underlying background image. In addition, it attempts to create bright colors (the sky) on top of

(a) (b)

Figure 7.24. Results of color correction process with two projectors: (a) the limited intensity capabilities of a single projector result in visible artifacts; (b) the contribution of a second projector reduces these effects. (*Displayed artwork courtesy of the Vatican Museum, Rome and the British Museum, London; images reprinted from [19] © IEEE; see Plate VI.*)

dark color pigments on the canvas. In Figure 7.24(a), a single projector is used. Intensity values that are too large are clipped and result in visible artifacts. Balancing the intensity load between two projectors reduces these artifacts clearly (Figure 7.24(b)).

Due to hardware acceleration of today's graphics cards, the color correction process can be easily performed in real time. Note that none of the photographs here have been retouched. Slight variations of color and brightness are due to different camera responses.

The color correction method has limitations that are mainly defined by the capabilities of the applied hardware. For example, the restricted resolution, brightness, contrast, minimum focus distance, and black-level of video projectors are issues that will certainly be improved by future generations. The XGA resolution and the brightness of 1100 ANSI lumen of the low-cost projectors that were used to create the results for this section was appropriate for small- and medium-sized paintings in a normally lit environment. An upscaling is possible by using more projectors, but downscaling would either result in a loss of effective resolution or in focus problems.

Black, for instance, is a color that cannot be projected. Instead, the environment light together with the black level of the projectors, illuminates areas that need to appear black. However, the human vision system adjusts well to *local contrast effects* which makes these areas appear much darker than they actually are. Even with little environment light, the high black level of video projectors causes this illusion to fail in extreme situations, such as the one shown in Figure 7.25. The development of video projectors indicates that a decrease of the black level and an increase of the contrast ratio can be expected in the future.

Light can also damage the artwork. Especially ultra violet (UV) and infrared (IR) radiation produced by the lamps of video projectors is critical. Commercially available UV/IR blocking filters can be mounted in front of the projectors' lenses to remove most of these unwanted rays while transmitting visible wavelengths. For the remaining visible light portion, a rule of thumb advises to illuminate valuable and delicate pieces permanently with no more than 100 lx–150 lx. The potential damage caused by light is cumulative (e.g., 1 hour with 1000 lx equals 1000 hour with 1 lx) and depends on the material and color of the painting and the wavelength (i.e., the color) of the light. A temporary illumination of a higher light intensity is not critical. During a 2–3 minute presentation, increased lighting is only temporary and such highlight situations usually only appear (if at all) for

Figure 7.25. Rembrandt's self-portrait: (a) copy of original painting as it looks today (illuminated under environment light); (b)–(d) various cleaning stages to remove the overpainted layers form 1935(d), 1950(c) and 1980(b) are projected onto (a). Only black and white photographs of these stages are available. The high black-level of the video projectors prevents the creation of a totally black color on the canvas. Extreme regions, such as overlaid hair and hat cannot appear completely black for this reason. (*Displayed artwork courtesy of the Museum het Rembrandthuis, Amsterdam; images reprinted from [19] © IEEE; see Plate VIII.*)

a short period of time (e.g., a few seconds) at varying locations on the canvas. Nevertheless, using the intensity adjustment described, the maximum light level can be constrained to be below an upper threshold. However, this might cause visible artifacts depending on the presented content, the painting, and the environment light (as described in Figure 7.22). Thus, it is important to reach a good balance between total illumination (projected light and environment light) over time and convincing presentation effects.

The method described currently considers only the intensity of the environment light E. This is adequate for regular white light sources but will result in artifacts if visible color shading is created on the canvas by the environment illumination. Without a modification to either the mathematical model or the rendering process, the environment light's color can be compensated by determining it with the aid of *colorimeters*, encoding this information in E, and passing it to the pixel shader.

The presented concept and techniques are applicable in combination with diffuse pictorial artwork, such as watercolor paintings, pen or ink drawings, sanguine sketches, or matte oil paintings. Extreme light and view-dependent effects, such as *non-Lambertian specular reflections*, *self-shadows*, *sub-surface scattering*, and inter-reflections that are created by brush strokes, paint material, or canvas textures cannot be handled with this method. Techniques to augment geometrically non-trivial (i.e., non-flat) textured surfaces are described in the following section.

7.6 Augmenting Geometrically Non-Trivial Textured Surfaces

In this section, we describe how a color- and geometry-corrected projection onto *arbitrarily shaped* and textured surfaces is possible in real time, and how calibration can be realized fully automatically, fast, and robustly [20]. Off-the-shelf components, like a consumer LCD video beamer, a CCD camcorder, and a personal computer comprising a TV card and a pixel-shading capable graphics board can be used.

7.6.1 Creating Virtual Projection Canvases

This concept combines camera feedback with *structured light projection* to gain information about the screen surface and the environment. Neither the geometry of the surface nor internal or external parameters of projector and camera have to be known for calibrating the system. This makes it extremely robust and easy to use which is crucial for some applications, such as home entertainment or ad-hoc stereoscopic visualizations in everyday environments.

A modular camera component has to be temporarily placed approximately at the sweet-spot of the observers—pointing at the screen surface. The projection unit (i.e., the projector) can be placed at an arbitrary location. Its light frustum must also cover the screen surface area.

During calibration, the camera mimics the *target perspective* which is the optimal viewing position for which the projection unit will be calibrated. The user can either define the *display area* by sketching the outlines of a *virtual projection canvas* over a portion of the camera image, or it is automatically derived from the margins of the camera field of view. Lens distortion of the camera is adjusted at the beginning to deliver images without radial distortion.

The system can then determine all parameters that are required for real-time geometric predistortion and color correction of video frames delivered by PAL/NTSC compliant devices like DVD players or game consoles, or generated by a rendering tool.

After the system has been calibrated, the camera module can be removed. From now on, the projector unit corrects incoming video signals geometrically and radiometrically in real time. If the corrected images are projected onto the non-trivial screen surface, they appear as being displayed onto a plain white canvas from target perspectives at or near the *observers' sweet-spot*. However, this projection canvas is completely virtual and does not exist in reality.

7.6.2 Vanishing Shapes

One goal is to geometrically predistort the input images (e.g., video frames or a rendered image) in such a way that they appear correct if projected onto a geometrically non-trivial surface and are observed from an area close to or at the target perspective.

Wide-field-of-view cameras are sometimes used in a sweet-spot position to calibrate multiple overlapping projectors [149]. Projecting pixels and capturing them with the camera results in a projector-to-camera pixel mapping. For performance reasons, only a subset of projector pixels are usually displayed and captured, while the mapping for the remaining ones are linearly interpolated. For arbitrarily shaped surfaces with fine geometric details, however, a high-resolution pixel correspondence has to be generated in an acceptable amount of time.

To realize this, *time-multiplexed line strip scanning* techniques as known from *structured light* three-dimensional range-finder systems can be adapted (Figures 7.26(c) and (d)). If the camera is placed at the observers' sweet-spot, it equals the target perspective. A variation of a column-row coded pattern projection method with phase shifting (similar to that proposed in [60]) is applied to compute a pixel *displacement map*, which is a look-up table that maps every camera pixel to the corresponding projector pixel.

Figure 7.26. (a) Projection onto a scruffy corner; (b) normal uncorrected projection; (c)–(d) line-strip scanning of surface; (e) geometry-corrected projection on virtual canvas; (f) final geometry and color-corrected image (*Movie footage: The Jackal, © 1997 Universal Pictures; images reprinted from [20] © IEEE; see Plate XIII.*)

To ensure a correct synchronization between projection and capturing during the scanning process, the camera's *latency* has to be known. Latency results from hardware image compression and data transfer between camera and receiving device. It is particularly high for consumer camcorders, since they do not target real-time image processing applications. The latency of the camera can be determined automatically at the beginning of the geometric calibration process. This is done by sending out sample patterns and measuring the maximum time until these patterns can be detected in the recorded camera images.

Due to the different resolutions of camera and projector and to their varying distances and perspectives to the screen surface, the displacement map does not represent a one-to-one pixel mapping. In fact, it might not be complete since surface portions can lie in shadow areas. Multiple projector units can be used in this case. Furthermore, different projected line strips might project on the same camera pixel. The average values are computed and stored in the displacement map to achieve a *sub-pixel precision.*

If projector and camera are calibrated, the entire three-dimensional geometry of the screen surface can be recovered using the pixel correspondences in the displacement map and triangulation. This is exactly how some three-dimensional scanners function. However, since both devices are not expected to be located at known positions, the displacement map allows us only to map each camera pixel from the target perspective into the perspective of the projector. This is sufficient and results in an undistorted perspective (Figure 7.26(e)) without having to know the three-dimensional shape of the screen surface.

To benefit from hardware-accelerated computer graphics, the displacement map can be converted into a texture map (realized with a 32bit/16bit P-buffer) that stores references for every projector pixel to its corresponding video pixels (stored in the R, G channels) and camera pixels (encoded in the B, A channels). This texture is then passed as a parameter to a modern pixel shader to implement a real-time image warping via a *pseudo pixel displacement mapping*. Today, pixel shaders are standard components of many consumer graphics cards and provide per-pixel operations. Besides the displacement texture map, several other parameter textures are passed to the pixel shader, such as the uncorrected input image itself. All that is necessary for a geometric image predistortion is to render a single two-dimensional rectangle into the entire frame buffer of the projector. This triggers the rasterization of every projector pixel through the pixel shader before it is displayed. The colors of incoming pixels are simply overwritten by new colors that result from the corresponding input image pixels. It is possible to find these pixels in the input image with the aid of the displacement texture map. This has the same effect as actually moving the input image pixels to new positions within the projector frame buffer. Note that colors are not just copied from input image pixels to projector pixels; they are also modified to enable a color correction, which will be discussed in the next section.

This allows *warping* of every pixel of the input image in real time without geometric information of the screen surface. The final projection can be perceived geometrically correct at or near the target perspective. Depending on the shape of the screen surface, a more or less extreme distortion will be sensed.

Even though the predistorted projected image can now be observed geometrically correct on a non-trivially shaped surface, its uncorrected colors will still be blended with the texture of the screen surface. In the next section, the color correction process will be explained.

7.6.3 Neutralizing Textures

If light strikes a surface, only a fraction of its original intensity and color is reflected back while the rest is absorbed. For *Lambertian surfaces* (completely diffuse surfaces), the amount and color of reflected light depends on several parameters, such as the surface's material color (M), the light color and intensity that leaves the source (I), as well as the distance (r) and the incidence angle (α) of light rays with respect to the surface, together called form factor (F). For perfectly diffuse surfaces, *Lambert's law* approximates the diffuse reflection of light for each spectral component (see Lighting and Shading in Section 2.4.1) with $R = IFM$, where $F = \cos(\alpha)/r^2$.

In addition to the light projected by a video projector, the environment light is blended with the surface in the same way. Assuming additive color mixing, Lambert's law can be extended to take this into account: $R = EM + IFM$, where E is the intensity and color of the environment light. The main difference between environment light and projected light is that the latter is controllable.

The goal is to neutralize this natural blending effect by projecting an image (I) in such a way that its blended version on the screen surface is perceived in its known original colors (R). Since only diffuse Lambertian screen surfaces are considered (most other surface types are improper for a video projection), one simply has to solve the equation for I: $I = (R - EM)/FM$.

Because information about internal and external parameters of the projector or camera is not required, each component (E, M, and F) cannot be determined individually. Rather than that, the products EM and FM can be measured while R is the given input image.

If the video projector displays a bright white image ($I = 1$) onto the screen surface within a dark environment ($E = 0$), then the camera captures an image that is proportional to FM. Further, if the projector is turned off ($I = 0$), an image of the screen surface that is captured under environment light is proportional to EM. These assumptions imply that the projector and camera are color- and intensity-adjusted, and that automatic brightness control, focus, and white-balancing are turned off. They are simple approximations, but allows to determine the required parameters robustly and without performing complicated measurements and using additional special purpose devices.

(a) (b)

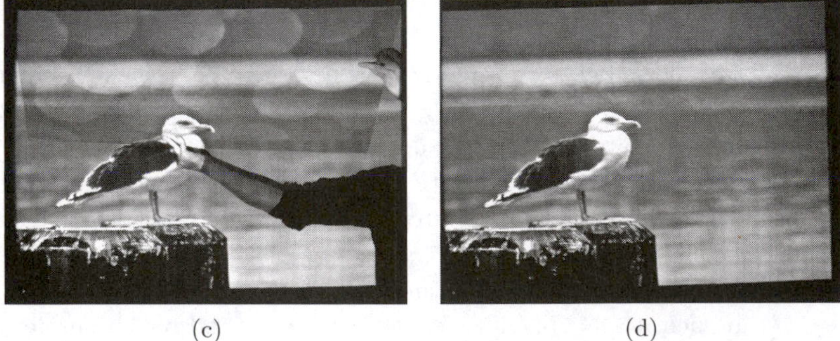

(c) (d)

Figure 7.27. (a) Projection onto a wallpapered pitched roof area; (b) projection with uncorrected colors; (c) color correction projected onto white piece of paper; (d) color corrected image on wallpaper. All projections are geometry corrected. (*Images reprinted from [20] © IEEE; see Plate XII.*)

Representing all these parameters as textures, the pixel shader can perform this computation in real time (>100 fps on a NVIDIA GeForce FX6800). As mentioned in the previous section, pixels of images taken from the camera view (EM and FM) as well as the input image R are warped to the view of the projector via pixel displacement mapping. This ensures a correct concatenation of corresponding pixels. The resulting image I is finally displayed from the perspective of the projector. These computations are performed on all three RGB color channels separately. In addition the pixel shader allows for fine-tuning of the output images by considering manually set color and brightness parameters (also for *gamma correction*). In addition, it clips out extreme intensity situations to avert visible artifacts.

Figure 7.27 illustrates an example of a geometry-corrected projection onto a wallpapered pitched roof area. If the input image is not color corrected, then the projected colors are blended (Figure 7.27(b)) with the colors of the screen surface. The texture of the wallpaper interferes with the video image which results in a disturbing effect. The color-corrected image (I) is partially shown in Figure 7.27(c) by projecting it onto a white cardboard. Blending I with the screen surface results in the image shown in Figure 7.27(d)—which is a close approximation to the original input image R. In this case, the screen surface becomes almost invisible.

Further examples of projections onto other surface types, such as window curtains and natural stone walls are presented in Section 8.8.

7.6.4 Limitations

Obviously, the geometry correction, but in particular the color correction, will fail if the screen surface's material is absorbing light entirely. This is also the case if only a part of the visible spectrum that needs to be displayed is completely absorbed even if other parts are reflected. Fortunately, such materials, like velvet, are comparatively rare in our everyday environments. In fact, most diffuse materials produce fairly acceptable results. If the surface is capable of reflecting a certain fraction of the desired light color, it is just a matter of how much incident light is required to produce the required output. If one projector is not capable of generating the necessary amount of light, multiple projectors can complement each other.

Additionally, there are several technical limitations that lower the current quality of this concept. One important factor is the *limited resolution* of consumer camcorders. If the camera has to be placed far away from a large screen surface to capture the entire display area, fine details on the surface cannot be detected and corrected. Higher-resolution cameras (e.g., mega-pixel digital cameras) can be used to provide better quality images. In fact, most camcorders already combine two devices in one: a high-resolution digital camera and a video camera that delivers a live video stream. They facilitate both a fast geometry correction and a high-quality color correction.

On the projector side, the limited resolution, the *low dynamic range*, and the *high black level* of consumer devices represent the main restrictions. A too low projector resolution causes too large pixel projections that cannot cover smaller pigments on the screen surface precisely. In particular, the black level cannot be controlled and, consequently, contributes to the environment light. Even in a completely dark room, the black level of a projector causes the screen surface to be visible all the time. As it is the case for a normal projection onto a regular canvas, the black level, and the environment light make it difficult to display dark colors. However, the human visual system adapts well to local contrast effects. Dark areas surrounded by brighter ones appear much darker than they actually are. Even though these problems will be solved with future-generation projectors, one general problem will remain: the limited depth focus of conventional projectors prevents them from displaying images on screen surfaces that are extremely curved. Since laser projectors, which are capable of focusing on non-planar surfaces, are still far too expensive for a consumer market, the use of multiple projectors is once again a promising solution to this problem.

Finally, one last issue remains for a projection onto non-planar surfaces: a single projector can cast shadows on the screen surface that appear as cuttings in the presented output image from the target perspective. These shadow areas, however, can be covered by other projectors that contribute from different directions. This will be explained in the next section.

7.6.5 Using Multiple Projectors

As already mentioned, the use of multiple projectors can enhance the final quality of the output image. They can complement each other to achieve *higher light intensities* and to *cancel out individual shadow regions.* Their output images can fully or partially overlap, or they can be completely independent of each other. In addition, they allow the covering of large screen surfaces with high-resolution image tiles. This leads to an overall resolution which cannot be achieved with a single projector. Such configurations are known as *tiled screen displays.* An example is shown in Figure 7.26, where two partially overlapping projectors generate a high-resolution 16:9 format.

Dual output graphics cards, for instance, can be used to run two projectors in synchronization (PCI-Express busses support two graphics cards and the synchronization of four projectors). During the geometry calibration, two displacement maps are generated sequentially, one for each projector. Consequently, pixels can be mapped from the camera view into the perspective of each projector in such a way that they are projected exactly to the same spot on the screen surface. Thus, for N projectors the individual light intensities add up to: $R = EM + I_1F_1M + I_2F_2M + \ldots + I_NF_NM$.

A *balanced load* among all projectors can be achieved by assuming that

$$I_i = I_1 = I_2 = \ldots = I_N.$$

This implies that

$$R = EM + I_i(F_1M + F_2M + \ldots + F_NM),$$

and one can solve for

$$I_i = (R - EM)/(F_1M + F_2M + \ldots + F_NM).$$

This is equivalent to the assumption that the total intensity arriving on the screen surface is virtually produced by a single high-capacity projector. Physically, however, it is evenly distributed among multiple low-capacity units. Although each projector sends the same output intensity, the potentially varying form factors cause different fractions to arrive at the screen surface. These fractions are additively mixed on the surface and lead to the final result

$$\begin{aligned} R &= EM + I_iF_1M + \ldots + I_iF_NM \\ &= EM + (R - EM)(F_1M + \ldots + F_NM)/(F_1M + \ldots + F_NM). \end{aligned}$$

As for a single projector, I_i is computed in real time by a pixel shader that receives the parameter textures EM and $F_1M \ldots F_NM$. The form-factor components F_iM can be determined in two ways, either by sequentially sending out a white image (I=1) from each projector and capturing each component one by one, or by capturing a single image that is proportional to $F_1M + \ldots + F_NM$ by sending each projector's maximum contribution simultaneously. Although the latter method is conceptually more compact, the first method prevents the camera's CCD/CMOS sensor from being *overmodulated*. Shadow regions caused by individual projectors are captured in the form factor components. Consequently, they are cancelled out automatically as a side effect. This, however, implies that the projectors are placed in a way that each surface portion can be reached by at least one projector. Smooth transitions among different contributions can be achieved with *cross-fading techniques* that are common for multi-projector setups.

7.7 Summary and Discussion

Video projectors offer many possibilities for augmented reality applications. While some of them are fairly well investigated, others are still unexplored. The unique ability to control the illumination on a per-pixel basis and to synchronize it to a rendering application allows us to solve existing problems and to create new applications. These "intelligent" light sources are sometimes referred to as *smart projectors*.

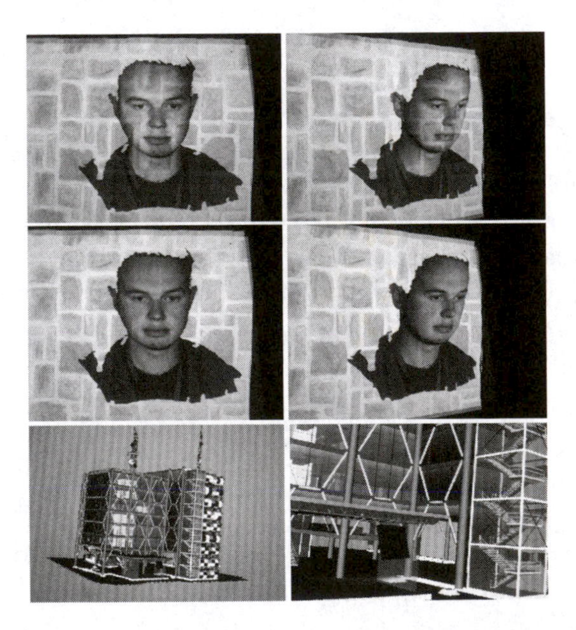

Figure 7.28. Geometrically corrected stereo-pairs of 3D scene projected onto a natural stone wall: (top) without radiometric compensation; (middle) with radiometric compensation; (bottom) radiometrically corrected stereoscopic walk-through projected onto a papered wall. (*Prototype realized by the Bauhaus-University Weimar; see Plate IX.*)

Radiometric, photometric, and geometric image compensation enables an ad-hoc and view-dependent (i.e., head-tracked) stereoscopic visualization on everyday surfaces, as shown in Figure 7.28.

Smart Projectors are video projectors enhanced with sensors to gain information about the environment. Although cameras are the most common sensor types used for smart projector implementations, other sensors, like tilt-sensors, have been used. Completely calibrated and mounted as a single camera-projector unit or realized with separated components, some smart projectors allow a dynamic elimination of shadows cast by the user [179], an automatic keystone correction on planar screens [178], or a manually aligned shape-adaptive projection on second-order quadric display surfaces [158], like cylinders, domes, ellipsoids, or paraboloids. For projection planes, multiple projector units can be automatically registered based on their *homographic relationships* that are determined with the aid of cameras [208]. In this case, camera feedback is also used for intensity blending and color matching [100, 101] of multiple projector contributions. Calibrated stereo cameras together with projectors allow us to directly scan the three-dimensional geometry of an arbitrary display surface. This enables an undistorted projection for a known head-tracked observer position [151]. Synchronized projector-camera pairs can encode structured light into color images that are imperceptible to the human observer [32]. This allows a simultaneous *acquisition of depth information* and presentation of graphical content. Radiometric compensation allows projection onto colored and textured surfaces.

Today's projectors still face technological limitations, such as high black level, relatively low resolution, dynamic range, and depth focus. These issues may be solved with future projector generations.

Projectors in the near future, for instance, will be compact, portable, and with the built-in awareness which will enable them to automatically create satisfactory displays on many of the surfaces in the everyday environment. Alongside the advantages, there are limitations, but we anticipate projectors being complementary to other modes of display for everyday personal use in the future and to have new application areas for which they are especially suited. LEDs are replacing lamps, and reflective, instead of *transmissive displays* (DLPs, LCOS), are becoming popular. Both lead to improved efficiency requiring less power and less cooling. DLP and LCOS projectors can display images at extremely high frame rates, currently 180 Hz and 540 Hz respectively, but lack video bandwidth. Several efforts are already in the making and are very promising.

A future *mobile projector* may double as a "flat panel" when there is no appropriate surface to illuminate, or ambient light is problematic. Super bright, sharp infinite focus laser projectors are also becoming widespread which may allow shape-adaptive projection without focus and ambient lighting problems. Other *multi-focal projection techniques* may also solve the fixed focus problems for conventional projection technologies. Finally *novel lamp designs*, especially those based on LEDs or lasers are creating smaller, lighter, efficient, and long-life solutions.

Two important problems in augmented reality, object identification and determining the pose of the displayed image with respect to the physical object, can be solved, for instance, by using photosensing RFID tags. This is being explored as a *radio frequency identification and geometry* (RFIG) discovery method [159].

8

Examples of Spatial AR Displays

This chapter provides a link between the previous, more technical, chapters and an application-oriented approach and describes several existing spatial AR display configurations. We first outline examples that utilize the projector-based augmentation concept in both a small desktop (e.g., the *Shader Lamps* approach) and a large immersive (e.g., the *Being There* project) configuration. In addition, an interactive extension, called *iLamps*, that uses hand-held projectors is described. Furthermore, several spatial optical see-through variations that support single or multiple users such as the *Extended Virtual Table* and the *Virtual Showcase* are explained. It is shown how these techniques can be combined with projector-based illumination techniques to present real and virtual environments consistently. A scientific workstation, the *HoloStation*, is described. It allows the combination of optical hologram records of fossils with interactive computer simulations. Finally, two implementations (*Augmented Paintings* and *Smart Projectors*) are presented; they use real-time radiometric correction and geometric warping to augment artistic paintings with multimedia presentations, as well as to make projector-based home entertainment and ad-hoc stereoscopic visualizations in everyday environments possible without artificial canvases.

Potential application areas for the display configurations described in this chapter are industrial design and visualization (e.g., Shader Lamps, iLamps, Extended Virtual Table, Smart Projector), scientific simulations (e.g., HoloStation), inertial design and architecture (e.g., Being There), digital storytelling and next-generation edutainment tools for museums (e.g., Virtual Showcase and Augmented Paintings), and home entertainment (e.g., Smart Projector). However, the interested reader can easily imagine further application domains, such as those in an artistic context.

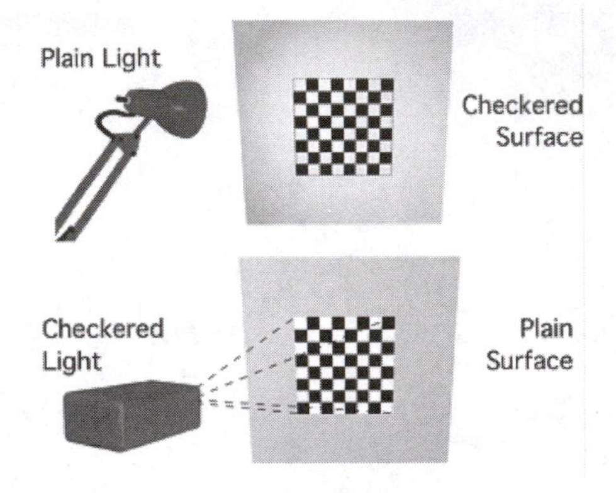

Figure 8.1. Concept of Shader Lamps. Physical textures (above) and Shader Lamp textures (below) (*Images reprinted from [155] © Springer-Verlag.*)

Another goal of this chapter is to show that spatial augmented reality display configurations can be used successfully and efficiently outside research laboratories. The Virtual Showcase, for instance, has been presented to more than 120,000 visitors of more than 11 exhibitions in museums, trade shows, and conferences. Unattended running times of four months and more are an indicator of the fact that it is possible to make the technology (soft and hardware) robust enough to be used by museums and in other public places.

8.1 Shader Lamps

Consider a special case of spatial augmented reality. The idea is to replace a physical object with its inherent color, texture, and material properties with a neutral object and projected imagery, reproducing the original appearance directly on the object. Furthermore, the projected imagery can be used to reproduce alternative appearances, including alternate shading, lighting, and even animation. The approach is to effectively "lift" the visual properties of the object into the projector and then reproject them onto a neutral surface. We use the phrase *Shader Lamps* to describe this mode of operation for projectors [155]. Consider the effect shown in Figure 7.1. The underlying physical object is a white diffuse vase. (The other objects such as the book and flowers are also real objects.) Can we make this white vase appear to look like it is made of marble, plastic or metal? Can we change the color or texture? The pictures show that the vase

(a) (b)

Figure 8.2. (a) Underlying physical model of the Taj Mahal; (b) the Taj Mahal enhanced with Shader Lamps. (*Images reprinted from [155] © Springer-Verlag.*)

can be effectively 'painted' by projecting an image with view-independent diffuse shading, textures, and intensity correction. The view-dependent effects such as specular highlights are generated for a given user location by modifying reflectance properties of the graphics model. The figure shows the appearance of a red plastic and a green metallic material on the clay vase.

Although, there have been other attempts at augmenting appearances of objects by projecting color or texture, those effects are very limited and have been achieved for only specific applications. The real challenge in realizing this as a new medium for computer graphics lies in addressing the problems related to complete illumination of non-trivial physical objects. The approach presented here offers a compelling method of visualization for a variety of applications including dynamic mechanical and architectural models, animated or "living" dioramas, artistic effects, entertainment, and even general visualization for problems that have meaningful physical shape representations. We present and demonstrate methods for using multiple Shader Lamps to animate physical objects of varying complexity, from a flower vase (Figure 7.1), to some wooden blocks, to a model of the Taj Mahal (Figure 8.2)

Object textures. In the simplest form, techniques such as Shader Lamps can be used to dynamically change the color of day-to-day objects, to add useful information in traditional tasks, or to insert instructional text and images. For example, engineers can mark the areas of interest, like drilling locations, without affecting the physical surface. As seen in the Taj Mahal example (Figure 8.2), we can render virtual shadows on scaled models. City planners can move around such blocks and visualize global effects in 3D on a tabletop rather than on their computer screen. Since 2001, Hiroshi Ishii et al. have started to look beyond flat tabletops to augment (morphable) terrains [140].

For stage shows, we can change not just the backdrops, but also create seasons or aging of the objects in the scene. Instead of beaming laser vector images haphazardly onto potentially non-planar surfaces and hoping they do not get too distorted, we can create shapes with laser displays on large buildings by calibrating the laser device with respect to a 3D model of the building. We can also induce apparent motion, by projecting animated texture onto stationary objects. In [157], we describe a collection of techniques to create believable apparent motion. Interesting non-photorealistic effects can also be generated using recently available real-time rendering algorithms [153].

Interactive surface probing. An interactive 3D touch-probe scanning system with closed-loop verification of surface reconstruction (tessellation) can be realized by continuously projecting enhanced images of the partial reconstruction on the object being scanned. This process will indicate useful information to the person while scanning, such as the required density of points, the regions that lack samples, and the current deviation of the geometric model from the underlying physical object.

8.1.1 Tracking, Detection and Control

So far we have presented techniques for a head-tracked moving user, but assumed that the projector and display surfaces are static (Figure 8.3 (a)). However, the geometric framework is valid even when the projector or display surface is movable (Figure 8.3(b)). Let us consider the three possibilities when one of the three components; the user, projector or display surface, is dynamic.

User tracking. User location is crucial for view-dependent effects. As we saw in the last chapter, with simple head tracking, a clay vase can appear to be made of metal or plastic. We can also render other view-dependent effects, such as reflections. The concept can be extended to some larger setups. Sculptors often make clay models of large statues before they create the molds. It may be useful for them to visualize how the geometric forms they have created will look with different materials or under different conditions in the context of other objects. By projecting guiding points or

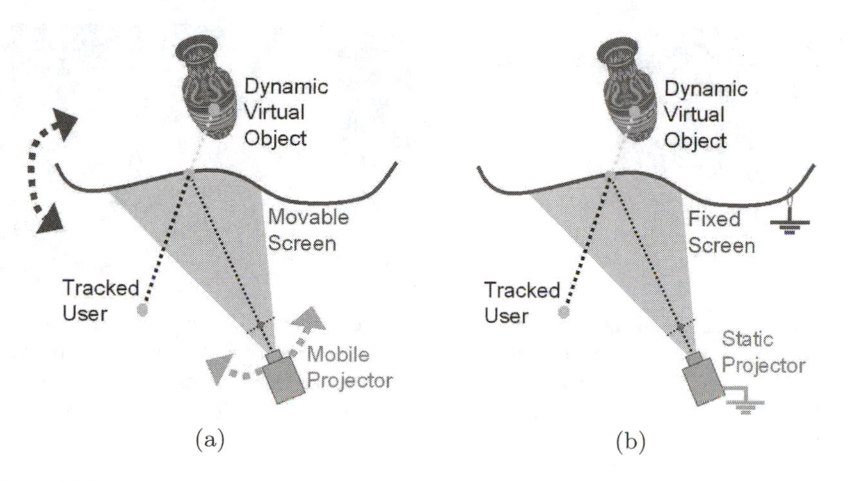

Figure 8.3. (a) In the geometric framework for applications discussed so far, the projector and display surface are assumed static; (b) however, the same geometric relationship can be expressed for dynamic projector and movable display surfaces.

lines (e.g. wire-frame), from the computer models, the sculptors can verify the geometric correctness of the clay models.

Image-based illumination can be very effectively used in movie studios where static miniature sets are painstakingly built and then updated with fine details. By projecting detailed textures on coarse geometry, the work can be completed with sophisticated photo-editing programs. Modification of such sets becomes much easier. The same idea can be further used to simplify the match-moving process as well. Match-moving involves registering synthetic 3D elements to a live shot. The difficulty arises in accurately determining the eleven camera parameters, intrinsics, and pose, at every frame of the shot. During post-processing, those exact camera parameters are needed for rendering the 3D elements. Using spatially augmented reality, we can instead project the synthetic 3D elements directly onto the miniature sets. If the synthetic element is a moving character, we can project simply the silhouette. In either case, the 3D elements will appear perspectively correct with appropriate visibility relationships with respect to the camera. Recall the important advantage of SAR over head-mounted augmented reality with respect to dynamic misregistration—reduced or no sensitivity to errors in rotation. The same idea can be exploited here. In fact, the projection of perspectively correct images on the miniature set requires only the 3D location of the camera optical center (3 parameters). It depends neither on the five internal parameters nor on the three rota-

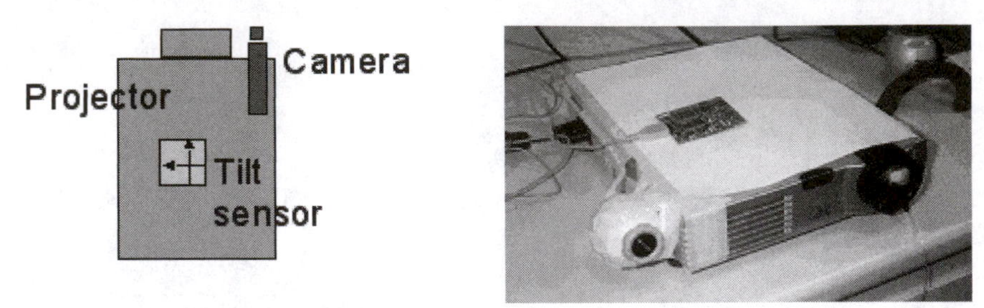

Figure 8.4. Extending the geometric framework to tracked projectors: the projector, augmented with a camera and tilt sensor, can automatically detect relative orientation with respect to a vertical planar screen; the three angles can be used for keystone-correction without the aid of any external devices. (*Images reprinted from [154] © IEEE.*)

tional parameters of the camera. The reduction in search space can greatly improve the match-moving registration quality.

8.1.2 Projector Tracking

Projectors are getting lighter and smaller; they are now *micro-mobile* and *ultra-portable*. What are some applications of movable projectors?

Changing pose. Using steerable, but otherwise fixed projectors, [208] and [139] have created interesting applications. In [208], Ruigang Yang et al. at the University of North Carolina (UNC) have described a planar seamless display using a set of projectors whose beams can be steered using a rotating mirror. For image generation, they use the same geometric framework and exploit the technique described in Section 5.1.5. In [139], Claudio Pinhanez describes the "Everywhere Display" projector to augment planar surfaces in offices with prewarped two-dimensional sprites. Raskar and Beardsley have created a so-called *self-correcting* projector [154]. The three rotational angles are computed automatically in near-real time as follows. The elevation and roll, which are angular displacements out of the horizontal plane are detected using gravity-based tilt sensors (Figure 8.4). The azimuth angle in the horizontal plane is computed using a rigidly attached camera which analyzes the projected image. No external fiducials, which can be fixed in the world coordinates to create a Euclidean frame of reference, are needed. The section below on *iLamps* describes more applications.

Changing internal parameters. The pin-hole projection model has been a powerful abstraction for representing geometric relationships. However, in addition to the internal and external parameters, the projection model can be improved by considering the optics, e.g., radial distortion, aperture, and focus issues, of a realistic projector. Majumder and Welch [102] at UNC have used optical properties, such as the focus, dynamic range, and color gamut of a projector to create novel effects, in what they call "Computer Graphics Optique." The idea is to exploit overlapping projectors to superimpose and add images rendered by separate graphics pipelines. For example, the effective color gamut of a displayed image can be improved beyond the range of a single projector.

8.1.3 Illuminating Dynamically Moving Objects

We can extend the framework in a different direction by allowing the display surface to move. There are two choices for the type of image generation. The virtual object may remain fixed in the world coordinates or appear to move with the display surface. If the display portal is used as a window into the virtual world, the desired view of the user should be maintained so that changes in display surface position remain transparent to the user. On the other hand, for example, in spatially augmented reality, the display surface is expected to retain its current state of shading and remain attached to associated virtual objects. In both cases, the geometric framework can be used with a minor modification. In the first case, the virtual object is not transformed and stays attached to the world coordinate system. In the second case, the virtual object is transformed in the same manner as the display surface.

A movable display portal that remains transparent to the user can be used, for example, for a magic lens system [7]. The user can move around the display surface and the virtual objects behind it become "visible." A movable and tracked white paper, when illuminated, can provide an x-ray vision of parts inside an engine or cables behind a wall.

We can also illuminate real objects so that the surface textures appear glued to the objects even as they move. To display appropriate specular highlights and other view-dependent effects, it may be easier to move the object in a controlled fashion rather than track one or more users. For example, in showroom windows or on exhibition floors, one can show a rotating model of the product in changing colors or with different features enhanced. When users can be assumed to be watching from near a sweet-spot, rotating along a fixed axis allows a single degree of freedom but enables multiple people to simultaneously look at the enhanced object without head-tracking.

Figure 8.5. Extending the geometric framework to tracked display surfaces: two projectors illuminate tracked white objects. (*Images reprinted from [6] © ACM.*)

A good example of a display with six-degree of freedom movement is a compelling 3D paint system built by Deepak Badhyopadhyay et al. [6] at UNC (Figure 8.5). A two-handed painting interface allows interactive shading of neutral colored objects. Two projectors illuminate the tracked object held in one hand while a paintbrush is held in the other hand. The painted-on appearance is preserved throughout six-degrees-of-freedom movement of the object and paintbrush. There is no intensity feathering, so regions in projector overlap may appear brighter. The intensity blending algorithm presented in Section 5.5.4 works in the presence of depth discontinuities, but it is too slow to work in real time. Robust intensity blending of two or more projectors on a movable display surface remains an interesting research problem.

Intensity blending without explicit feathering can be used for some new applications. Consider an interactive clay modelling system as a 3D version of "connect-the-dots" to provide feedback to a modeler. For example, two synchronized projectors could successively beam images of the different parts of the intended 3D model in red and green. A correct positioning of clay will be verified by a yellow illumination. On the other hand, incorrect position of the surface will result in two separate red and green patterns. After the shape is formed, the same Shader Lamps can be used to guide the painting of the model or the application of a real material with matching reflectance properties.

8.2 Being There

A compelling example of spatial augmentation is the walk-through of virtual human-sized environments like those built by Kok-Lim Low et. al. in

Figure 8.6. Spatially augmenting large environments: (Top left) virtual model; (Top right) physical display environment constructed using styrofoam blocks; (Bottom row) augmented display—note the view dependent nature of the display, the perspectively correct view through the hole in the wall and the windows. (*Images courtesy of Kok-Lim Low [93]; see Plate X.*)

the *Being There* project at UNC [93]. Instead of building an exact detailed physical replica for projection, the display is made of simplified versions. For example, primary structures of building interiors and mid-sized architectural objects (walls, columns, cupboards, tables, etc.), can usually be approximated with simple components (boxes, cylinders, etc.). As seen in Figure 8.6, the display is made of construction styrofoam blocks. The main architectural features that match the simplified physical model retain 3D auto-stereo, but the other details must be presented by projecting view-dependent images. Nevertheless, the experiment to simulate a building interior is convincing and provides a stronger sense of immersion when compared to SID, as the user is allowed to really walk around in the virtual environment. However, strategic placement of projectors to allow complete illumination and avoiding user shadows is critical.

8.3 iLamps: Mobile Projectors

Most AR systems have continued to use static projectors in a semi-permanent setup, one in which there may be a significant calibration process prior to using the system. Often, the systems are specifically designed for a particular configuration. But the increasing compactness and cheapness of projectors is enabling much more flexibility in their use than is found currently. For example, portability, and cheapness open the way for clusters of projectors which are put into different environments for temporary deployment, rather than a more permanent setup. As for hand-held use, projectors are a natural fit with cell phones and PDAs. Cell phones provide access to the large amounts of wireless data which surround us, but their size dictates a small display area. An attached projector can maintain compactness while still providing a reasonably-sized display. A hand-held cell-phone–projector becomes a portable and easily-deployed information portal.

These new uses will be characterized by opportunistic use of portable projectors in arbitrary environments. The research challenge is how to create *Plug-and-disPlay* projectors which work flexibly in a variety of situations. This requires generic application-independent components, in place of monolithic and specific solutions.

Consider a basic unit which is a projector with an attached camera and tilt sensor. Single units can recover 3D information about the surrounding environment, including the world vertical, allowing projection appropriate to the display surface. Multiple, possibly heterogeneous, units are deployed in clusters, in which case the systems not only sense their external environment but also the cluster configuration, allowing self-configuring seamless large-area displays without the need for additional sensors in the environment. We use the term *iLamps* to indicate intelligent, locale-aware, mobile projectors.

8.3.1 Object Adaptive Projection

We describe object augmentation using a hand-held projector, including a technique for doing mouse-style interaction with the projected data. Common to some previous approaches, we do object recognition by means of fiducials attached to the object of interest. Our fiducials are piecodes, colored segmented circles like the ones in Figure 8.7, which allow thousands of distinct color-codings. As well as providing identity, these fiducials are used to compute camera pose (location and orientation) and, hence, pro-

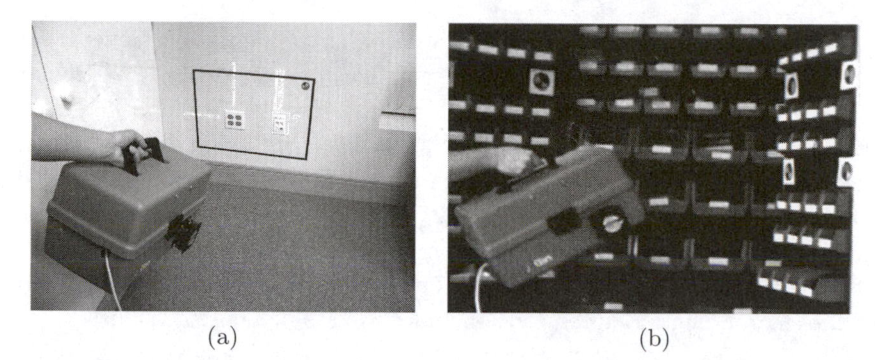

(a) (b)

Figure 8.7. Context-aware displays: (a) augmentation of an identified surface; (b) guidance to a user-requested object in storage bins. (*Images reprinted from [158] © ACM.*)

jector pose since the system is fully calibrated. We use four coplanar points in known position in a homography-based computation for the pose of the calibrated camera. The points are obtained from the segments of a single piecode, or from multiple piecodes, or from one piecode plus a rectangular frame. For good results, augmentation should lie within or close to the utilized points. With projector pose known relative to a known object, content can be overlaid on the object as required.

Advantages of doing object augmentation with a projector rather than by annotated images on a PDA include (1) the physical size of a PDA puts a hard limit on presented information; (2) a PDA does augmentation in the coordinate frame of the camera, not the user's frame, and requires the user to context-switch between the display and physical environment; (3) a PDA must be on the user's person while a projector can be remote; (4) projection allows a shared experience between users. Eye-worn displays are another important augmentation technique, but they can cause fatigue, and there are stringent computational requirements because of the tight coupling between user motion and the presented image (e.g., a user head rotation must be matched precisely by a complementary rotation in the displayed image). Projection has its own disadvantages—it is poor on dark or shiny surfaces and can be adversely affected by ambient light; it does not allow private display. But a key point is that projector-based augmentation naturally presents to the user's own viewpoint, while decoupling the user's coordinate frame from the processing. This helps in ergonomics and is easier computationally.

A hand-held projector can use various aspects of its context when projecting content onto a recognized object. We use proximity to the object to determine level-of-detail for the content. Other examples of context for content control would be gestural motion, history of use in a particular spot, or the presence of other devices for cooperative projection. The main uses of object augmentation are (1) information displays on objects, either passive display, or training applications in which instructions are displayed as part of a sequence (Figure 8.7(a)); (2) physical indexing in which a user is guided through an environment or storage bins to a requested object (Figure 8.7(b)); (3) indicating electronic data items which have been attached to the environment. Related work includes the Magic Lens [7], Digital Desk [202], computer augmented interaction with real-world environments [160], and Hyper mask [120].

Mouse-style interactions with augmentation data. The most common use of projector-based augmentation in previous work has been straightforward information display to the user. A hand-held projector has the additional requirement over a more static setup of fast computation of projector pose, so that the augmentation can be kept stable in the scene under user motion. But a hand-held projector also provides a means for doing mouse-style interactions—using a moving cursor to interact with the projected augmentation, or with the scene itself.

Consider first the normal projected augmentation data; as the projector moves, the content is updated on the projector's image plane, so that the projected content remains stable on the physical object (Figure 8.8). Now, assume we display a cursor at some fixed point on the projector image plane, say at the center pixel. This cursor will move in the physical scene

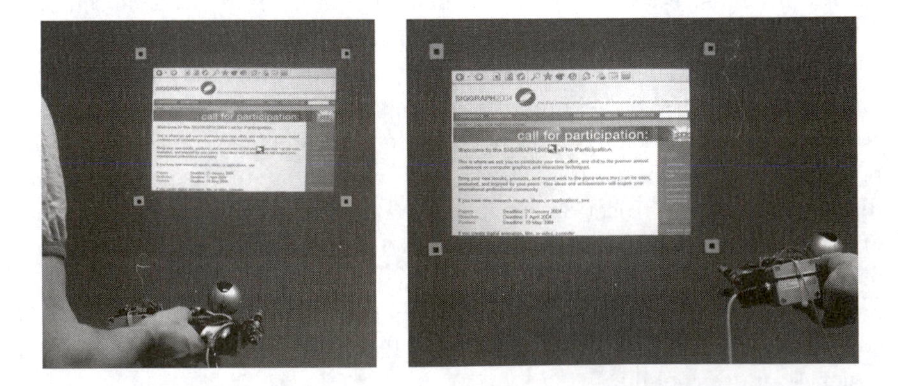

Figure 8.8. Absolute stabilization projects to a fixed location, regardless of projector position. (*Images reprinted from [159] © ACM.*)

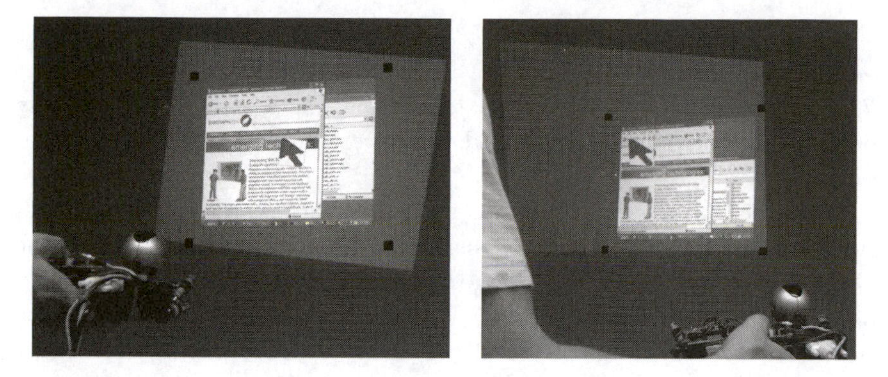

Figure 8.9. During interactive projection, a moving cursor can be tracked across a projection that is static on the display surface to do mouse-like interaction with projected data. In these images, the background pixels are intentionally rendered in red to show the extent of projection. The pointer remains at the center of the projector field while displayed desktop remains stable within the surface features. (*Images reprinted from [159] © ACM.*)

in accordance with the projector motion. By simultaneously projecting the motion-stabilized content and the cursor, we can emulate mouse-style interactions in the scene (Figure 8.9). For example, we can project a menu to a fixed location on the object, track the cursor to a menu item (by a natural pointing motion with the hand-held projector), and then press a button to select the menu item. Alternatively, the cursor can be used to interact with the physical scene itself, for example doing cut-and-paste operations with the projector indicating the outline of the selected area and the camera capturing the image data for that area. In fact, all the usual screen-based mouse operations have analogs in the projected domain.

8.4 The Extended Virtual Table

Virtual reality provides the user with a sense of spatial presence (visual, auditory, or tactile) inside computer-generated synthetic environments.

Opaque, head-mounted displays (HMDs) and surround-screen (spatially immersive) displays such as CAVEs, and domed displays are VR devices that surround the viewer with graphics by filling the user's field of view. To achieve this kind of immersion, however, these devices encapsulate the user from the real world, thus making it difficult or even impossible in many cases to combine them with habitual work environments.

Other, less immersive display technology is more promising to support seamless integration of VR into everyday workplaces. *Table-like display devices* and *wall-like projection systems* allow the user to simultaneously perceive the surrounding real world while working with a virtual environment.

This section describes an optical extension for workbench-like single-sided or multiple-sided (that is, L-shaped) projection systems called the *Extended Virtual Table* [10]. A large beam splitter mirror is used to extend both viewing and interaction space beyond the projection boundaries of such devices. The beam splitter allows an extension of exclusively virtual environments and enables these VR display devices to support AR tasks. Consequently, the presented prototype features a combination of VR and AR. Because table-like display devices can easily be integrated into habitual work environments, the extension allows the linkage of a virtual with a real workplace, such as a table-like projection system, with a neighboring real workbench.

8.4.1 Physical Arrangement

The Extended Virtual Table (xVT) consists of a virtual and a real workbench (Figure 8.10) and supports active stereoscopic projection, head-tracking of a single user, and six-degrees-of-freedom (6-DOF) tracking of input devices.

A large front surface beam splitter mirror separates the virtual workbench from the real workspace. With the bottom leaning onto the projection plane, the mirror is held by two strings that are attached to the ceiling.

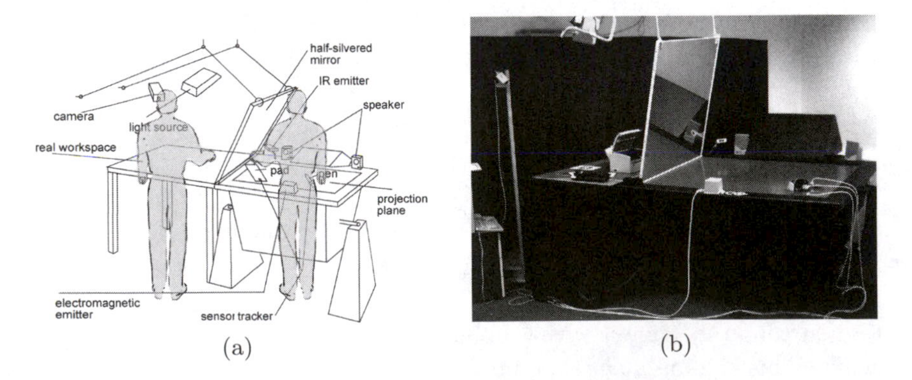

Figure 8.10. (a) Conceptual sketch; (b) photograph of the xVT prototype. (*Images reprinted from [10] © MIT Press.*)

The length of the strings can be adjusted to change the angle between the mirror and the projection plane or to allow an adaptation to the workbench's slope. A light source is adjusted in such a way that it illuminates the real workbench, but does not shine at the projection plane.

In addition, the real workbench and the walls behind it are covered with a black awning to absorb light that otherwise would be diffused by the wallpaper beneath it and would cause visual conflicts if the mirror were used in a see-through mode. Note that a projector-based illumination (see Chapter 7) would solve this problem.

Finally, a camera is used to continuously capture a video stream of the real workspace, supporting an optical tracking of paper markers that are placed on top of the real workbench.

8.4.2 General Functioning

Users can either work with real objects above the real workbench or with virtual objects above the virtual workbench. Elements of the virtual environment, which are displayed on the projection plane, are spatially defined within a single world coordinate system that exceeds the boundaries of the projection plane, covering also the real workspace.

The mirror plane splits this virtual environment into two parts that cannot be visible simultaneously to the user. This is due to the fact that only one part can be displayed on the projection plane (which serves as a secondary screen in the AR mode and as a primary screen in the VR mode). The user's viewing direction is continuously determined to support an intuitive visual extension of the visible virtual environment. If, on the one hand, the user is looking at the projection plane, the part of the environment that is located over the virtual workbench is displayed. If, on the other hand, the user is looking at the mirror, the part of the environment located over the real workbench is transformed, displayed, and reflected in such a way that it appears as a continuation of the other part in the mirror.

Using the information from the head-tracker, the user's viewing direction is approximated by computing the single line of sight that originates at his or her point of view and points towards his or her viewing direction. The plane the user is looking at (that is, the projection plane or mirror plane) is the one that is first intersected by this line of sight. If the user is looking at neither plane, no intersection can be determined, and nothing needs to be rendered at all.

In case the user is looking at the mirror, the part of the virtual environment behind the mirror has to be transformed in such a way that, if displayed and reflected, it appears stereoscopically and perspectively correct at the right place behind the mirror.

As described in Chapter 6, a slight modification to the common fixed function rendering pipeline allows us to use a simple affine transformation matrix that reflects the user's viewpoint (that is, both eye positions that are required to render the stereo images) and to inversely reflect the virtual environment over the mirror plane.

If the graphical content from the viewer's averting side of the mirror is inversely reflected to the opposite side and is then rendered from the viewpoint that is reflected, the projected virtual environment will not appear as a reflection in the mirror. The user will see the same scene that he or she would perceive without the mirror if the projection plane were large enough to visualize the entire environment. This effect is due to the neutralization of the computed inverse reflection by the physical reflection of the mirror, as discussed in Chapter 6. and can be done without increasing the computational rendering cost.

The mirror-plane parameters can be determined within the world coordinate system in multiple ways: an electromagnetic tracking device, for example, can be used to support a three-point calibration of the mirror plane. An optical tracking system can be used to recognize markers that are temporarily or permanently attached to the mirror. Alternatively, because the resting points of the mirror on the projection plane are known and do not change, its angle can be simply measured mechanically.

Note, that all three methods can introduce calibration errors caused by either tracking distortion (electromagnetic or optical) or human inaccuracy.

To avoid *visual conflicts* between the projection and its corresponding reflection especially for areas of the virtual environment whose projections are close to the mirror, a graphical clipping plane can be rendered that exactly matches the mirror plane. Visual conflicts arise if virtual objects spatially intersect the side of the user's viewing frustum that is adjacent to the mirror, because, in this case, the object's projection optically merges into its reflection in the mirror. The clipping plane culls away the part of the virtual environment that the user is not looking at. Note that the direction of the clipping plane can be reversed, depending on the viewer's viewing direction while maintaining its position. The result is a small gap between the mirror and the outer edges of the viewing frustum in which no graphics are visualized. This gap helps to differentiate between projection and reflection and, consequently, avoids visual conflicts. Yet, it does not allow virtual objects that are located over the real workbench to reach

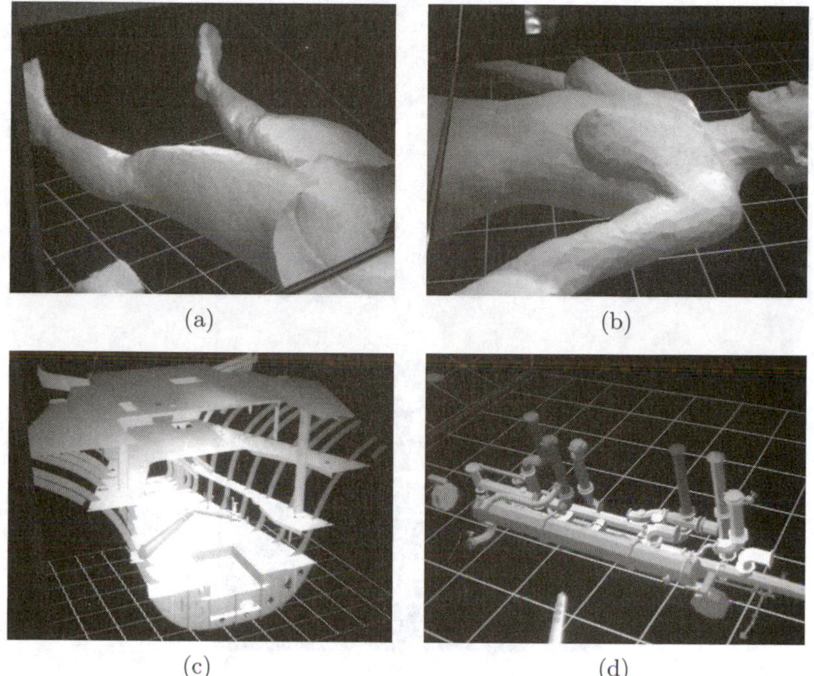

Figure 8.11. (a) and (b) A large coherent virtual content; (c) and (d) two independent virtual contents viewed in the mirror or on the projection plane. The real workspace behind the mirror (left side) is not illuminated. (*Images reprinted from [10] © MIT Press.*)

through the mirror. The clipping plane can be activated or deactivated for situations in which no, or minor, visual conflicts between reflection and projection occur to support a seamless transition between both spaces.

If the real workspace behind the beam splitter mirror is not illuminated, the mirror acts like a full mirror and supports a non-simultaneous visual extension of an exclusively virtual environment (that is, both parts of the environment cannot be seen at the same time). Figure 8.11 shows a large coherent virtual scene whose parts can be separately observed by either looking at the mirror or at the projection plane.

Figure 8.12 shows a simple example in which the beam splitter mirror is used as an optical combiner. If the real workspace is illuminated, both the real and the virtual environment are visible to the user, and real and virtual objects can be combined in the AR manner. Note that the ratio of intensity of the transmitted light and the reflected light depends upon the

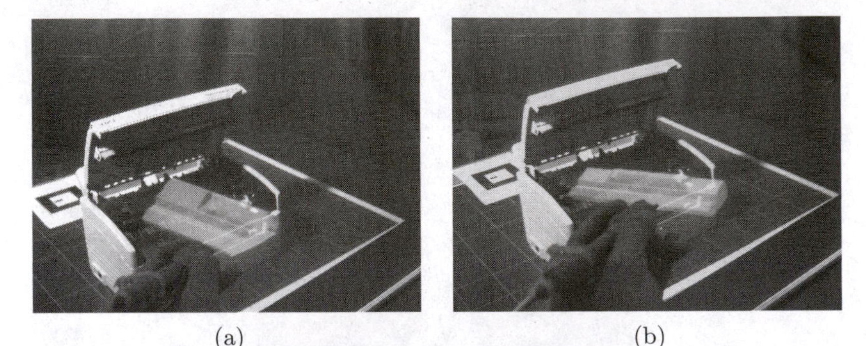

(a) (b)

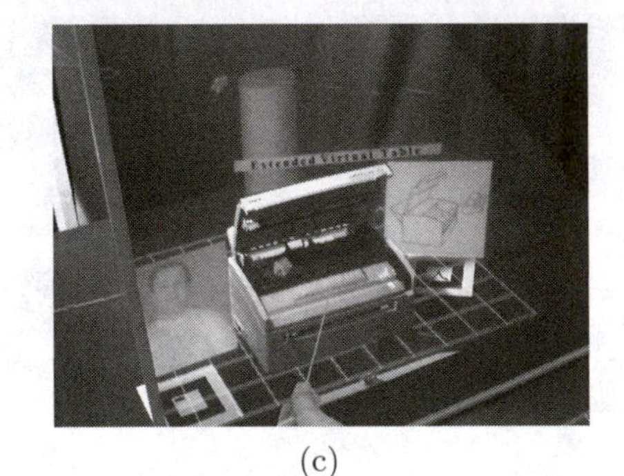

(c)

Figure 8.12. (a) and (b) A virtual object is pushed through the mirror; (c) a clipping plane is attached to the mirror—ray casting and optical tracking within an augmented real environment, a virtual cartridge is mounted into a real printer. (*Images reprinted from [10] © MIT Press.*)

angle between the beam splitter and the projection plane. Acute angles highlight the virtual content, and obtuse angles let the physical objects shine through brighter.

Virtual objects can be exchanged between both sides of the mirror in different ways, in the see-through mode as well as in the opaque mode. For example, they can be picked with a tracked stylus-like interaction device, either directly or from a distance over the virtual workbench, or indirectly over the real workbench. Virtual objects can then be pushed or pulled through the mirror (see Figure 8.12), either directly or indirectly.

As illustrated in Figure 8.12(c), a virtual laser beam that is cast from the stylus through the mirror is used to select and manipulate (that is, to move and place) virtual objects behind the mirror plane. This ray tool allows for interaction with virtual objects on a remote basis and offers an indirect object placement by "beaming" the object back and forth along the ray.

Virtual-virtual and real-virtual *object collision detection* is applied over the real and the virtual workbench to simulate a realistic interaction behavior between objects. This fundamental collision detection capability, for instance, enables us to implement gravity-based automatic object placement methods. Real-virtual object occlusion is implemented by rendering black "phantoms," which are geometric placeholders for the real objects, into the depth buffer. The application of projector-based illumination (Chapter 7), instead of a simple analog light source, also allows for a correct virtual-real object occlusion. This is not shown in Figure 8.12.

8.4.3 A Comparison with Head-Mounted Displays

The Extended Virtual Table has three major drawbacks: mobility, direct interaction with augmented real objects, and single-user application.

Stable and precise *long-range tracking* will be available in the near future, enabling AR applications using HMDs and head-mounted projector displays to be highly mobile. Nevertheless, the intention of the xVT as a spatial AR display is to combine two table-like workplaces in which the users focus on the workspace above the workbenches. For this, neither long-range tracking nor a high degree of mobility is required.

Head-mounted displays also offer direct interaction with augmented real objects that are in arm's reach of the user. In the case of the xVT, the mirror represents a physical barrier for the user's hands and the input devices and, to a large degree, prevents direct interaction with superimposed objects. It is possible to either directly interact with real objects on the real workbench and with virtual objects on the virtual workbench, or indirectly interact with virtual objects above the real workbench through the mirror.

Although HMDs provide an individual image plane for each participant of a multiple user session, users of large projection systems have to share the same image plane. Thus, multiple-user scenarios are difficult to realize with such technology. Because the xVT also uses a single large projection plane, it has the same problem. However, some projection solutions exist that simultaneously support two and more users. The problems of multi-user use is addressed with the following display example.

8.5 The Virtual Showcase

The *Virtual Showcase* [9] has the same form-factor as a real showcase making it compatible with traditional museum displays (Figure 8.13). Real

Figure 8.13. Illustration of Virtual Showcase applications in museums.

scientific and cultural artifacts are placed inside the Virtual Showcase allowing their three-dimensional graphical augmentation. Inside the Virtual Showcase, virtual representations and real artifacts share the same space providing new ways of merging and exploring real and virtual content. The virtual part of the showcase can react in various ways to a visitor enabling intuitive interaction with the displayed content. These interactive showcases represent a step towards *ambient intelligent landscapes*, where the computer acts as an intelligent server in the background and visitors can focus on exploring the exhibited content rather than on operating computers.

In contrast to the Extended Virtual Table, Virtual Showcases offer the possibility to simultaneously support multiple head-tracked viewers and to provide a seamless surround view on the augmented real environment located within the showcase.

8.5.1 Physical Arrangement

A Virtual Showcase consists of two main parts, a convex assembly of beam splitters mirrors and one or multiple graphics secondary displays. Head-tracking and 6-DOF tracking of interaction devices are also used.

Many different versions of Virtual Showcases have been developed (see Figure 8.14 and various figures in Chapters 4 and 6). Some of them use a large projection display and curved, cone-shaped mirror optics, and others use multiple monitors in addition to planar pyramid-shaped mirror optics.

(a) (b) (c) (d) (e)

Figure 8.14. Different Virtual Showcase variations: (a) single-user display; (b)–(d) multi-user display; (e) seamless surround view.

These variations serve as illustrative examples to explain several of the rendering techniques described in Chapter 6.

While the pyramid-shaped prototypes simultaneously support up to four viewers, the cone-shaped prototypes provide a seamless surround view onto the displayed artifact for fewer viewers. Optional components of the Virtual Showcase are additional video projectors that are used for a *pixel-precise illumination* of the real content (e.g., to create consistent occlusion and illumination effects) and video cameras to receive feedback from the inside of the display. Techniques that use these components are described in Chapter 7.

8.5.2 General Functioning

By assigning each viewer to an individual mirror, the pyramid-like Virtual Showcase prototype can support up to four observers simultaneously. Looking through the mirror optics allows the users to see the reflection of the corresponding screen portion at the same time, and within the same spatial space as the real object inside the showcase.

Since convex mirror assemblies tessellate the screen space into individual mirror reflection zones which do not intersect or overlap, a single pixel that is displayed on the secondary screen appears (for one viewpoint) exactly once as a reflection. Consequently, a definite one-to-one mapping between the object space and the image space is provided by convex mirror assemblies.

Observed from a known viewpoint, the different images optically merge into a single consistent mirror image by reflecting the secondary screen, whereby this image space visually equals the image of the untransformed image space geometry.

Building Virtual Showcases from one single mirror sheet, instead of using multiple planar sections reduces the calibration problem to a single registration step and consequently decreases the error sources. In addition, the edges of adjacent mirror sections (which can be annoying in some applications) disappear. It provides a seamless surround view onto the displayed artifact. However, using curved mirror optics requires a curvilinear of the image before it is displayed. How to warp the image for curved beam splitter mirrors is also described in Chapter 6.

8.5.3 A Platform for Augmented Reality Digital Storytelling

Museums are facing more and more competition from theme parks and other entertainment institutions. Some museums have learned that their visitors do not only want to be educated, but also to be entertained. Consequently they are investigating new edutainment approaches. Technologies such as multimedia, virtual reality, and augmented reality, in combination with appropriate interaction and storytelling techniques, may become the new *presentation forms* of museums in the future.

Interactive *digital storytelling* techniques are being used in combination with new media forms, such as virtual reality and augmented reality.

The technological progress that is being made within these areas allows for the shifting of interactive digital storytelling more and more into the third dimension and into the physical world. One of the main advantages of this transition is the possibility to communicate information more effectively with digital means by telling stories that can be experienced directly within a real environment or in combination with physical objects. The user experience is thus transformed from relating different pieces of information to one another to "living through" the narrative. The perceptual quality and the unique aura of a real environment (e.g., a historical site) or object (e.g., an ancient artifact) cannot be simulated by today's technology. Thus it is not possible to substitute them with virtual or electronic copies without them losing their flair of originality. This circumstance can be a crucial reason for using augmented reality as a technological basis for interactive digital storytelling.

To use the Virtual Showcase as a digital storytelling platform [14], five major components can be identified: *content generation*, authoring, presentation, *interaction*, and *content management*.

Conventional content types (such as three-dimensional models, animations, and audio, etc.) can be generated, authored, and presented with

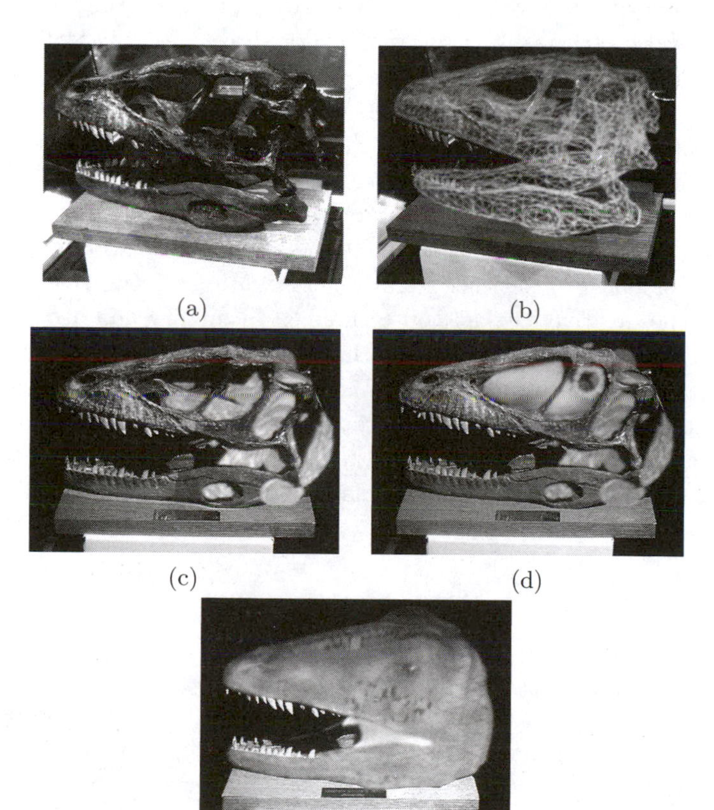

Figure 8.15. Storytelling with the Virtual Showcase: (a) the physical skull of Deinonychus is placed inside the display; (b) a scanned skull geometry is registered to the real counterpart; (c) different muscle groups are augmented; (d) the paranasal air sinuses and the bony eye rings are integrated into the skull; (e) the skin is superimposed on the skull. (*Images reprinted from [13] © IEEE.*)

spatial AR displays, such as the Virtual Showcase; these are the same as the content types that are essential for traditional or VR digital storytelling. More interesting, however, are the unconventional content types that are specific to augmented reality, or even to spatial augmented reality. For example, static and dynamic illumination of the real objects is an unconventional content that can be created, authored, managed, interacted with, and presented within a digital story.

In an early case study, the Virtual Showcase was used to communicate the scientific findings in dinosaur soft-tissue reconstruction [13] of a leading paleontologist to novice museum visitors in an interesting and exciting way. Several layers of soft tissue were overlaid over the physical skull and animated to illustrate their functions (Figure 8.15).

Projector-based illumination was used to create consistent occlusion effects between muscles and bones. Text labels and audio narrations (a verbal explanation by the palaeontologists) were presented in sync to the animations, and two users were supported simultaneously.

At the time of the writing of this book, more than two hundred thousand visitors have experienced this demonstration during several public events. The positive feedback on technological quality and acceptance of such an approach that was gathered throughout informal user studies gives an indication of the potential of augmented reality in areas such as edutainment. Some of these exhibits required the display to run unattended for as long as three months, demonstrating that the technology can be made robust enough to be used in public places, such as museums.

8.6 The HoloStation

A hologram is a photometric emulsion that records interference patterns of coherent light. The recording itself stores the amplitude, wavelength, and phase information of light waves. In contrast to simple photographs, which can record only amplitude and wavelength information, holograms can reconstruct complete optical wavefronts. This results in the captured scenery having a three-dimensional appearance that can be observed from different perspectives. Two main types of optical holograms exist: transmission and reflection holograms. To view a transmission hologram, the light source and observer must be on opposite sides of the holographic plate. The light is transmitted through the plate before it reaches the observer's eyes. Those portions of the emulsion not recorded or illuminated remain transparent. To view a reflection hologram, the light source and observer must be on the same side of the holographic plate. The light reflects from the plate toward the observer's eyes. As with transmission holograms, unrecorded or non-illuminated portions of the emulsion remain transparent. These two hologram types have generated a spectrum of variations. Although some holograms can be reconstructed only with laser light, others can be viewed only under white light. Rainbow holograms, one of the most popular white-light transmission hologram types, diffract each wavelength of the light through a different angle. This lets viewers observe the recorded scene from different horizontal viewing positions but also makes the scene appear in different colors when observed from different points. In contrast to rainbow holograms, white-light reflection holograms can provide full parallax and display the recorded scene in a consistent color for different viewing positions. Monochrome holograms are most common, but high-quality full-color holograms do exist. Today, computer graphics and raster displays offer a megapixel resolution and the interactive rendering of megabytes of data. Holograms, however, provide a terrapixel resolution and are able to present an information content in the range of terrabytes in real time. Both are dimensions that will not be reached by computer

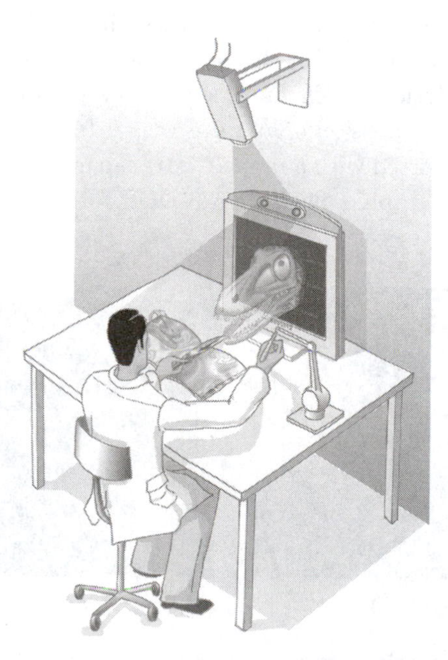

Figure 8.16. Illustration of a holographic workstation—the HoloStation. (*Images reprinted from [18] © IEEE.*)

graphics and conventional displays within the next years, even if Moore's law proves to hold in the future.

Chapter 7 discussed how to utilize projector-based illumination for reconstructing optical holograms in such a way that they can be augmented with computer graphics [18]. The subsequent sections here describe several desktop configurations, called *HoloStations*, which use this concept (Figure 8.16).

8.6.1 Physical Arrangement

The desktop prototypes shown in Figure 8.17 consist entirely of off-the-shelf components, including either an autostereoscopic lenticular-lens sheet display with integrated head-finder for wireless user tracking (Figure 8.17(a)) or a conventional CRT screen with active stereo glasses, wireless infrared tracking, and a touch screen for interaction (Figure 8.17(b)).

Both prototypes use LCD projectors for reconstructing the holographic content. A single PC with a dual-output graphics card renders the graphical content on the screen and illuminates the holographic plate on the video projector. In both cases, the screen additionally holds further front layers—glass protection, holographic emulsion, and optional beam splitter mirror (used for transmission holograms only). Interaction with the graphical content is supported with a mouse, a transparent touch screen mounted in front of the holographic plate, or a 6-DOF force-feedback device.

(a) (b)

Figure 8.17. HoloStation prototypes: (a) autostereoscopic version with head-finder and force-feedback; (b) active stereoscopic version with infrared tracking and touch-screen interaction; the video beamers are not visible from the outside.

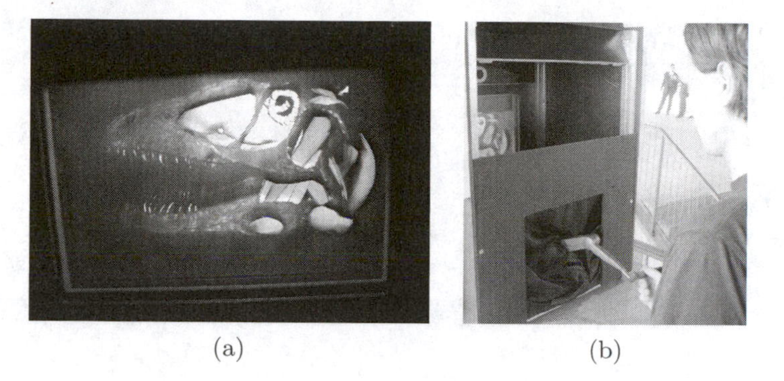

(a) (b)

Figure 8.18. Force-feedback interaction with an augmented hologram.

8.6.2 General Functioning

Stereoscopics or autostereoscopics are integrated into holograms as described in Chapter 7. In addition, haptic interaction devices allow us to feel computer-generated three-dimensional content by simulating *force feedback*. Depending on the position of the 6-DOF stylus-like device, force vectors and magnitudes are calculated dynamically based on pre-defined material parameters of the corresponding holographic or graphical content.

This enables the user to virtually touch and feel the hologram and the integrated virtual models (Figure 8.18). For both content types, a three–dimensional geometrical model is required to determine surface intersections with the stylus which lead to the proper force computations. While the holographic content is static, the graphical content can be deformed by the device in real time. In the examples shown in Figure 8.18, using the force feedback device allows the touching and feeling of bones and muscles with different material parameters. While bones feel stiff, muscles and air sinus feel soft. In addition, the soft tissue can be pushed in under increasing pressure, and it expands back when released.

The ability to control the reconstruction of a hologram's object wave allows the seamless integration into common *desktop-window environments*. If the holographic emulsion that is mounted in front of a screen is not illuminated, it remains transparent. In this case, the entire screen content is visible and an interaction with software applications on the desktop is possible. The holographic content (visible or not) is always located at a fixed spatial position within the screen/desktop reference frame. An application that renders the graphical content does not necessarily need to be displayed in full screen mode (as in the previous examples), but

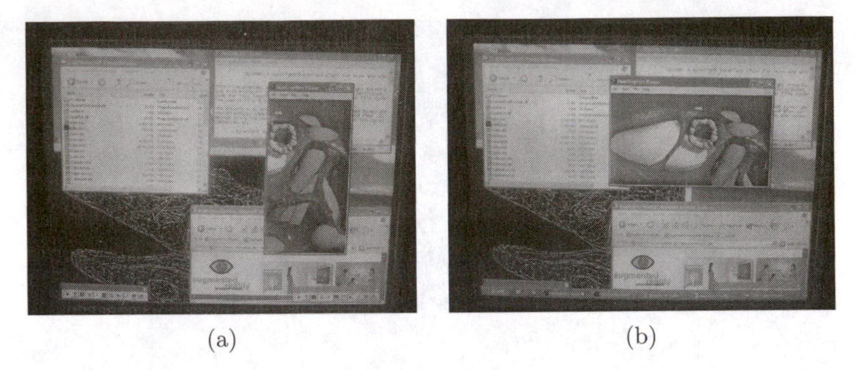

(a) (b)

Figure 8.19. A holographic window in different states.

can run in a *windows mode* covering an arbitrary area on the desktop behind the emulsion. If position and dimensions of the graphics window are known, the projector-based illumination can be synchronized to bind the reference wave to the portion of the emulsion that is located directly on top of the underlying window. Thereby, all the rendering techniques that are described in Chapter 7 (partial reconstruction and intensity variations) are constrained to the window's boundaries. The remaining portion of the desktop is not influenced by the illumination, or by the graphical or holographic content.

In addition, the graphical content can be rendered in such a way that it remains registered with the holographic content even if the graphical window is moved or resized. This simple, but effective, technique allows a seamless integration of holograms into common desktop environments. We can temporarily minimize the "holographic window" or align it over the main focus while working in other applications. Figure 8.19 shows a holographic window in different states on a desktop together with other applications. It displays an optical (monochrome) white-light reflection hologram of a dinosaur skull with integrated graphical three-dimensional soft tissues. A stereoscopic screen was used in this case, because autostereoscopic displays (such as lenticular screens or barrier displays) do not yet allow an undisturbed view of non-interlaced two-dimensional content.

8.6.3 HoloGraphics for Science, Industry, and Education

A combination of interactive computer graphics and high-quality optical holograms represents an alternative that can be realized today with off-the-shelf consumer hardware. Several commercial online services already offer uncomplicated and inexpensive ways to create color holograms from a set of images or video clips. With this technology, users can create holograms with almost any content, even outdoor scenes.

Archaeologists, for example, already use optical holograms to archive and investigate ancient artifacts. Scientists can use *hologram copies* to perform their research without having access to the original artifacts or settling for inaccurate replicas. They can combine these holograms with interactive computer graphics to integrate real-time simulation data or perform experiments that require direct user interaction, such as packing reconstructed soft tissue into a fossilized dinosaur skull hologram. In addition, specialized interaction devices can simulate haptic feedback of holographic and graphical content while scientists are performing these interactive tasks. An entire collection of artifacts will fit into a single album of holographic recordings, while a light-box-like display such as that used for viewing X-rays can be used for visualization and interaction. This approach has the potential for wide application in other industries. In the automotive industry, for example, complex computer models of cars and components often lack realism or interactivity. Instead of attempting to achieve high visual quality and interactive frame rates for the entire model, designers could decompose the model into sets of interactive and static elements. The sys-

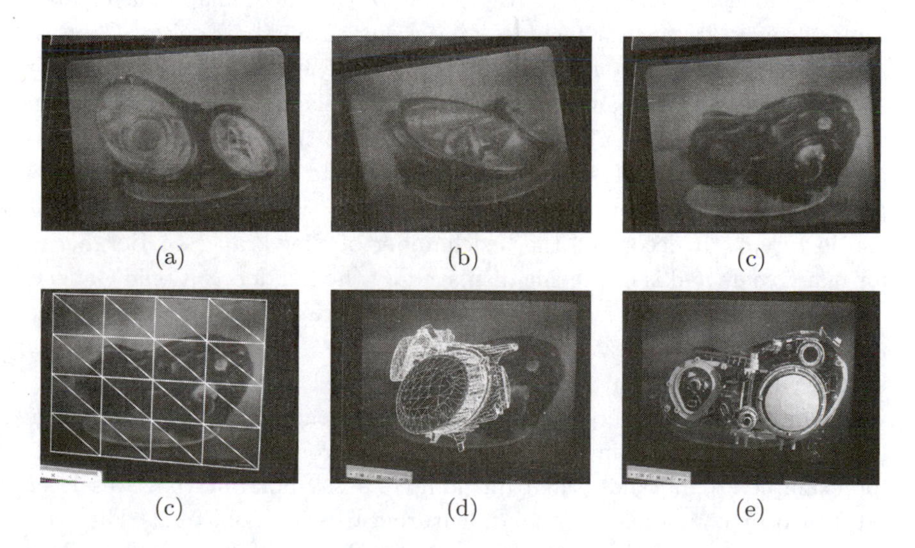

Figure 8.20. (a)–(c) Digital multiplex hologram of real head-light observed from different perspectives; (d) registration of image plane; (e)–(f) augmentation with interactive CAD models.

tem could record physical counterparts of static elements in a hologram with maximum realism and release computational resources to render the interactive elements with a higher quality and increased frame rate. Multiplexing the holographic content also lets users observe and interact with the entire model from multiple perspectives (Figure 8.20).

Augmenting optical holograms in museums with animated multimedia content lets exhibitors communicate information about the artifact with more excitement and effectiveness than text labels offer. Such displays can also respond to user interaction. Because wall-mounted variations require little space, museums can display a larger number of artifacts. Clearly, holograms or other replicas cannot substitute for original artifacts because viewing those originals is the main reason patrons visit a museum. If, however, a unique artifact is unavailable or too fragile to be displayed, a hologram offers an enticing alternative by showing the artifact as a high-quality, three-dimensional image that, combined with computer graphics, lets users experience it interactively.

8.7 Augmented Paintings

Working high above the floor of the Sistine Chapel in the Vatican of Rome, Michelangelo Buonarroti painted, between 1509 and 1512, some of the finest pictorial images of all time. On the ceiling of the papal chapel, he created a masterpiece fresco that includes nine scenes from the Book of Genesis. Among them is the famous Creation of Adam scene, showing God touching Adam's hand. In 1510, an initial study led Michelangelo to draw the Adam figure as a sanguine on a piece of paper. Today, this early drawing is displayed by the British Museum in London. Around 1518, Jacopo Pontormo painted yet another scene from the Book of Genesis, called Joseph and Jacob in Egypt. It decorated the bedchamber of Pier Francesco Borgherini for many years and is now being displayed by the London National Gallery. This oil painting made headlines after art historians discovered incredible underdrawings underneath the top paint layer. The underdrawings show that the artist had decided to flip the background design after starting the work, and he simply overpainted his initial approach. Such underdrawings have been found in many Renaissance and Baroque paintings. In 1634, for example, Rembrandt van Rijn painted a self-portrait that was later retouched by one of his students to feature a Russian nobleman. The original portrait was hidden under layers of paint for more than three hundred years, until it was uncovered recently by art restorers. It was sold for nearly seven million British pounds.

Figure 8.21. Illustration of an Augmented Paintings installation in a museum. (*Image reprinted from [19] © IEEE.*)

Pictorial artwork, such as these examples, can tell interesting stories. The capabilities of museums to communicate this and other information, however, are clearly limited. Text legends and audio guides can mediate facts, but offer little potential for presenting visual content, such as embedded illustrations, pictures, animations, and interactive elements.

Edutainment is becoming an important factor for museums. By applying new media technologies, such as computer graphics, virtual reality, and augmented reality, exhibit-oriented information might be communicated more effectively, but certainly in a more exciting way (Figure 8.21).

Chapter 7 has described methods of how to augment flat textured surfaces, such as paintings, with any kind of projected computer graphics [19], and outlines a possible display configuration as well as imaginable presentation and interaction techniques for such an approach.

8.7.1 Physical Arrangement

Besides ceiling mounted projectors, as outlined in Figure 8.21, a *lectern-like setup* (Figure 8.22) is also possible. In this case, a dual-output PC, two projectors, stereo speakers, and a touchpad are integrated into the same frame.

The projectors are directed towards the painting on the opposite wall, which is covered by the transparent film-material described in Chapter 7. A down-scaled replica of the painting is placed underneath the touchpad, which allows *synchronized pointing* or *drag-and-drop interactions*.

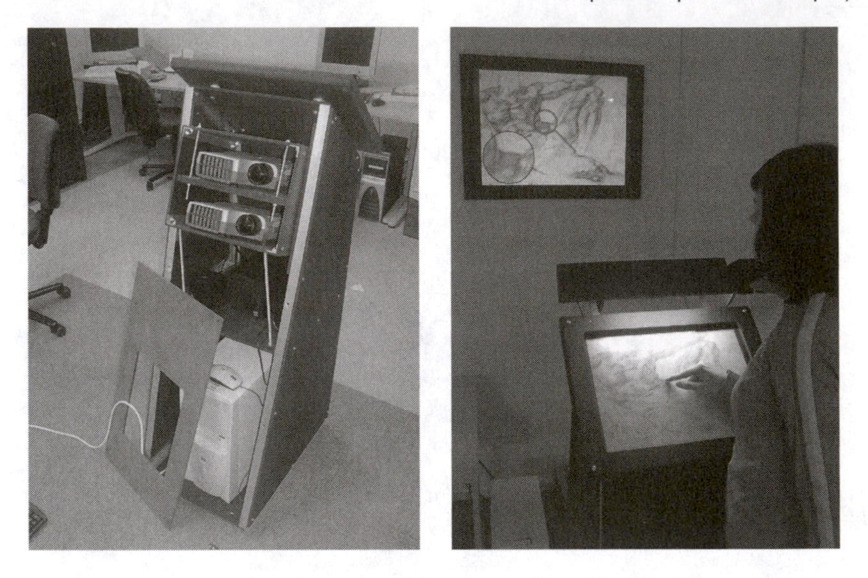

Figure 8.22. Lectern-like display configuration with integrated video-beamers; PC, and touchpad. (*Right image reprinted from [19] © IEEE.*)

Note, that the frame has to be attached (e.g., bolted) to the ground to prevent small movements of the projectors while interacting with the touchpad, which would cause misregistrations and *visual artifacts* of the calibrated graphical overlays on the physical painting.

8.7.2 General Functioning

The basic color correction process that is described in Chapter 7 allows visualization of all sorts of information and effects. This opens the potential for applying a wide range of established and new *presentation tools and techniques*. They can be categorized into six main classes:

1. *Inlay objects*—e.g., textual information, images, videos, arrow annotations;

2. *Lens effects*—e.g., magnification, X-ray, toolglasses and magic lenses;

3. *Focus effects*—e.g., blurring, decolorization, hiding, highlighting;

4. *Three-dimensional effects*—e.g., walk or flythrough, object observation;

5. *Modification effects*—e.g., drawing style modification, reillumination, color restoration;

6. *Audio effects* (e.g., verbal narrations, music, or sound that is synchronized to the visual effects).

Figure 8.23 illustrates a variety of examples. *Embedded information windows* (Figure 8.23(a)), such as the semi-transparent text panel or the opaque image/video panel are dynamic—in contrast to simple physical text labels—and can be directly linked to corresponding portions of the artwork (e.g., via arrow annotations, etc.). The *magnifying glass* in Figure 8.23(b) is interactively controlled by the touchpad integrated into the lectern and allows zooming in on interesting areas to identify brush stroke and hatching

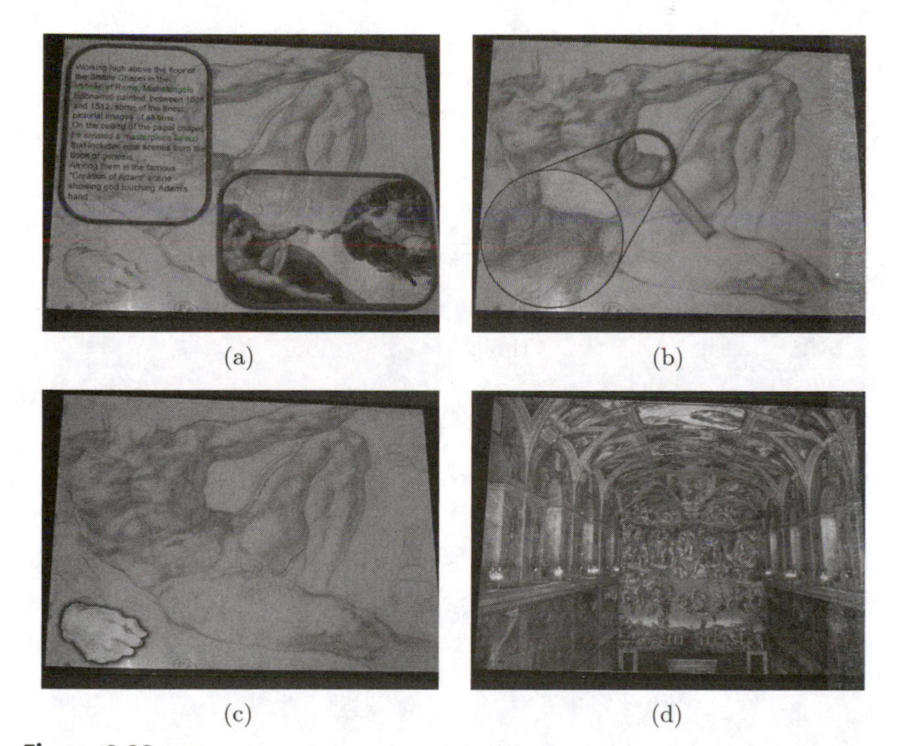

Figure 8.23. Examples of presentation techniques: (a) inlay text and image; (b) magnification; (c) focus through highlighting and decolorization; (d) three-dimensional flythrough through the Sistine chapel. Note, that in images (a)–(c) the Adam drawing itself is not projected. (*Displayed artwork courtesy of the Vatican Museum, Rome and the British Museum, London; images reprinted from [19] © IEEE; see Plate XI.*)

styles of the artist. Usually, museum visitors are restricted from getting close enough to recognize such details. To draw the observer's attention to a certain image portion, it can be brought into the foreground while the remaining part is eclipsed. In Figure 8.23(c), the *focus area* is highlighted while the rest is decolorized. This is achieved technically by projecting the inverse colors of the background image. In this case, the colors of the pigments on the canvas are physically cancelled out and appear in grayscale. Figure 8.23(d) shows a *three-dimensional flythrough* through the Sistine chapel, building a spatial correlation to surrounding paintings and the environment. This allows a virtual relocation of the observer to remote places.

Stereoscopic rendering (and optional head-tracking technology) allows an immersive experience. Although the transparent film preserves the po-

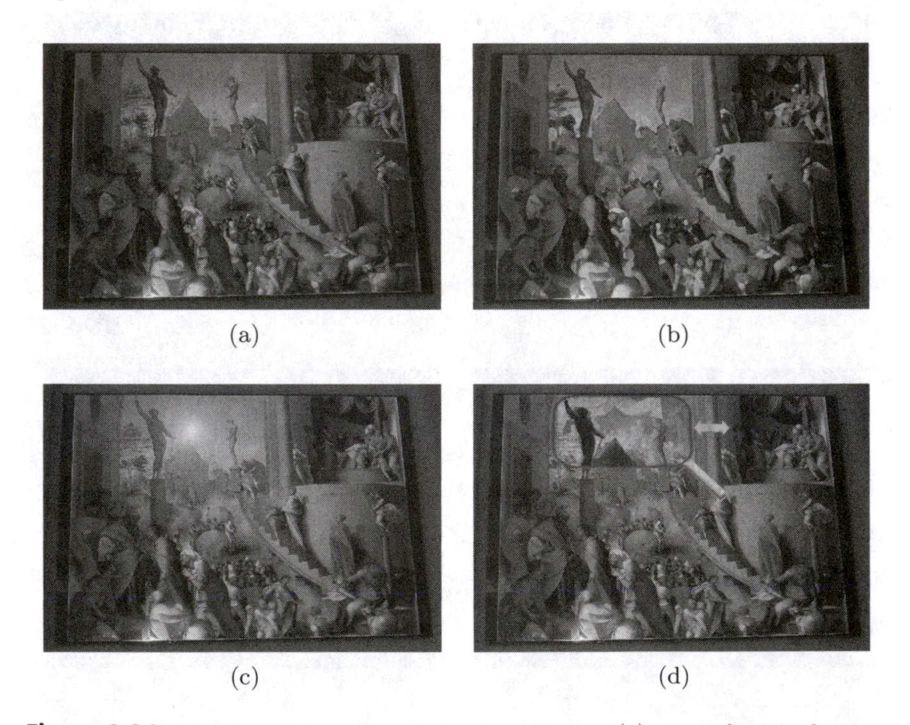

Figure 8.24. Pontormo's Joseph and Jacob in Egypt: (a) copy of original painting illuminated under environment light; (b) modification of painting style from oil on wood to watercolor on paper via a 2D artistic filter; (c) reillumination and lens flare; (d) registered visualization of underdrawings (infrared recordings are black and white) and explanation. (*Displayed artwork courtesy of the National Gallery, London; images reprinted from [19] © IEEE; see Plate XVI.*)

larization of light, the underlying canvas does not. Consequently, only active or anaglyphic stereo solutions can be used. A mechanical (half opaque/half transparent) shutter wheel rotating with 120 Hz in front of two video beamers can also be used. A light sensor measures the position of the wheel and triggers the infrared synchronization signal of the shutter glasses. In combination with conventional LCD projectors, this is a simple and cost efficient alternative to active stereo, compared to CRT projectors.

Figure 8.24 illustrates further presentation examples, such as modifications of the *painting style* via two-dimensional *artistic filters*, visualization of the *underdrawings*, and scene modification through *reillumination*.

For non-interactive presentations, it is possible to pregenerate the entire content. In this case, well-established content creation and authoring tools (e.g., digital imaging, three-dimensional modelling, video editing, and audio mixing packages) can be used. They already provide techniques such as animations, image filtering, roll-over effects, etc., as well as professional graphical user interfaces. For interactive presentations, however, the generated content has to be managed by the presentation software linked to an interactive framework.

If no user interaction is required, an embedded AVI player can be used to map video frames into input images on the fly. This represents a direct interface to the display-specific player framework that comprises rendering and color correction, and allows users to benefit from the capabilities of established content creation and authoring packages, such as Photoshop, 3D Studio Max, Maya, or Premiere.

8.7.3 Using Pictorial Artwork as Information Displays

Using pictorial artwork as information displays opens another door to *embedded multimedia content* in museums and art galleries. The method described previously is simple, seamless, cost and space efficient, robust, and compatible. These are all important factors for museum operators, content creators, and museum visitors. The presented techniques allow users to think of a wide variety of presentation and interaction tools that have not yet been explored. Dynamic view management and label placement are only two examples. As for other areas, the development of efficient interaction techniques and devices for such displays presents an interesting challenge. Digital storytelling techniques and content creation/management tools will be the link between specialized presentation techniques and display technology and standardized software platforms that can be operated by media artists and designers.

8.8 Smart Projectors

The invention of television radically shaped the 20th century. Today, it is the number one entertainment medium for most of us. New and innovative display technologies are changing their image, now faster than ever before. While the biggest step for cathode ray tube technology was the transistor from black-and-white to color, liquid crystal and plasma displays are currently transforming TVs into large flat panel screens. Clearly there is an emerging trend towards *large screen displays*. Falling prices are leading to a booming market for home entertainment technology. However, there are still physical limitations related to all of these technologies that result in constraints for maximum screen size, display size, refresh rate, or power consumption.

Video projectors are another display type that is just on the cusp of conquering the entertainment market; they have experienced an enormous metamorphosis during the last decade. Their cost/performance development is comparable to the development of personal computers. Video projectors have one vital advantage over other display technologies; they can generate images that are much larger than the devices themselves without being constrained by most of the limitations mentioned previously. But this unique ability comes at a price. We have to give up living space to set up artificial projection canvases that must be as large as the images to be displayed. Home theaters, for instance, can allocate entire rooms. In many situations, temporary or stationary canvases also harm the ambience of environments like our living rooms or historic sites equipped with projector-based multimedia presentations.

Smart Projectors, however, do not require artificial canvases. Instead, they allow a correct projection onto arbitrary existing surfaces, such as wallpapered walls or window curtains [20].

8.8.1 Physical Arrangement and General Functioning

Smart Projectors are video projectors that are enhanced with sensors to gain information about the environment. Although cameras are the most common sensor types used for smart projector implementations, other sensors, like *tilt sensors* have been used. Completely calibrated and mounted as a single camera-projector unit or realized with separated components, some smart projectors allow a dynamic elimination of shadows cast by the user, an automatic keystone correction on planar screens, or a manually aligned shape-adaptive projection on second-order quadric display

(a) (b)

Figure 8.25. The Smart Projector concept: (a) scanning the projection surface; (b) displaying a geometry and color corrected image. (*Images reprinted from [20] © IEEE.*)

surfaces, like cylinders, domes, ellipsoids, or paraboloids. For projection planes, multiple projector units can be automatically registered based on their homographic relationships that are determined with the aid of cameras. In this case, camera feedback is also used for intensity blending and color matching of multiple projector contributions. Calibrated stereo cameras together with projectors allow us to directly scan the three-dimensional geometry of an arbitrary display surface. This enables an undistorted projection for a known head-tracked observer position.

All of these approaches require the calibration of camera(s) and projector(s) at one point to determine their intrinsic (focal length, principle point, skew angle, aspect ratio, field of view) and *extrinsic*(position and orientation) parameters. Although some systems are capable of displaying geometrically predistorted images for a known observer position onto non-planar surfaces scanned or modelled beforehand, these surfaces are still fairly simple, like adjacent even walls. Surfaces with fine geometric details, such as natural stone walls or window curtains would represent an overkill for real-time predistortion realized by processing a high-resolution three-dimensional model.

Per-pixel color correction becomes feasible in real time with the new capabilities of recent graphics chips. This has been described for planar and geometrically non-trivial but textured, surfaces in Chapter 7.

Also described in Chapter 7, this smart projector concept combines camera feedback with structured light projection to gain information about the screen surface and the environment. The modular camera component can be detached from the projection unit of the smart projector for calibration. It has to be temporarily placed approximately at the sweet-spot of the observers, pointing at the screen surface (Figure 8.25(a)). The projection unit can be placed at an arbitrary location. Its light frustum must also cover the screen surface area. The system will then automatically determine all parameters that are required for real-time geometric predistortion and color correction. After the system has been calibrated, the camera module can be removed. From now on, the projector unit corrects incoming video signals geometrically and photometrically. If the corrected images are projected onto the non-trivial screen surface, they appear as being displayed onto a plain white canvas from the camera's target perspectives at or near the observer's sweet-spot. However, this projection canvas is completely virtual and does not exist in reality (Figure 8.25(b)). The details on calibration and image correction are explained in Chapter 7.

A compact device, such as the one sketched in Figure 8.25 has not yet been built. Instead, off-the-shelf components, like a consumer LCD video beamer, a CCD camcorder, and a personal computer comprising a TV card and a pixel-shading-capable graphics board are used for realizing first proof-of-concept prototypes.

8.8.2 Embedded Entertainment with Smart Projectors

Video projectors will play a major role for future home entertainment and edutainment applications, ranging from movies and television, to computer games, and multimedia presentations. Smart video projectors have the potential to sense the environment and adapt to it. This promotes a seamless embedding of display technology into our everyday life.

The example described previously can function without an artificial canvas (Figure 8.26) and consequently leaves a bit more freedom to us in the decision on how to arrange our living space. Recent improvements in computer graphics hardware have made this possible. Modern graphics chips support a per-pixel displacement which makes the computation intensive processing of three-dimensional geometry for image predistortion unnecessary. In addition, a color correction is supported on a pixel-precise basis. Display surfaces do not have to be plain and white, and video images do not have to be rectangular (as illustrated in Figure 8.27, for instance). Instead, many existing surfaces can be temporarily converted to a display screen by projecting color and geometry corrected images onto them. Most important for consumer applications is, that the calibration of such a device is fast, fully automatic, and robust, and that the correction of video signals can be achieved in real time. Neither geometry information nor projector

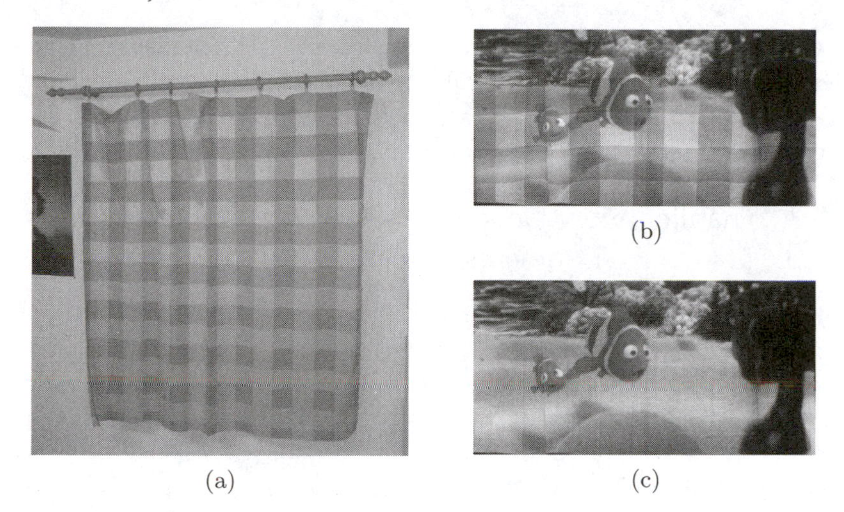

Figure 8.26. (a) Projection onto a curtain; (b) uncorrected colors; (c) corrected colors. Both projections are geometry corrected (*Movie footage: Finding Nemo, © 2003 Pixar Animation Studios; images reprinted from [20] © IEEE.*)

or camera parameters need to be known. Instead, the entire calibration and correction is done on a per-pixel level.

Future hardware improvements will pave the way for further developments with smart projectors. Graphics chips, for instance, will be more powerful than ever before. Next-generation projectors will continue to enhance quality factors, like brightness, resolution, dynamic range, and black

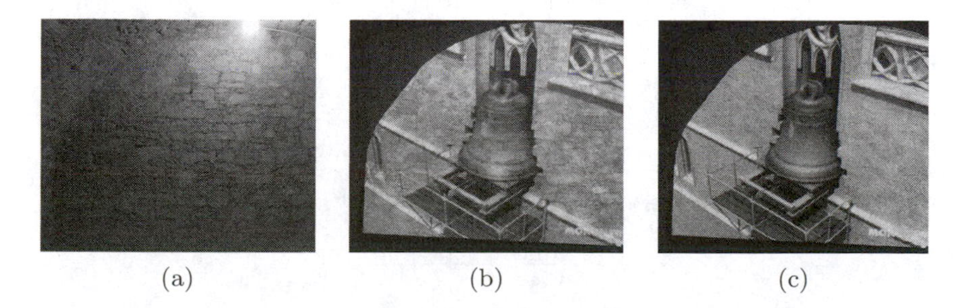

Figure 8.27. (a) Stenciled projection onto a natural stone wall inside the vaults of a castle; (b) color-uncorrected; (c) color-corrected. Both projections are geometry corrected (*Displayed content: "The Recovery of Gloriosa", © 2004 Bennert-Monumedia GmbH; images reprinted from [20] © IEEE.*)

level. Prices will decrease with an increasing market. The development of digital cameras and camcorders will behave in a similar way. Thanks to an ongoing miniaturization, all of these components could soon be integrated into compact devices that are as inconspicuous as light bulbs.

8.8.3 Enabling View-Dependent Stereoscopic Projection in Real Environments

As mentioned in Chapter 7, view-dependent stereoscopic projection can be enabled on ordinary surfaces (geometrically complex, colored, and textured) within everyday environments. However, several problems have to be solved to ensure a consistent disparity-related depth perception. As illustrated in Figure 8.28, we can summarize that a consistent and view-dependent stereoscopic projection onto complex surfaces is enabled by four main components: geometric warping, radiometric compensation, multi-focal projection, and multi-projector contributions. Implementing these components as hardware-accelerated dynamic pixel shaders achieves interactive frame rates. Radiometric compensation is essential for a consistent disparity-related depth perception of stereoscopic images projected onto colored and textured surfaces. Multi-projector techniques (see Chapter 5) enable intensity and color blending, or shadow removal for such con-

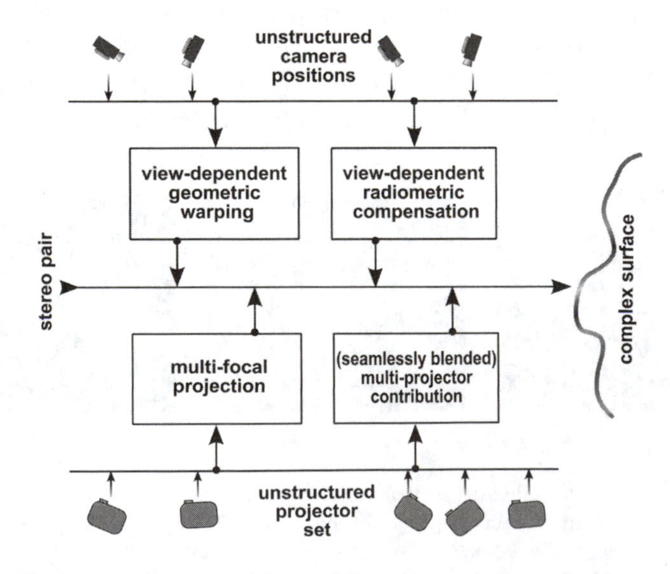

Figure 8.28. Main components for enabling stereoscopic projection on everyday surfaces.

Figure 8.29. Visualizing registered structures of (top) an adjacent room; (center) stairways and lower building level for architectural applications in a realistic environment; (bottom) envisioned application example in an architectural domain. (*Prototype realized by the Bauhaus-University Weimar, 2005; see Plate XV.*)

figurations. Pixel-precise image-based or projective texture-based warping technique (e.g., such as explained in Chapter 5) provide the required precision for radiometrically compensated projections of stereo pairs onto geometrically and radiometrically complex surfaces. They provide view-dependent rendering at interactive frame rates and support the correction of non-linear projector distortions. Partially unfocussed projections affect disparity-related depth perception and can make the correct fusion of stereo pairs difficult. Since conventional projectors provide only a single focal plane, it is physically impossible to generate sharp images on surfaces with extreme depth variations. Multi-focal projection methods that offer an efficient image composition from multiple projection units represent novel software solutions to this problem. As outlined in the architectural examples shown in Figure 8.29, such techniques do not offer only new possibilities for augmented reality and virtual reality. They also allow a merger of both technologies and give some applications the possibility to benefit from the conceptual overlap of AR and VR. An ad-hoc system configuration that is not constrained to laboratory conditions will increase the practicability of such a tool.

8.9 Summary and Discussion

In this chapter, we have presented numerous examples of spatial AR displays and their applications in different domains, such as museums, home entertainment, science, and industry. Some of them are optical see-through configurations that utilize mirror beam or transparent screen combiners and, in combination with projector-based illumination, achieve a high degree of consistency between real and virtual environments. Others use projectors to augment physical objects directly without additional optical elements.

From an application point of view, there is one major difference between optical see-through augmentations and projector-based augmentations. An optical see-through display allows the visualization of virtual objects as if they are floating in thin air (e.g., next to a physical object). A projector-based augmentation is restricted to a projection onto the real object's surfaces even when stereoscopic rendering is used. This prevents us from displaying virtual objects next to real ones if no underlying projection surface is given. However, projector-based augmentation and illumination are well suited for modifying the real object's surface appearance, such as shading, color, texture, or visibility properties. Thus, spatial optical see-through and projector-based techniques go hand in hand. They complement each other and lead to efficient, consistent, and realistic spatial AR presentations.

The robustness, attractiveness, and efficiency of technology are essential for a successful application of spatial AR in public places, such as museums. Some of the display configurations that have been described in this chapter tell the first success stories and provide a promising perspective for upcoming installations.

9

The Future

For non-mobile tasks, novel approaches have taken augmented reality beyond traditional eye-worn or hand-held displays, thus, enabling new application areas. These spatial approaches exploit components such as video projectors, optical elements, holograms, radio frequency tags, and tracking technologies. Due to the decrease in cost of these devices and graphics resources, there has been a considerable interest in using such augmented reality systems in universities, labs, museums, and in the art community. New technology will not only open new possibilities for SAR, but also for other display concepts, such as hand-held and head-attached displays.

9.1 Displays

9.1.1 Projectors

Projectors of the near future will be compact, portable, and with the built-in awareness which will enable them to automatically create satisfactory displays on many of the surfaces in our everyday environment. Alongside the advantages, there are limitations, but we anticipate projectors being complementary to other modes of display for everyday personal use in the future. We expect that there will be new application areas for which they are especially suited. LEDs are replacing lamps. Reflective, instead of transmissive displays, (DLPs, LCOS) are becoming popular. Both lead to improved efficiency requiring less power and less cooling. DLP and LCOS projectors can display images at extremely high frame rates, currently 180 Hz and 540 Hz respectively, but they lack video bandwidth. Efforts to increase video bandwidth are already in the making and are very promising. For example, Symbol Technologies [184] has demonstrated a small laser projector (two tiny steering mirrors for vertical and horizontal deflection) and has even built a hand-held three-dimensional scanner based on such a projector. Siemens has built a mini-beamer attachment for mobile-phones [171]. Cam3D has built a Wedge display where a projector can be converted into a flat panel display by projecting images at the bottom of a wedge shaped glass [189]. A future mobile projector may function as a as flat panel when there is no appropriate surface to illuminate or when ambient light is problematic. Super bright, sharp infinite focus laser projectors are also becoming widespread [75]. They may allow shape-adaptive projection without focus and ambient lighting problems. In addition suitable input devices are also being developed; e.g., Canesta [28] has built a projected laser pattern on which one can type. The finger movement is detected by IR sensing. Finally novel lamp designs, especially those based on LEDs or lasers are creating smaller, lighter, and more efficient long-life solutions.

9.1.2 Autosteroscopic Displays

Parallax displays are display screens (e.g., LCD displays) that are overlaid with an array of light directing elements. Depending on the observer's location, the emitted light that is presented by the display is directed in a way that it appears to originate from different parts of the display while changing the viewpoint. If the light is directed to both eyes individually, the observer's visual system interprets the different light information to be emitted by the same spatial point. Examples of parallax displays are parallax barrier displays (Figure 2.18(a)) that use a controllable array of light-blocking elements (e.g., a light-blocking film or liquid crystal barriers [136]) in front of a CRT screen. Depending on the observer's viewpoint, these light-blocking elements are used to direct the displayed stereo images to the corresponding eyes. Other examples are lenticular sheet displays (Figure 2.18(b)) that use an array of optical elements (e.g., small cylindrical or spherical lenses) to direct the light for a limited number of defined viewing zones.

Parallax displays can be designed and mass produced in a wide range of sizes and can be used to display photorealistic images. Many different types are commercially available and used in applications ranging from desktop screens to cell phone displays. We believe that the application of parallax displays will be the next logical and feasible step for SAR solutions towards autostereoscopy.

In the future, a stronger combination of computer graphics and holography (as explained in Chapter 4) can also be expected. Any 3D rendered image can, in principle, be transformed into a holographic interference pattern basically by using Fourier transformations. These, however, presuppose large computational, network, and storage capacities. For instance, a prerendered or recorded movie-length holographic video could require a petabyte (1 million gigabytes) of storage. Holographic storage itself, which uses lasers to record data in the three dimensions of a clear crystalline medium may be a future solution to the problem that non-static holograms require a massive amount of data. While computer-generated static images can already be transformed into large digital holograms using holographic printers [80], interactive or real-time electronic holography [96] with an acceptable quality (size, resolution, and color) would still require the invention of more advanced light modulators, faster computers with a higher bandwidth, and better compression techniques. However, the combination of high quality optical holograms and interactive computer graphics is already feasible with off-the-shelf hardware today [17].

9.1.3 Flexible Displays

Organic light emitting diodes [68], for instance, may replace the crystalline LEDs that are currently being used to build the miniature displays for HMDs. OLEDs promise to produce cheap and very high-resolution full-color matrix displays that can give head-attached displays a technological advance. In contrast to normal LEDs, OLEDs are made from plastic compounds rather than from semi-conducting elements, such as silicon or gallium, etc. Like LEDs, OLEDs glow when voltage is applied. Two main classes exist today: small molecule OLEDs and polymer OLEDs. While small molecule OLEDs are built up by depositing molecules of the compound onto the display itself under very low pressure, polymer OLEDs have the active molecules suspended in a liquid-like pigment in paint. It can be printed onto displays using ink jets, screen printing, or any of the various contact techniques used for ordinary inks. Small molecule OLED displays are limited in size, but they may be suitable for head-mounted displays. Polymer OLEDs can be used to build large scale displays, such as 500 inch displays or larger. Resolution approaching 300 dpi is also possible, approaching the quality of ink on paper. They may become more interesting for spatial AR approaches.

OLEDs have some general advantages:

- Because OLED displays are fluorescent and don't require backlights, they will need far less power than LCD screens (currently LCDs require three times more power than OLEDs);

- OLED layers can be made much thinner than LCD layers (about a thousand times thinner than a human hair);
- In contrast to LCDs, OLEDs don't use polarized light filters. The displayed images on OLEDs can be viewed from a much wider angle;
- OLEDs have a much wider working temperature range than LCDs;
- OLEDs can be printed onto large-scale and flexible display surfaces of any shape and (almost) any material.

However, the OLED compounds degrade over time (especially when they come in contact with oxygen or water) limiting the maximum lifetime of a display. Different colors degrade at different rates making the color balance change. These problems may be solved in the future.

Light emitting polymers [26] provide the opportunity for the fabrication of large, flexible, full-color, fast emissive displays with a high-resolution, a wide viewing angle, and a high durability. In LEP technology, a thin film of light-emitting polymer is applied onto a glass or plastic substrate coated with a transparent electrode. A metal electrode is evaporated on top of the polymer. The polymer emits light when the electric field between the two electrodes is activated. The response time of LEPs is ultra-fast (sub-microsecond) and is unaffected by temperature. Consequently, they may support high enough frame rates for active stereoscopic rendering. Light emission occurs at low voltage, and (unlike LCD or plasma displays) it can be fabricated on a single sheet of glass. Also, because it can be made of flexible plastic substrates, it is not only extremely difficult to break, but can also be molded into different shapes and contours.

The advantages of LEPs can be summarized as follows:

- Since a low voltage is required, LEPs need little power;
- The response time of LEPs is very high, potentially allowing active stereoscopic rendering;
- LEPs can be fabricated on transparent glass or plastic surfaces of any shape;
- LEPs provide a high contrast (currently between 3–5 times higher than LCDs).

Transparent LEPs may present other future alternatives for spatial AR configurations.

Electronic papers (or E-Ink) [206] also has potential as an AR display technology of the future. Here an electric charge moves magnetic colored capsules within the "paper" either toward or away from the surface in order to form an image. The capsules retain their positions until another charge is applied. The ink simply resides on the display while an image is not changing; therefore, it consumes no power. Philips and other companies in the field are working on color as well as bendable applications. The main advantage of electronic paper is that it does not require any power, as long as the displayed image does not change. Large color screen configurations exist that have been built from smaller electronic paper tiles.

Several approaches exist that allow a projection of two-dimensional images onto a screen composed of fog [135] or modified air [74]. Such fog/air displays suffer from distortion caused by turbulence but allow through-the-screen interactions since they lack a physical screen boundary.

Solid-state displays [39] generate visible photons (i.e., light) within a transparent host material by exciting optically active ions with special energy sources. Ions at a known three-dimensional position within the host materials can be excited by crossing energy beams, such as infrared lasers, ultraviolet sources of radiation, or electron beams.

Examples of suitable host materials are various gases, crystals, and electro-active polymers. The host material must provide several properties:

- In its initial state, it must be transparent;
- It must emit visible light in its excited state;
- Its inner structure must be homogeneous;
- It must have a refraction index similar to air to avoid distortion.

If it were possible to use air as the host material, then this approach would represent the "Holy Grail," not only for spatial augmented reality displays, but for three-dimensional display technology in general. Unfortunately this is not yet feasible. In addition, conceptual problems, such as the *ghost voxel problem* (when energy beams that are used to create voxels intersect with other beams and create unwanted voxels) have to be solved.

In the short run, especially high-resolution bright and flexible projection devices, high-performance and cost efficient rendering hardware, reliable, precise and wireless tracking technology, and advanced interaction techniques and devices will pave the way for forthcoming alternative AR configurations. In the long run, new display concepts, such as autostereoscopy and holography will replace goggle-bound stereoscopic displays (at least for non-mobile applications). However, the underlying technology must be robust and flexible, and the technology that directly interfaces to users should adapt to humans, rather than forcing users to adapt to the technology. Therefore, human-centered and seamless technologies, devices, and techniques will play a major role for augmented reality of the future. That this is feasible can be demonstrated based on the example of one early SAR display, the Virtual Showcase [9]. Since its invention in 2000, several prototypes have been exhibited to an approximate total number of more than 25,000–30,000 visitors/users (a user-study and picture gallery are available through http://www.spatialar.com/). Some events were technically supervised short-term exhibitions (1–7 days) at conferences and trade shows; others have been non-supervised long-term exhibitions (3–4 months) in museums. These uses have shown that SAR technology can be robust, attractive, and efficient enough to be applied successfully outside research laboratories.

9.2 Supporting Elements

New hardware for tagging, registering, or linking physical objects to the digital world are emerging. AR has a clear connection with ubiquitous computing and the wearable computer domain. Let us look at the new sensors and input devices.

9.2.1 Electronic Tags

Small electronic components, or tags, that can respond to radio frequency, ultrasound, or light are being used to simplify the task of locating and identifying objects. Electronic tags have several advantages compared to optical tags.

Adding distinctive visual markers for tracking is not always practical. Optical tags require sufficient area and are more prone to wear and tear. The number of distinguishable identifiers in optical tags remains small, while electronic tags allow a very long binary sequence to be used. Electronic tags are also not affected by ambient illumination, and in many cases do not require line of sight to the tags. Authoring also becomes easy with electronic tags because data can be stored locally. On the other hand, electronic tags are expensive and battery-operated tags require frequent maintenance. The added complexity increases the cost, and radio frequency or ultrasound tags are difficult to precisely locate.

Multiple research groups are looking at adding intelligence to objects and, in some cases, building human interactions around them. *Smart-Its* provides interconnected smart devices that can be attached to everyday items [64]. *SenseTable* tracks and augments the area around sensing tablets on a tabletop using a projector [134]. Intelligent furniture has also been explored [125].

Some systems use active RF tags that respond to laser pointers. The *FindIT* flashlight uses a one-way interaction and an indicator light on the tag to signal that the desired object has been found [98]. Other systems use a two-way interaction, where the tag responds back to the PDA using a power-hungry protocol like 802.11 or X10 ([132], respectively [161]). CoolTown [188] uses beacons that actively transmit devices references although without the ability to point and without visual feedback.

Radio frequency identification tags (RFID) allow querying of the identification code stores on the tag using a radio frequency reader. Powered radio-frequency tags use a battery that is about the size of a watch-battery, have a lifetime of a few years, and have a cost of a few dollars. In contrast, passive RFID tags are unpowered, can be as small as a grain of rice, and cost only about ten cents. Prices of both are dropping, but the price differential will remain. The size and cost properties are such that RFID is showing signs of being adopted as a mass-deployment technology.

Current commercial applications include embedding of RFID tags in packaging for inventory control, non-contact access control, and ear tags for livestock. Despite the interest in RFID, the available functionality is very limited; an RF-reader broadcasts a request, and in-range tags collect energy from the RF pulse and reply. Since only in-range tags respond, this can be used for coarse location sensing.

A collection of sensing and interaction techniques for mobile devices is described in Hinckley [62]. Augmentation of physical world objects has been primarily achieved via eye-worn or head-mounted displays [5] or hand-held screens. Screen-based augmentation using PDA, camera, and RFID tags is described in [198] and[160].

9.2.2 Location Sensing

Location sensing systems such as the Olivetti Active Badge [195] and Xerox PARCTAB [197] recover location and have been used for passive tracking,

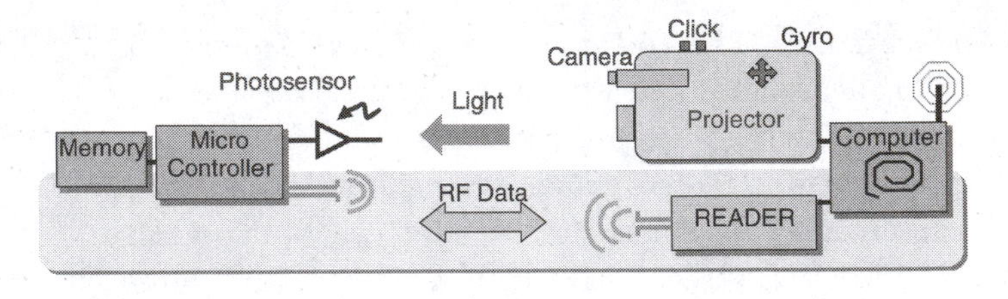

Figure 9.1. RFIG: communication between the photosensing tag (left) and hand-held projector system (right). (*Images reprinted from [159] © ACM.*)

but not for an interactive system. In the Cricket project [187], the system recovers the pose of a hand-held device using installed RF and ultrasound beacons, and uses the computed location and orientation for projected augmentation.

Two important problems in augmented reality, object identification and determining the pose of the displayed image with respect to the physical object, can be solved by using photosensing RFID tags (Figure 9.1). This is being explored as a radio frequency identification and geometry (RFIG) discovery method [159]. Figure 9.2 shows a current application in a warehouse scenario where two employees can locate products that are about to expire using the RF channel to send the query and an optical channel

Figure 9.2. Warehouse scenario with RFIG tags. A user directs a hand-held projector at tagged inventory, with communication mediated by two channels—RF and photo-sensing on the tags. The user sees a projection of the retrieved tag information collocated with the physical objects and performs a desktop-like interaction with the projection. A second user performs similar operations, without conflict in the interaction because the projector beams do not overlap. (*Image reprinted from [159] © ACM.*)

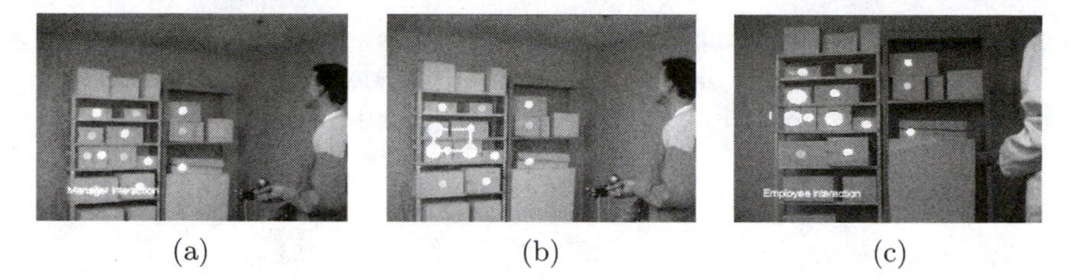

(a) (b) (c)

Figure 9.3. Warehouse scenario implementation: (a) manager locates items about to expire (marked in white circles); (b) annotates some of those items (marked in larger white circles); (c) employee retrieves and views the same annotations from a different projector view. (*Images reprinted from [159] © ACM; see Plate XIV.*)

to locate the corresponding wireless tags. The hand-held projector finally displays the appropriate information overlaid on those objects (Figure 9.3).

A distinguishing factor between RFIG and related systems is in the parallel nature of the interaction. Laser-pointer systems [98, 132] require a user to identify a target object and direct the pointer at it to initiate interaction. However, accurate pointing is difficult when the tags become visually imperceptible, and multiple tags can only be dealt with serially. In contrast, a RFIG system uses a casually-directed projector to interact with all tagged objects within the projector beam and can select all objects or a subset for further interaction. The computed location can be used to support sophisticated geometric operations.

9.3 Summary and Discussion

In this chapter, we have presented numerous examples of emerging technologies and how they are likely to impact spatial augmented reality systems. We discussed innovative optics for displays, new materials such as light emitting polymers, promising developments in sensor networks including those using photosensors and the excitement surrounding radio frequency identification tags.

Trends in miniaturization are changing the way projector-based augmented reality systems will be built. The embedding of computation power and data in everyday environments is now becoming feasible as cost and power problems diminish. Tag-based environments will provide real-time feedback to AR systems and simplify problems in authoring, recognition, and location sensing. However, these hardware components involve many challenges, and it is difficult to predict which technologies will be prevalent. Nevertheless, emerging prototypes that we have described demonstrate innovative ways to interact with such ambient intelligence.

A

Calibration of a Projector (or a Camera)

A.1 Source Code for Calibration of a Projector (or a Camera)

Input: set of corresponding 3D points `points.{X,Y,Z}` and 2D pixels `pixels.{u,v}`
Output of Step 3: Perspective 3 by 4 projection matrix, `ProjMV`
Output of Step 4: OpenGL 4×4 matrices, `ProjMat`, `ModelViewMat`

```
void calculateMatrices()
{
// Step 1: Fill Matrix that covers the constraining equations
  Matrix lhs(2 * nPointsFound, 12); // lhs=LeftHandSide
  for (int i = 0; i < nPointsFound; i++) {
    // odd rows
    lhs(2 * i + 1, 1)  =  points[i].X;
    lhs(2 * i + 1, 2)  =  points[i].Y;
    lhs(2 * i + 1, 3)  =  points[i].Z;
    lhs(2 * i + 1, 4)  =  1;
    lhs(2 * i + 1, 5)  =  0;
    lhs(2 * i + 1, 6)  =  0;
    lhs(2 * i + 1, 7)  =  0;
    lhs(2 * i + 1, 8)  =  0;
    lhs(2 * i + 1, 9)  = -pixels[i].u * points[i].X;
    lhs(2 * i + 1, 10) = -pixels[i].u * points[i].Y;
    lhs(2 * i + 1, 11) = -pixels[i].u * points[i].Z;
    lhs(2 * i + 1, 12) = -pixels[i].u;

    // even rows
    lhs(2 * i + 2, 1)  =  0;
    lhs(2 * i + 2, 2)  =  0;
```

```
    lhs(2 * i + 2, 3)  =  0;
    lhs(2 * i + 2, 4)  =  0;
    lhs(2 * i + 2, 5)  =  points[i].X;
    lhs(2 * i + 2, 6)  =  points[i].Y;
    lhs(2 * i + 2, 7)  =  points[i].Z;
    lhs(2 * i + 2, 8)  =  1;
    lhs(2 * i + 2, 9)  = -pixels[i].v * points[i].X;
    lhs(2 * i + 2, 10) = -pixels[i].v * points[i].Y;
    lhs(2 * i + 2, 11) = -pixels[i].v * points[i].Z;
    lhs(2 * i + 2, 12) = -pixels[i].v;
  }

// Step 2: Find u-vector corresponding to smallest singular
// value (S)(=Solution)
  DiagonalMatrix D(12);
  Matrix U(12, 12);

  // lhs.t() denotes matrix transpose
  SVD(lhs.t() * lhs, D, U);

  // Column containing smallest sing. value
  int smallestCol = 1;

  // find smallest
  for (int j = 1; j < 13; j++)
    if ((D(smallestCol) * D(smallestCol)) > (D(j) * D(j)))
      smallestCol = j;
  ColumnVector S = U.Column(smallestCol);

// Step 3: write 12x1-Vector S as 3x4 Matrix (row-wise)
  Matrix ProjMV(3, 4);
  for (int k = 0; k < 12; k++)
    ProjMV}((k / 4) + 1,(k % 4) + 1) = S(k + 1);

// Step 4: decompose ProjMV in Proj- and ModelView-matrices}
  double scale =
    sqrt(ProjMV.SubMatrix(3, 3, 1, 3).SumSquare());
  ProjMV /= scale;

  //ProjMV /= ProjMV(3, 4);
  if (ProjMV(3, 4) > 0) ProjMV * = -1;

  ColumnVector Q1 = (ProjMV.SubMatrix(1, 1, 1, 3)).t();
  ColumnVector Q2 = (ProjMV.SubMatrix(2, 2, 1, 3)).t();
  ColumnVector Q3 = (ProjMV.SubMatrix(3, 3, 1, 3)).t();
```

```
  double q14 = ProjMV(1, 4);
  double q24 = ProjMV(2, 4);
  double q34 = ProjMV(3, 4);

  double tz = q34;
  double tzeps = 1;
  if (tz > 0) tzeps = -1;

  tz = tzeps * q34;
  RowVector r3 = tzeps * Q3.t();
  double u0 = (Q1.t() * Q3).AsScalar();
  double v0 = (Q2.t() * Q3).AsScalar();
  double a = crossNorm(Q1, Q3);
  double b = crossNorm(Q2, Q3);
  RowVector r1 = tzeps * (Q1.t() - (u0 * Q3.t())) / a;
  RowVector r2 = tzeps * (Q2.t() - (v0 * Q3.t())) / b;
  double tx = tzeps * (q14 - u0 * tz) / a;
  double ty = tzeps * (q24 - v0 * tz) / b;

  // create Rotation Matrix and Translation Vector
  Matrix RotMatrix(3,3);
  RotMatrix = r1 & r2 & r3;
  ColumnVector t(3);
  t << tx << ty << tz;

// Step 5: Expand found matrices to 4 x 4 matrices
  // Projection
  Matrix IntMat(4, 4);
  IntMat(1, 1) = -a;
  IntMat(1, 2) =  0;
  IntMat(1, 3) = -u0;
  IntMat(1, 4) =  0;
  IntMat(2, 1) =  0;
  IntMat(2, 2) = -b;
  IntMat(2, 3) = -v0;
  IntMat(2, 4) =  0;
  IntMat(3, 1) =  0;
  IntMat(3, 2) =  0;
  IntMat(3, 3) =     -(gfFarPlane + gfNearPlane) /
                      (gfFarPlane - gfNearPlane);
  IntMat(3, 4) = -2 * gfFarPlane * gfNearPlane /
                      (gfFarPlane - gfNearPlane);
  IntMat(4, 1) =  0;
  IntMat(4, 2) =  0;
  IntMat(4, 3) = -1;
  IntMat(4, 4) =  0;
```

```
  // Rotation & Translation
  Matrix ExtMat(4, 4);
  RowVector nulleins(4);
  nulleins << 0.0 << 0.0 << 0.0 << 1.0;
  ExtMat = ((r1 & r2 & r3) | t) & nulleins;

// Step 6: Set matrices as Current MV/Proj-matrices}
  for (int l = 0; l < 16; l++) {
    MVMat[l] = ExtMat((l % 4) + 1, (l / 4) + 1);
    ProjMat[l] = IntMat((l % 4) + 1, (l / 4) + 1);
  }

  // ProjMat has to be multiplied by VP^-1
  mat16dMult( invViewport, ProjMat, ProjMat );

// Step 7: Save matrices to file ("IntMat0.dat"/"ExtMat0.dat")
  ofstream lfInt, lfExt;
  lfInt.open("IntMat0.dat", ios::out);
  lfExt.open("ExtMat0.dat", ios::out);
  for (i = 0; i < 16; i++) {
    lfInt << IntMat(i % 4 + 1, i / 4 + 1)
          << ((i % 4 - 3)?" ":"\n");
    lfExt << ExtMat(i % 4 + 1, i / 4 + 1)
          << ((i % 4 - 3)?" ":"\n");
  }
}
```

A.2 Quadric Image Transfer

Vertex Shader code in Cg to implement Quadric Image Transfer for projecting on quadric curved surfaces.

```
 vertout main(appin IN,
             uniform float4x4 modelViewProj,
             uniform float4 constColor,
             uniform float3x3 A,
             uniform float3x3 E,
             uniform float3 e)
{
  vertout OUT;
  float4 m1 = float4(IN.position.x,
                     IN.position.y,
                     IN.position.z,
```

```
                          1.0f);
  float4 m, mi;
  float3 m2, mp;
  float scale;

  m = mul(modelViewProj, m1);
  m2.x = m.x/m.w;
  m2.y = m.y/m.w;
  m2.z = 1;
  scale = mul(m2, mul(E,m2));
  mp = mul(A,m2) + sqrt(scale) * e;
  mi.x = m.w * (mp.x) / (mp.z);
  mi.y = m.w * (mp.y) / (mp.z);
  mi.zw = m.zw;

  OUT.position = mi;
  // Use the original per-vertex color specified
  OUT.color0 = IN.color0;

  return OUT;
}
```

B

OpenGL's Transformation Pipeline Partially Re-Implemented

B.1 General Definitions

```
typedef double MATRIX_4X4 [16];     //   4x4 homogeneous matrix in
                                    //   column-first order
MultMatrix(m);                      //   multiplies matrix m onto
                                    //   current matrix stack

nv (a,b,c);                         //   normalize vector [a,b,c]
xvv(a,b,c,d,e,f,g,h,i)              //   cross product:
                                    //   [g,h,i]=[a,b,c]x[d,e,f]
```

B.2 Projection Functions

```
void Frustum(double l, double r, double b, double t, double n,
             double f)
//parameters OpenGL convention
{
    MATRIX_4X4 m;

    m[0]=(2*n)/(r-l); m[4]=0.0;          m[8]=(r+l)/(r-l); m[12]=0.0;
    m[1]=0.0;          m[5]=(2*n)/(t-b); m[9]=(t+b)/(t-b); m[13]=0.0;
    m[2]=0.0;          m[6]=0.0;          m[10]=-(f+n)/(f-n);
    m[14]=-(2*f*n)/(f-n);
    m[3]=0.0;          m[7]=0.0;          m[11]=-1.0;       m[15]=0.0;

    MultMatrix(m);
}
```

```
void Perspective(double fovy, double aspect, double zNear,
                 double zFar)
//parameters in OpenGL convention
{
    MATRIX_4X4 m;
    double f;

    f=1.0/tan(rad(fovy)/2.0);
    m[0]=f/aspect; m[4]=0.0; m[8]=0.0;   m[12]=0.0;
    m[1]=0.0;      m[5]=f;   m[9]=0.0;   m[13]=0.0;
    m[2]=0.0;      m[6]=0.0; m[10]=(zFar+zNear)/(zNear-zFar);
    m[14]=(2*zFar*zNear)/(zNear-zFar);
    m[3]=0.0;      m[7]=0.0; m[11]=-1.0; m[15]=0.0;

    MultMatrix(m);
}
```

B.3 Transformation Functions

```
void Scale(double x, double y, double z)
//parameters in OpenGL convention
{
    MATRIX_4X4 m;

    m[0]=x;   m[4]=0.0; m[8]=0.0;  m[12]=0.0;
    m[1]=0.0; m[5]=y;   m[9]=0.0;  m[13]=0.0;
    m[2]=0.0; m[6]=0.0; m[10]=z;   m[14]=0.0;
    m[3]=0.0; m[7]=0.0; m[11]=0.0; m[15]=1.0;

    MultMatrix(m);
}
void Translate(double x, double y, double z)
//parameters in OpenGL convention
{
    MATRIX_4X4 m;

    m[0]=1.0; m[4]=0.0; m[8]=0.0;  m[12]=x;
    m[1]=0.0; m[5]=1.0; m[9]=0.0;  m[13]=y;
    m[2]=0.0; m[6]=0.0; m[10]=1.0; m[14]=z;
    m[3]=0.0; m[7]=0.0; m[11]=0.0; m[15]=1.0;

    MultMatrix(m);

}
```

```
void Rotate(double a, double i, double j, double k)
//parameters in OpenGL convention
{
    MATRIX_4X4 m;

    nv(&(i),&(j),&(k));
    double  cosa=cos(rad(a)), sina=sin(rad(a)), cosa1=1.0-cosa,
        n11=i*i*cosa1, n22=j*j*cosa1, n33=k*k*cosa1,
        n1=i*sina, n2=j*sina, n3=k*sina,
        n12=i*j*cosa1, n13=i*k*cosa1, n23=j*k*cosa1;

    m[0]=cosa+n11; m[4]=n12-n3;   m[8]=n13+n2;    m[12]=0;
    m[1]=n12+n3;   m[5]=cosa+n22; m[9]=n23-n1;    m[13]=0;
    m[2]=n13-n2;   m[6]=n23+n1;   m[10]=cosa+n33; m[14]=0;
    m[3]=0.0;      m[7]=0.0;      m[11]=0.0;      m[15]=1.0;

    MultMatrix(m);

}
\end{verbtim}
\end{small}

\section{Camera Control Function}

\begin{small}
\begin{verbatim}
void LookAt (double eyex, double eyey, double eyez,
             double centerx, double centery, double centerz,
             double upx, double upy, double upz)
//parameters in OpenGL convention
{
   MATRIX_4X4 m;
   double F[3],UP[3],s[3],u[3];

   F[0]=centerx-eyex; F[1]=centery-eyey; F[2]=centerz-eyez;
   nv (&(F[0]),&(F[1]),&(F[2]));

   UP[0]=upx; UP[1]=upy; UP[2]=upz;
   nv (&(UP[0]),&(UP[1]),&(UP[2]));

   xvv (F[0] , F[1] , F[2], UP[0], UP[1], UP[2],
        &(s[0]),&(s[1]),&(s[2]));
   nv (&(s[0]),&(s[1]),&(s[2]));
```

```
   xvv (s[0] , s[1] , s[2], F[0] , F[1] , F[2],
        &(u[0]),&(u[1]),&(u[2]));
   nv (&(u[0]),&(u[1]),&(u[2]));

  m[0]=s[0];  m[4]=s[1];  m[8]=s[2];   m[12]=0.0;
  m[1]=u[0];  m[5]=u[1];  m[9]=u[2];   m[13]=0.0;
  m[2]=-F[0]; m[6]=-F[1]; m[10]=-F[2]; m[14]=0.0;
  m[3]=0.0;   m[7]=0.0;   m[11]=0.0;   m[15]=1.0;

  MultMatrix(m);
  Translate(-eyex,-eyey,-eyez);
}
```

B.4 Additional Functions

```
void Project2Plane(double a, double b, double c, double d,
                   double x, double y, double z)
//[a,b,c,d] : plane parameters (normal, distance to origin)
//[x,y,z] : projection center
//defines a matrix that projects every point from [x,y,z]
// onto an arbitrary plane [a,b,c,d]
{
  double dot;
  dot = a*x+b*y+c*z+d ;

  MATRIX_4X4 m;

  m[0]=dot-x*a; m[4]=0.f-x*b; m[8]=0.f-x*c;  m[12]=0.f-x*d;
  m[1]=0.f-y*a; m[5]=dot-y*b; m[9]=0.f-y*c;  m[13]=0.f-y*d;
  m[2]=0.f-z*a; m[6]=0.f-z*b; m[10]=dot-z*c; m[14]=0.f-z*d;
  m[3]=0.f-a;   m[7]=0.f-b;   m[11]=0.f-c;   m[15]=dot-d;

  MultMatrix(m);
}


void Reflect(double a, double b, double c, double d)
//[a,b,c,d] : plane parameters (normal, distance to origin)
//defines a matrix that reflects every point over an arbitrary
//plane [a,b,c,d]
{
     MATRIX_4X4 m;

    //set up reflection matrix
    m[0]=1-2*a*a; m[4]=-2*a*b;  m[8]=-2*a*c;    m[12]=-2*a*d;
    m[1]=-2*a*b;  m[5]=1-2*b*b; m[9]=-2*b*c;    m[13]=-2*b*d;
    m[2]=-2*a*c;  m[6]=-2*b*c;  m[10]=1-2*c*c; m[14]=-2*c*d;
```

```
    m[3]=0;         m[7]=0;         m[11]=0;        m[15]=1;

    MultMatrix(m);

}
```

Bibliography

[1] M. Agrawala, A. Beers, B. Froehlich, P. Hanrahan, I. MacDowall, and M Bolas. "The Two-User Responsive Workbench: Support for Collaboration Through Individual Views of a Shared Space." In *Proceedings of SIGGRAPH 97, Computer Graphics Proceedings, Annual Conference Series*, edited by Turner Whitted, pp. 327–332. Reading, MA: Addison Wesley, 1997.

[2] J. Allard, V. Gouranton, G. Lamarque, E. Melin, and B. Raffin. "Softgenlock: Active Stereo and GenLock for PC Clusters." In *Proceedings of Immersive Projection Technology and Eurographics Virtual Environments Workshop 2003 (IPT/EGVE'03)*, pp. 255–260. New York: ACM Press, 2003.

[3] The Augmented Reality Homepage. Available from World Wide Web (http://www.augmented-reality.org/), 2004.

[4] R. T. Azuma. "A Survey of Augmented Reality." *Presence: Teleoperators and Virtual Environments.* 6:4 (1997), 355–385.

[5] Ronald Azuma, Yohan Baillot, Reinhold Behringer, Steven Feiner, Simon Julier, and Blair MacIntyre. "Recent Advances in Augmented Reality." *IEEE Computer Graphics and Applications* 21:6 (2001), 34–47.

[6] D. Bandyopadhyay, R. Raskar, and H. Fuchs. "Dynamic Shader Lamps: Painting on Real Objects." In *International Symposium on Augmented Reality (ISAR)*, pp. 89–102. New York: ACM Press, 2001.

[7] Eric A. Bier, Maureen C. Stone, Ken Pier, William Buxton, and Tony D. DeRose. "Toolglass and Magic Lenses: The See-Through Interface." In *Proceedings of SIGGRAPH 93, Computer Graphics Proceedings, Annual Conference Series*, edited by James T. Kajiya, pp. 73–80. New York: ACM Press, 1993.

[8] O. Bimber, L. M. Encarnação, and D. Schmalstieg. *Augmented Reality with Back-Projection Systems using Transflective Surfaces. Computer Graphics Forum (Proceedings of EUROGRAPHICS 2000 - EG'2000* 19:3 (2000), 161–168.

[9] O. Bimber, B. Fröhlich, D. Schmalstieg, and L. M. Encarnação. *The Virtual Showcase. IEEE Computer Graphics & Applications* 21:6 (2001), 48–55.

[10] O. Bimber, L. M. Encarnação, and P. Branco. "The Extended Virtual Table: An Optical Extension for Table-Like Projection Systems." *Presence: Teleoperators and Virtual Environments* 10:6 (2001), 613–631.

[11] O. Bimber "Interactive Rendering for Projection-Based Augmented Reality Displays." Ph.D. Diss., University of Technology Darmstadt, 2002. Available from World Wide Web (http://elib.tu-darmstadt.de/diss/000270/)

[12] O. Bimber and B. Fröhlich. "Occlusion Shadows: Using Projected Light to Generate Realistic Occlusion Effects for View-Dependent Optical See-Through Displays." In *Proceedings of International Symposium on Mixed and Augmented Reality (ISMAR'02)*, pp. 186–195. Los Alamitos, CA: IEEE Press, 2002.

[13] O. Bimber, S. M. Gatesy, L. M. Witmer, R. Raskar and L. M. Encarnação. "Merging Fossil Specimens with Computer-Generated Information." *IEEE Computer* 35:9 (2002), 45–50.

[14] O. Bimber, L. M. Encarnação and D. Schmalstieg. "The Virtual Showcase as a New Platform for Augmented Reality Digital Storytelling." In *Proceedings of Eurographics Workshop on Virtual Environemnts IPT/EGVE'03*, pp. 87–95. New York: ACM Press, 2003.

[15] O. Bimber, A. Grundhöfer, G. Wetzstein, G., and S. Knödel. "Consistent Illumination within Optical See-Through Augmented Environments." In *Proceedings of IEEE/ACM International Symposium on Mixed and Augmented Reality (ISMAR'03)*, pp. 198–207. Los Alamitos, CA: IEEE Press, 2003.

[16] O. Bimber, B. Fröhlich, D. Schmalstieg, and L. M. Encarnação. "Real-Time View-Dependent Image Warping to Correct Non-Linear Distortion for Curved Virtual Showcase Displays." *Computers and Graphics - The International Journal of Systems and Applications in Computer Graphics* 27:4 (2003), 512–528.

[17] O. Bimber and R. Raskar. *Modern Approaches to Augmented Reality.* Conference Tutorial Eurographics, 2004.

[18] O. Bimber "Combining Optical Holograms with Interactive Computer Graphics." *IEEE Computer* 37:1 (2004), 85–91.

[19] O. Bimber, F. Coriand, A. Kleppe, E. Bruns, S. Zollmann, and T. Langlotz. "Superimposing Pictorial Artwork with Projected Imagery." *IEEE MultiMedia* 12:1 (2005), 16–26.

[20] O. Bimber, A. Emmerling, and T. Klemmer. "Embedded Entertainment with Smart Projectors." *IEEE Computer*, 38:1 (2005), 48–55.

[21] J. F. Blinn and M. E. Newell. "Texture and Reflection in Computer Generated Images." *Communications of the ACM* 19 (1976), 542–546.

[22] S. Boivin and A. Gagalowicz. "Image-Based Rendering of Diffuse, Specular and Glossy Surfaces from a Single Image." In *Proceedings of SIGGRAPH 2001, Computer Graphics Proceedings, Annual Conference Series*, edited by E. Fiume, pp. 107–116. Reading, MA: Addison-Wesley, 2001.

[23] R. P. Brent. *Algorithms for Minimization without Derivatives.* Engelwood Cliffs, NJ: Prentice-Hall, 1973.

[24] D. E. Breen, R. T. Whitaker, E. Rose, and M. Tuceryan. "Interactive Occlusion and Automatic Object Placement for Augmented Reality." *Computer Graphics Forum (Proceedings of EUROGRAPHICS '96)* 15:3 (1996), C11–C22.

[25] D. C. Brown. "Close-Range Camera Calibration." *Photogrammetric Engineering* 37:8 (1971), 855–866.

[26] J. H. Burroughes. "Light Emitting Polymers." *Nature* (1990) 347.

[27] P. J. Burt and E. H. Adelson. "A Multiresolution Spline with Applications to Image Mosaic." *ACM Trans. on Graphics* 2:4 (1983), 217–236.

[28] Canesta Corporation. Miniature Laser Projector Keyboard. Available from World Wide Web (http://www.canesta.com/), 2002

[29] Shenchang Eric Chen and Lance Williams. "View Interpolation for Image Synthesis." In *Proceedings of SIGGRAPH 93, Computer Graphics Proceedings, Annual Conference Series*, edited by James T. Kajiya, pp. 1279–288. New York: ACM Press, 1993.

[30] W. Chen, H. Towles, L. Nylard, G. Welch, and H. Fuchs. "Toward a Compelling Sensation of Telepresence: Demonstrating a Portal to a Distant (Static) Office." In *Proceedings of IEEE Visualization*, pp. 327–333. Los Alamitos, CA: IEEE Press, 2000.

[31] Y. Chen, D. Clark, A. Finkelstein, T. Housel, and K. Li. "Automatic Alignment of High Resolution Multi-Projector Displays using an Un-Calibrated Camera." In *Proceedings of IEEE Visualization*, pp. 125–130. Los Alamitos, CA: IEEE Press, 2000.

[32] D. Cotting, M. Naef, M. Gross, H. and Fuchs. "Embedding Imperceptible Patterns into Projected Images for Simultaneous Acquisition and Display." In *Proceedings of IEEE/ACM International Symposium on Mixed and Augmented Reality (ISMAR'04)*, pp. 100–109. Los Alamitos, CA: IEEE Press, 2004.

[33] A. Criminisi, I. Reid, and A. Zisserman. "Duality, Rigidity and Planar Parallax." In *Proceedings of the European Conference on Computer Vision*, Lecture Notes in Computer Science, 1407, pp. 846–861. Berlin: Springer-Verlag, 1998.

[34] G. Cross and A. Zisserman. "Quadric Surface Reconstruction from Dual-Space Geometry." In *Proceedings of 6th International Conference on Computer Vision(Bombay, India)*, pp. 25–31. Los Alamitos, CA: IEEE Press, 1998.

[35] C. Cruz-Neira, D. Sandin, T. DeFanti. "Surround-Screen Projection-Based Virtual Reality: The Design and Implementation of the CAVE." In *Proceedings of SIGGRAPH 93, Computer Graphics Proceedings, Annual Conference Series*, edited by James T. Kajiya, pp. 135–142. New York: ACM Press, 1993.

[36] M. Czernuszenko, D. Pape, D. Sandin, T. DeFanti, L. Dawe, and M. Brown. "The ImmersaDesk and InfinityWall Projection-Based Virtual Reality Displays." *Computer Graphics*, 31:2 (1997) 46–52.

[37] P. E. Debevec, Yizhou Yu, and George D. Borshukov. "Efficient Viewdependent Image-Based Rendering with Projective Texture-Mapping." In *Eurographics Rendering Workshop*, pp. 105–116. Berlin: Springer-Verlag, 1998.

[38] J. Dorsey, F. X. Sillion, and D. P. Greenberg. "Design and Simulation of Opera Lighting and Projection Effects." In *Proceedings of SIGGRAPH 95, Computer Graphics Proceedings, Annual Conference Series*, edited by Robert Cook, pp. 41–50. Reading, MA: Addison Wesley, 1995.

[39] E. A. Downing, L. Hesselink, J. Ralston, and R. A. Macfarlane. "Three-Color, Solid-State Three-Dimensional Display." *Science* 273 (1996), 1185–1189.

[40] F. Dreesen, and G. von Bally. "Color Holography in a Single Layer for Documentation and Analysis of Cultural Heritage." In *Optics within Life Sciences (OWLS IV)*, pp. 79–82. Berlin: Springer Verlag 1997.

[41] F. Dreesen, H. Deleré, and G. von Bally. "High Resolution Color-Holography for Archeological and Medical Applications." In *Optics within Life Sciences (OWLS V)*, pp. 349–352. Berlin: Springer Verlag, 2000.

[42] G. Drettakis, F. Sillion. "Interactive Update of Global Illumination using a Line-Space Hierarchy." In *Proceedings of SIGGRAPH 97, Computer Graphics Proceedings, Annual Conference Series*, edited by Turner Whitted, pp. 57–64. Reading, MA: Addison Wesley, 1997.

[43] Evans and Sutherland Web Page. Available from World Wide Web (http://www.es.com/), 2002.

[44] O. Faugeras. *Three-Dimensional Computer Vision: A Geometric Viewpoint.* Cambridge, MA: MIT Press, 1993.

[45] G. E. Favalora, D. M. Hall, M. Giovinco, J. Napoli, and R. K. Dorval, "A Multi-Megavoxel Volumetric 3-D Display System for Distributed Collaboration." In *Proceedings of IEEE GlobeCom 2000 Workshop, Applications of Virtual Reality Technologies for Future Telecommunication Systems.* Los Alamitos, CA: IEEE Press, 2000.

[46] S. Feiner, B. MacIntyre, Marcus Haupt, and Eliot Solomon. "Windows on the World: 2D Windows for 3D Augmented Reality." In *Proceedings of ACM Symposium on User Interface Software and Technology*, pp. 145–155. New York: ACM Press, 1993.

[47] R. Fernando and M. Kilgard. *Cg Tutorial: The Definitive Guide to Programmable Real-Time Graphics.* Reading, MA: Addison Wesley Professional, 2003.

[48] M. Figl, W. Birkfellner, C. Ede, J. Hummel, R. Hanel, F. Watzinger, F. Wanschitz, R. Ewers, and H. Bergmann. "The Control Unit for a Head Mounted Operating Microscope used for Augmented Reality Visualization in Computer Aided Surgery." In *Proceedings of International Symposium on Mixed and Augmented Reality (ISMAR'02)*, pp. 69–76. Los Alamitos, CA: IEEE Press, 2002.

[49] G. W. Fitzmaurice. "Situated Information Spaces and Spatially Aware Palmtop Computer." *CACM* 35:7 (1993), pp. 38–49.

[50] A. Fournier, A. S. Gunawan, and C. Romanzin. "Common Illumination between Real and Computer Generated Scenes." In *Proceedings of Graphics Interface'93*, pp. 254–262. Wellesley, MA: A K Peters, Ltd., 1993.

[51] J. Fründ, C. Geiger, M. Grafe, and B. Kleinjohann. "The Augmented Reality Personal Digital Assistant." In *Proceedings of International Symposium on Mixed Reality.* Los Alamitos, CA: IEEE Press, 2001.

[52] H. Fuchs, S. M. Pizer, L. C. Tsai, and S. H. Bloombreg. "Adding a True 3-D Display to a Raster Graphics System." *IEEE Computer Graphics and Applications* 2:7 (1982), 73–78.

[53] J. Gausemeier, J. Fruend, C. Matysczok, B. Bruederlin, and D. Beier". Development of a Real Time Image Based Object Recognition Method for Mobile AR-Devices." In *Proceedings of International Conference on Computer Graphics, Virtual Reality, Visualisation and Interaction in Africa*, pp. 133–139. New York: ACM Press, 2003.

[54] C. Geiger, B. Kleinjohann, C. Reimann, and D. Stichling. "Mobile Ar4All." In *Proceedings of IEEE and ACM International Symposium on Augmented Reality*, pp. 181–182. Los Alamitos, CA: IEEE Press, 2001.

[55] S. Gibson and A. Murta. "Interactive Rendering with Real-World Illumination." In *Proceedings of 11th Eurographics Workshop on Rendering, Brno, Czech Republic, June 2000*, pp. 365–376. Berlin: Springer-Verlag, 2000.

[56] G. Goebbels, K. Troche, M. Braun, A. Ivanovic, A. Grab, K. v. Lübtow, R. Sader, F. Zeilhofer, K. Albrecht, and K. Praxmarer. "ARSyS-Tricorder - Entwicklung eines Augmented Reality Systems für die intraoperative Navigation in der MKG Chirurgie." In *Proceedings of 2. Jahrestagung der Deutschen Gesellschaft für Computer- und Roboterassistierte Chirurgie e.V., Nürnberg*, 2004.

[57] C. M. Goral, K. E. Torrance, D. P. Greenberg, and B. Battaile. "Modeling the Interaction of Light between Diffuse Surfaces." *Computer Graphics, (Proc. SIGGRAPH '84)* 18:3 (1984), 212–222.

[58] Steven J. Gortler, Radek Grzeszczuk, Richard Szeliski, and Michael F. Cohen. "The Lumigraph." In *Proceedings of SIGGRAPH 96, Computer Graphics Proceedings, Annual Conference Series*, edited by Holly Rushmeier, pp. 43–54, Reading, MA: Addison Wesley,

[59] M.D. Grossberg, H. Peri, S. K. Nayar, and P. Belhumeur. "Making One Object Look Like Another: Controlling Appearance Using a Projector-Camera System." In *Proc. of IEEE Conference on Computer Vision and Pattern Recognition (CVPR '04)*, vol. 1, pp. 452–459. Los Alamitos, CA: IEEE Press, 2004.

[60] J. Guehring. "Dense 3D Surface Acquisition by Structured Light using Off-the-Shelf Components." In *Proceedings of SPIE: Videometrics and Optical Methods for 3D Shape Measuring* 4309, (2001), 220–231.

[61] P. Heckbert and P. Hanrahan. "Beam Tracing Polygonal Objects." *Computer Graphics (Proc. SIGGRAPH '84)* 18:3 (1984), 119–127.

[62] Ken Hinckley, Jeff Pierce, Mike Sinclar, and Eric Horvitz. "Sensing Techniques for Mobile Interaction." *ACM UIST CHI Letters* 2:2, (2000), 91–100.

[63] M. Hirose, T. Ogi, S. Ishiwata, and T. Yamada, T. "A Development of Immersive Multiscreen Displays (CABIN)." In *Proceedings of the VRSJ 2nd Conference*, pp. 137–140, 1997.

[64] Lars Erik Holmquist, Friedemann Mattern, Bernt Schiele, Petteri Alahuhta, Michael Beigl, and Hans-W. Gellersen. "Smart-Its Friends: A Technique for Users to Easily Establish Connections between Smart Artefacts." In *Proceedings of the 3rd International Conference on Ubiquitous Computing*, Lectures Notes in Computer Science 2201, pp. 273–291. Berlin: Springer-Verlag, 2001.

[65] H. Hoppe. "Progressive Meshes." In *Proceedings of SIGGRAPH 96, Computer Graphics Proceedings, Annual Conference Series*, edited by Holly Rushmeier, pp. 99–108. Reading, MA: Addison Wesley, 1996.

[66] H. Hoppe. "View-Dependent Refinement of Progressive Meshes." In *Proceedings of SIGGRAPH 97, Computer Graphics Proceedings, Annual Conference Series*, edited by Turner Whitted, pp. 189–197. Reading, MA: Addison Wesley, 1997.

[67] H. Hoppe. "Smooth View-Dependent Level-of-Detail Control and its Application to Terrain Rendering." In *Proceedings of IEEE Visualization'98*, pp. 35–42. Los Alamitos, CA: IEEE Press, 1998.

[68] W. E. Howard. "Organic Light Emitting Diodes (OLED)." In *OLED Technology Primer*. Available from World Wide Web (http://www.emagin.com/oledpri.htm), 2001. .

[69] HowardModels. Web Site. Available from World Wide Web (http://www.howardweb.com/model/), 2000.

[70] H. Hua, C. Gao, L. Brown, N Ahuja, and J. P. Rolland. "Using a Head-Mounted Projective Display in Interactive Augmented Environments." In *Proceedings of IEEE and ACM International Symposium on Augmented Reality 2001*, pp. 217–223. Los Alamitos, CA: IEEE Press, 2001.

[71] G. Humphreys and P Hanrahan. "A Distributed Graphics System for Large Tiled Displays." In *Proceedings of IEEE Visualization '99*, pp. 215–233. Los Alamitos, CA: IEEE Press, 1999.

[72] Raymond L. Idaszak, Jr. W. Richard Zobel, and David T. Bennett. Alternate Realities, now Elumens. Systems, Methods and Computer Program Products for Converting Image Data to Nonplanar Image Data. US Patent 6,104,405, 1997.

[73] M. Inami, N. Kawakami, D. Sekiguchi, Y. Yanagida, T. Maeda, and S. Tachi. "Visuo-Haptic Display Using Head-Mounted Projector." In *Proceedings of IEEE Virtual Reality 2000*, pp. 233–240. Los Alamitos, CA: IEEE Press, 2000.

[74] IO2 Technology. "The Heliodisplay." Available from World Wide Web (http://www.io2technology.com/), 2005.

[75] Jenoptik AG. Laser Projector. Available from World Wide Web (http://www.jenoptik.com/), 2002.

[76] J. T. Kajiya. "The Rendering Equation." *Computer Graphics (Proc. SIGGRAPH '86)* 20:4 (1986), 143–151.

[77] William A. Kelly, Lee T. Quick, Edward M. Sims, and Michael W. Tackaberry. "Polygon Fragmentation Method of Distortion Correction in Computer Image Generating Systems." US Patent 5,319,744, General Electric Company, USA, 1994. .

[78] R. Kijima and T. Ojika. "Transition between Virtual Environment and Workstation Environment with Projective Head-Mounted Display." In *Proceedings of IEEE Virtual Reality Annual International Symposium*, pp. 130–137. Los Alamitos, CA: IEEE Press, 1997.

[79] K. Kiyokawa, Y. Kurata, and H. Ohno. "An Optical See-Through Display for Mutual Occlusion of Real and Virtual Environments." In *Proceedings of IEEE & ACM ISAR 2000*, pp. 60–67. Los Alamitos, CA: IEEE Press, 2000.

[80] M. A. Klug. "Scalable Digital Holographic Displays." In *IS&T PICS 2001: Image Processing, Image Quality, Image Capture Systems Conference*, pp. 26–32. Springfield, VA: IS&T, 2001.

[81] K. C. Knowlton. "Computer Displays Optically Superimpose on Input Devices." *Bell Systems Technical Journal* 53:3 (1977) 36–383.

[82] J. S. Kollin, S. A. Benton, and M. L. Jepsen. "Real-Time Display of 3-D Computer Holograms by Scanning the Image of an Acousto-Optic Mudulator." In *Proceedings of SPIE, Holographic Optics II: Principle and Applications*, 1136, pp. 178–185. Bellingham, WA: SPIE, 1989.

[83] J. Kollin. "A Retinal Display For Virtual-Environment Applications." In *Proceedings of SID International Symposium, Digest Of Technical Papers* pp. 827. San Jose, CA: Society for Information Display, 1993.

[84] W. Krueger and B. Fröhlich. "The Responsive Workbench." *IEEE Computer Graphics and Applications* 14:3 (1994), 12–15.

[85] W. Krueger, C. A. Bohn, B. Fröhlich, H. Schüth, W. Strauss, and G. Wesche. "The Responsive Workbench: A Virtual Work Environment." *IEEE Computer* 28:7 (1995), 42–48.

[86] Laser Magic Productions. "Holograms, Transparent Screens, and 3D Laser Projections." Available from World Wide Web (http://www.laser-magic.com/transscreen.html), 2003.

[87] V. Lepetit, L. Vacchetti, D. Thalmann, D. and P. Fua. "Real-Time Augmented Face." In *Proceedings of International Symposium on Mixed and Augmented Reality (ISMAR'03)*, pp. 346–347. Los Alamitos, CA: IEEE Press, 2003.

[88] Marc Levoy and Pat Hanrahan. "Light Field Rendering." In *Proceedings of SIGGRAPH 96, Computer Graphics Proceedings, Annual Conference Series*, edited by Holly Rushmeier, pp. 31–42. Reading, MA: Addison Wesley, 1996.

[89] J. R. Lewis. "In the Eye of the Beholder." *IEEE Spectrum* (May 2004), 16–20.

[90] P. Lindstrom, D. Koller, W. Ribarsky, L. Hughes, N. Faust, and G. Turner. "Realtime, Continuous Level of Detail Rendering for Height Fields." In *Proceedings of SIGGRAPH 96, Computer Graphics Proceedings, Annual Conference Series*, edited by Holly Rushmeier, pp. 109–118. Reading, MA: Addison Wesley, 1996.

[91] H. C. Longuet-Higgins. "A Computer Algorithm for Reconstructing a Scene from Two Projections." *Nature* 293 (1981), 133–135.

[92] C. Loscos, G. Drettakis, and L. Robert. "Interactive Virtual Relighting of Real Scenes." *IEEE Transactions on Visualization and Computer Graphics* 6:3 (2000) 289–305.

[93] K. Low, G. Welch, A. Lastra, and H. Fuchs. "Life-Sized Projector-Based Dioramas." In *ACM Symposium on Virtual Reality Software and Technology, Banff, Alberta, Canada, November 2001*, pp. 161–168. New York: ACM Press, 2001.

[94] M. Lucente. "Interactive Computation of Holograms using a Look-Up Table." *Journal of Electronic Imaging* 2:1 (1993), 28–34.

[95] M. Lucente and A. Tinsley. "Rendering Interactive Images." In *Proceedings of SIGGRAPH 95, Computer Graphics Proceedings, Annual Conference Series*, edited by Robert Cook, pp. 387–397. Reading, MA: Addison Wesley, 1995.

[96] M. Lucente. "Interactive Three-Dimensional Holographic Displays: Seeing the Future in Depth." In *Computer Graphics (SIGGRAPH Special issue on Current, New, and Emerging Display Systems)*, 31:2 (1997), 63–67.

[97] P. Lyon. "Edge-Blending Multiple Projection Displays On A Dome Surface To Form Continuous Wide Angle Fields-of-View." In *Proceedings of 7th I/ITEC*, pp. 203–209. Arlington, VA: NTSA, 1985.

[98] Hongshen Ma and Joseph A. Paradiso. "The FindIT Flashlight: Responsive Tagging Based on Optically Triggered Microprocessor Wakeup." In *Proceedings of the 4th International Conference on Ubiquitous Computing*, Lecture Notes in Computer Science 2498, pp. 160–167. Berlin: Springer-Verlag, 2002.

[99] A. Majumder, Z. He, H. Towles, and G. Welch. "Color Calibration of Projectors for Large Tiled Displays." In *Proceedings of IEEE Visualization 2000*, pp. 102–108. Los Alamitos, CA: IEEE Press, 2000.

[100] A. Majumder, Z. He, H. Towles, and G. Welch. "Achieving Color Uniformity Across Multi-Projector Displays." In *Proceedings of IEEE Visualization 2000*, pp. 117–124. Los Alamitos, CA: IEEE Press, 2000.

[101] A. Majumder and R. Stevens. "Color Non-Uniformity in Projection Based Displays: Analysis and Solutions." *IEEE Transactions on Visualization and Computer Graphics* 10:2 (2004), 100–111.

[102] A. Majumder and G. Welch. "Computer Graphics Optique: Optical Superposition of Projected Computer Graphics." In *Proceedingas of the Fifth Immersive Projection Technology Workshop, in conjunction with the Seventh Eurographics Workshop on Virtual Environments, Stuttgart, Germany* pp. 81–88. Vienna: Springer-Verlag, 2001.

[103] D. Marr and T. Poggio. "A Computational Theory of Human Stereo Vision." *Proceedings of the Royal Society of London* 204 (1979), 301–328.

[104] T. H. Massie, and J. K. Salisbury. "The PHANTOM Haptic Interface: A Device for Probing Virtual Objects." *Proceedings of the ASME Dynamic Systems and Control Division* 55:1 (1994), 295–301.

[105] N. Max. "SIGGRAPH '84 Call for Omnimax Films." *Computer Graphics* 16:4 (1982) 208–214.

[106] S. McKay, S. Mason, L. S. Mair, P. Waddell, and M. Fraser. "Membrane Mirror-Based Display For Viewing 2D and 3D Images." In *Proceedings of SPIE* 3634 (1999) 144–155.

[107] S. McKay, S. Mason, L. S. Mair, P. Waddell, and M. Fraser. "Stereoscopic Display using a 1.2-M Diameter Stretchable Membrane Mirror." In *Proceedings of SPIE* 3639 (1999), 122–131.

[108] L. McMillan and G. Bishop. "Plenoptic Modeling: An Image-Based Rendering System." In *Proceedings of SIGGRAPH 95, Computer Graphics Proceedings, Annual Conference Series*, edited by Robert Cook, pp. 39–46. Reading, MA: Addison Wesley, 1995.

[109] P. Milgram and F. Kishino. "A Taxonomy of Mixed Reality Visual Displays." *IEICE Transactions on Information Systems E77-D* 12 (1994), 1321–1329.

[110] P. Milgram, H. Takemura et al. "Augmented Reality: A Class of Displays on the Reality-Virutality Continuum." In *Proceedings of SPIE: Telemanipulator and Telepresence Technologies* 2351 (1994), 282–292.

[111] D. Mitchell and P. Hanrahan. "Illumination from Curved Reflectors." *Computer Graphics (Proc. SIGGRAPH '92)* 26:2 (1992), 283–291.

[112] M. Moehring, C. Lessig, and O. Bimber. "Video See-Through AR on Consumer Cell-Phones." In *Proceedings of ISMAR'04*. Los Alamitos, CA: IEEE Press, 2004. See also "Optical Tracking and Video See-Through AR on Consumer Cell Phones." In *Proceedings of Workshop on Virtual and Augmented Reality of the GI-Fachgruppe AR/VR*, pp. 252–253. Aachen, Germany: Shaker-Verlag, 2004.

[113] T. Möller and B. Trumbore. "Fast, Minimum Storage Ray-Triangle Intersection." *journal of graphics tools* 2:1 (1997), 21–28.

[114] T. Naemura, T. Nitta, A. Mimura, and H. Harashima. "Virtual Shadows – Enhanced Interaction in Mixed Reality Environments." In *Proceedings of IEEE Virtual Reality (IEEE VR'02)*, pp. 293–294. Los Alamitos, CA: IEEE Press, 2002.

[115] E. Nakamae, K. Harada, T. Ishizaki, and T. Nishita. "A Montage Method: The Overlaying of Computer Generated Images onto Background Photographs." *Computer Graphics (Proc. SIGGRAPH 86)* 20:4 (1986), 207–214.

[116] Shree K. Nayar, Harish Peri, Michael D. Nayar, and Peter N. Belhumeur. "A Projection System with Radiometric Compensation for Screen Imperfections." In *Proc. of International Workshop on Projector-Camera Systems*, Los Alamitos, CA: IEEE Press, 2003.

[117] J. Neider, T. Davis, and M. Woo. *OpenGL Programming Guide.* Reading, MA: Addison-Wesley Publ., 1993.

[118] J. Neider, T. Davis, and M. Woo. *OpenGL Programming Guide. Release 1*, pp. 157–194. Reading, MA: Addison Wesley, 1996.

[119] W. Newman and R. Sproull. *Principles of Interactive Computer Graphics.* New York: McGraw-Hill, 1973.

[120] Franck Nielsen, Kim Binsted, Claudio Pinhanez, Tatsuo Yotsukura, and Shigeo Morishima. "Hyper Mask - Talking Head Projected onto Real Object." *The Visual Computer* 18:2 (2002), 111–120, 2002.

[121] S. Noda, Y. Ban, K. Sato, and K. Chihara. "An Optical See-Through Mixed Reality Display with Realtime Rangefinder and an Active Pattern Light Source." *Transactions of the Virtual Reality Society of Japan* 4:4 (1999), 665–670.

[122] E. Ofek and A. Rappoport. "Interactive Reflections on Curved Objects." In *Proceedings of SIGGRAPH 98, Computer Graphics Proceedings, Annual Conference Series*, edited by Michael Cohen, pp. 333–342. Reading, MA: Addison Wesley, 1998.

[123] E. Ofek. "Interactive Rendering of View-Dependent Global Lighting Phenomena." Ph.D. Diss., Hebrew University (Israel), 1999.

[124] T. Ogi, T. Yamada, K. Yamamoto, and M. Hirose. "Invisible Interface for Immersive Virtual World." In *Proceedings of the Immersive Projection Technology Workshop (IPT'01)*, pp. 237–246. Vienna: Springer-Verlag, 2001.

[125] O. Omojola, E. R. Post, M. D. Hancher, Y. Maguire, R. Pappu, B. Schoner, P. R. Russo, R. Fletcher, and N. Gershenfeld. "An Installation of Interactive Furniture." *IBM Systems Journal* 39:3/4 (2000), 861–879.

[126] Origin Instruments Corporation. Available from World Wide Web (http://www.orin.com/), 2002.

[127] J. O'Rourke. *Art Gallery Theorems and Algorithms.* New York: Oxford University Press, 1987.

[128] Panoram Technologies, Inc. Available from World Wide Web (http://www.panoramtech.com/), 2002.

[129] T. E. Parks. "Post Retinal Visual Storage." *American Journal of Psychology* 78 (1965), 145–147.

[130] J. Parsons and J. P. Rolland. "A Non-Intrusive Display Technique for Providing Real-Time Data within a Surgeons Critical Area of Interest." In *Proceedings of Medicine Meets Virtual Reality'98*, pp. 246–251. Amsterdam: IOS Press, 1998.

[131] W. Pasman and C. Woodward. "Implementation of an Augmented Reality System on a PDA." In *Proceedings of IEEE and ACM International Symposium on Mixed and Augmented Reality*, pp. 276–277. Los Alamitos, CA: IEEE Press, 2003.

[132] Shwetak N. Patel and Gregory D. Abowd. "A 2-Way Laser-Assisted Selection Scheme for Handhelds in a Physical Environment." In *Proceedings of the 5th International Conference on Ubiquitous Computing*, Lecture Notes in Computer Science 2864, pp. 200–207. Berlin: Springer-Verlag, 2003.

[133] E. D. Patrick, A. Cosgrove, J. Slavkovic, A. Rode, T. Verratti, and G. Chiselko. "Using a Large Projection Screen as an Alternative to Head-Mounted Displays for Virtual Environments." In *Proceedings of Conference on Human Factors in Computer Systems*, pp. 479–485. New York: ACM Press, 2000.

[134] James Patten, Hiroshi Ishii, and Gian Pangaro. "Sensetable: A Wireless Object Tracking Platform for Tangible User Interfaces." In *Conference on Human Factors in Computing Systems*, pp. 253–260. New York, ACM Press, 2001.

[135] L. D. Paulson. "Displaying Data in Thin Air." *IEEE Computer* (March 2004), 19.

[136] K. Perlin, S. Paxia, and J. S. Kollin. "An Autostereoscopic Display."In *Proceedings of SIGGRAPH 2000, Computer Graphics Proceedings, Annual Conference Series*, edited by Kurt Akeley, pp. 1319–326. Reading, MA: Addison-Wesley, 2000.

[137] C. Petz and M. Magnor. "Fast Hologram Synthesis for 3D Geometry Models using Graphics Hardware." In *Proceedings of SPIE'03, Practical Holography XVII and Holographic Materials IX* 5005 (2003), 266–275.

[138] C. Pinhanez. "The Everywhere Displays Projector: A Device to Create Ubiquitous Graphical Interfaces." In *Proceedings of the 3rd International Conference on Ubiquitous Computing*, Lecture Notes in Computer Science 2201, pp. 315–331. Berlin: Springer-Verlag, 2001.

[139] C. Pinhanez. "Using a Steerable Projector and a Camera to Transform Surfaces into Interactive Displays." In *Conference on Human Factors in Computing Systems*, pp. 369–370. New York: ACM Press, 2001.

[140] B. Piper, C. Ratti, and H Ishii. "Illuminating Clay: A 3-D Tangible Interface for Landscape Analysis." In *Conference on Human Factors in Computing Systems*, pp. 355–362. New York: ACM Press, 2002.

[141] Polhemus. Available from World Wide Web (http://www.polhemus.com/), 2002.

[142] T. Poston and L. Serra. "The Virtual Workbench: Dextrous VR." In *Proceedings of Virtual Reality Software an Technology (VRST'94)*, pp. 111–121. River Edge, NJ: World Scientific Publishing, 1994.

[143] PowerWall. Available from World Wide Web (http://www.lcse.umn.edu/research/powerwall/powerwall.html), June 2002.

[144] W. H. Press, S. A. Teukolsky, W. T. Vetterling, and B. P. Flannery. *Numerical Recipes in C - The Art of Scientific Computing*, Second edition, pp. 412–420. Cambridge, UK: Cambridge University Press, 1992.

[145] Homer L. Pryor, Thomas A. Furness, and E. Viirre. "The Virtual Retinal Display: A New Display Technology Using Scanned Laser Light." In *Proceedings of Human Factors and Ergonomics Society, 42nd Annual Meeting*, pp. 1570–1574. Santa Monica, CA: HFES, 1998.

[146] Pyramid Systems (now owned by FakeSpace). Available from World Wide Web (http://www.fakespacesystems.com/), 2002.

[147] R. Raskar, M. Cutts, G Welch, and W. Stürzlinger. "Efficient Image Generation for Multiprojector and Multisurface Displays." In *Proceedings of the Ninth Eurographics Workshop on Rendering*, pp. 139–144. Berlin: Springer-Verlag, 1998.

[148] R. Raskar, G. Welch, M. Cutts, A. Lake, L. Stesin, and H. Fuchs. "The Office of the Future: A Unified Approach to Image-Based Modeling and Spatially Immersive Displays." In *Proceedings of SIGGRAPH 98, Computer Graphics Proceedings, Annual Conference Series*, edited by Michael Cohen, pp. 179–188. Reading, MA: Addison Wesley, 1998.

[149] R. Raskar, G Welch, and H. Fuchs. "Seamless Projection Overlaps Using Image Warping and Intensity Blending." In *Fourth International Conference on Virtual Systems and Multimedia*. Amsterdam: IOS Press, 1998.

[150] R. Raskar, G. Welch, and H. Fuchs. "Spatially Augmented Reality." In *Proceedings of First IEEE Workshop on Augmented Reality (IWAR'98)*, pp. 63–72. Los Alamitos, CA: IEEE Press, 1998.

[151] R. Raskar, M. Brown, Y. Ruigang, W. Chen, G. Welch, H. Towles, B. Seales, and H. Fuchs. "Multiprojector Displays using Camera-Based Registration." In *Proceedings of IEEE Visualization*, pp. 161–168. Los Alamitos, CA: IEEE Press, 1999.

[152] R. Raskar, G. Welch, and W- C. Chen. "Table-Top Spatially Augmented Reality: Bringing Physical Models to Life with Projected Imagery." In *Proceedings of Second International IEEE Workshop on Augmented Reality (IWAR'99), San Francisco, CA*, pp. 64–71. Los Alamitos, CA: IEEE Press, 1999.

[153] R. Raskar. "Hardware Support for Non-Photorealistic Rendering." In *SIGGRAPH Eurographics Workshop on Graphics Hardware*, pp. 41–46. New York: ACM Press, 2001.

[154] R. Raskar and P. Beardsley. "A Self Correcting Projector." In *IEEE Computer Vision and Pattern Recognition (CVPR)*, pp. 89–102. Los Alamitos, CA: IEEE Press, 2001.

[155] R. Raskar, G. Welch, K. L. Low, and D. Bandyopadhyay. "Shader Lamps: Animating Real Objects with Image-Based Illumination." In *Proceedings of Eurographics Rendering Workshop*, pp. 89–102. Berlin: Springer-Verlag, 2001.

[156] R. Raskar, J. van Baar, and X. Chai. "A Lost Cost Projector Mosaic with Fast Registration." In *Fifth Asian Conference on Computer Vision*, pp. 161–168. Tokyo: Asian Federation of Computer Vision Societies, 2002.

[157] R. Raskar, R. Ziegler, and T. Willwacher. "Cartoon Dioramas in Motion." In *Proceedings of Int. Symp on Non-Photorealistic Animation and Rendering*, pp. 7–ff. New York; ACM Press, 2002.

[158] R. Raskar, J. van Baar, P. Beardsly, T. Willwacher, S. Rao, and C. Forlines. "iLamps: Geometrically Aware and Self-Configuring Projectors." *Transactions on Graphics (Proc. SIGGRAPH '03)* 22:3 (2003), 809–818.

[159] Ramesh Raskar, Paul Beardsley, Jeroen van Baar, Yao Wang, Paul Dietz, Johnny Lee, Darren Leigh, and Thomas Willwacher. "Rfig Lamps: Interacting with a Self-Describing World via Photosensing Wireless Tags and Projectors." *Transactions on Graph. (Proc. SIGGRAPH 04)* 23:3 (2004), 406–415.

[160] Jun Rekimoto, Brygg Ullmer, and Haro Oba. "DataTiles: A Modular Platform for Mixed Physical and Graphical Interactions." In *CHI 2001*, pp. 269–276. New York: ACM Press, 2001.

[161] Matthias Ringwald. "Spontaneous Interaction with Everyday Devices Using a PDA. In *Proceedings of the 4th International Conference on Ubiquitous Computing*, Lecture Notes in Computer Science 2498. Berlin: Springer-Verlag, 2002.

[162] J. P. Rolland, and T. Hopkins. "A Method of Computational Correction for Optical Distortion in Head-Mounted Displays." Technical Report. No. TR93-045, UNC Chapel Hill, Department of Computer Science, 1993.

[163] J. Rolland, H. Rich, and H. Fuchs. "A Comparison of Optical and Video See-Through Head-Mounted Displays." In *Proceedings of SPIE: Telemanipulator and Telepresence Technologies*, pp. 293–307. Bellingham, WA: SPIE, 1994.

[164] J. P. Rolland, Y. Baillot, and A. A. Goon. "A Survey of Tracking Technology for Virtual Environments." In *Fundamentals of Wearable Computers and Augmented Reality*, Chapter 3, pp. 67–112. Mahwah, NJ: Lawrence Erlbaum Assoc. Inc., 2001.

[165] D. Schmalstieg, A. Fuhrmann, G. Hesina, Zs. Szalavári, L. M. Encarnação, M. Gervautz, and W. Purgathofer. "The Studierstube Augmented Reality Project" *Presence: Teleoperators and Virtual Environments* 11:11 (2002), 32–54..

[166] C. Schmandt. "Spatial Input/Display Correspondence in a Stereoscopic Computer Graphics Workstation." *Computer Graphics (Proc. SIGGRAPH '83)* 17:3 (1983), 253–261.

[167] B. Schwald, H. Seibert, and T. A. Weller. "A Flexible Tracking Concept Applied to Medical Scenarios Using an AR Window." In *Proceedings of International Symposium on Mixed and Augmented Reality (ISMAR'02)*, pp. 261–262. Los Alamitos, CA: IEEE Press, 2002.

[168] M. Segal, C. Korobkin, R. van Widenfelt, J. Foran, and P. Haeberli. "Fast Shaddows and Lighting Effects Using Texture Mapping." *Computer Graphics (Proc. SIGGRAPH '92)* 26:2 (1992) 249–252.

[169] A. Shashua and S. Toelg. "The Quadric Reference Surface: Theory and Applications." *International Journal of Computer Vision (IJCV)* 23:2 (1997), 185–198.

[170] H. Shum and R. Szeliski. "Panoramic Image Mosaics." Technical Report MSR-TR-97-23, Microsoft Research, 1997.

[171] Siemens. Siemens Mini Beamer, 2002. On World Wide Web (http://w4.siemens.de/en2/html/press//newsdeskarchive/2002/ - foe02121b.html), 2002.

[172] M. Spanguolo. "Polyhedral Surface Decomposition Based on Curvature Analysis." In *Modern Geometric Computing for Visualization*, pp. 57–72. Berlin: Springer Verlag, 1992.

[173] D. Starkey and R. B. Morant. "A Technique for Making Realistic Three-Dimensional Images of Objects." *Behaviour Research Methods and Instrumentation* 15:4 (1983), 420–423.

[174] G. Stetten, V. Chib, D. Hildebrand, and J. Bursee. "Real Time Tomographic Reflection: Phantoms for Calibration and Biopsy." In *Proceedings of IEEE/ACM International Symposium on Augmented Reality (ISMAR'01)*, pp. 11–19. Los Alamitos, CA: IEEE Press, 2001.

[175] G. Stoll, M. Eldridge, D. Patterson, A. Webb, S. Berman, R. Levy, C. Caywood, M. Taveira, S. Hunt, and P. Hanrahan. "Lightning-2: A High-Performance Subsystem for PC Clusters." In *Proceedings of SIGGRAPH 2001, Computer Graphics Proceedings, Annual Conference Series*, edited by E. Fiume, pp. 141–148. Reading, MA: Addison-Wesley, 2001.

[176] D. Stricker, W. Kresse, J. F. Vigueras-Gomez, G. Simon, S. Gibson, J. Cook, P. Ledda, and A. Chalmers. *Photorealistic Augmented Reality*, Conference tutorial ISMAR 2003 (Second International Symposium on Mixed and Augmented Reality). Los Alamitos, CA: IEEE Press, 2003.

[177] W. Stuerzlinger. "Imaging All Visible Surface." In *Proceedings Graphics Interface*, pp. 115–122. Wellesley, MA: A K Peters, Ltd., 1999.

[178] R. Sukthankar, R. Stockton, and M. Mullin, M. "Automatic keystone Correction for Camera-Assisted Presentation Interfaces." In *Proceedings of International Conference on Multimodal Interfaces*, pp. 607–614. Berlin: Springer-Verlag, 2000.

[179] S. Sukthankar, T.J. Cham, and G. Sukthankar. "Dynamic Shadow Elimination for Multi-Projector Displays." In *Proceedings of IEEE Computer Vision and Pattern Recognition*, Vol. 2, pp. 151–157. Los Alamitos, CA: IEEE Press, 2001.

[180] A. Sullivan. "3-Deep: New Displays Render Images You Can Almost Reach Out and Touch." IEEE Spectrum (April 2005), 22–27.

[181] R. Surati. *Scalable Self-Calibrating Display Technology for Seamless Large-Scale Displays.* PhD Diss., Massachusetts Institute of Technology, 1999.

[182] I. E. Sutherland. "The Ultimate Display." *Proceedings of IFIP'65*, pp. 506–508, 1965.

[183] I. E. Sutherland. "A Head Mounted Three Dimensional Display." *Proceedings of the Fall Joint Computer Conference (AFIPS)* 33:1 (1968) 757–764.

[184] Symbol. Laser Projection Display. Available from World Wide Web (http://www.symbol.com/products/oem/lpd.html), 2002.

[185] Richard Szeliski. "Video Mosaics for Virtual Environments." *IEEE Computer Graphics and Applications* 16:2 (1996), 22–30.

[186] R. Szeliski and H. Shum. "Creating Full View Panoramic Mosaics and Environment Maps." In *Proceedings of SIGGRAPH 97, Computer Graphics Proceedings, Annual Conference Series*, edited by Turner Whitted, pp. 251–258. Reading, MA: Addison Wesley, 1997.

[187] S. Teller, J. Chen, and H. Balakrishnan. "Pervasive Pose-Aware Applications and Infrastructure." *IEEE Computer Graphics and Applications* 23:4 (2003), 14–18.

[188] The CoolTown Project. Available from World Wide Web (http://www.cooltown.com/research/), 2001.

[189] A. Travis, F. Payne, J. Zhong, and J. Moore. "Flat Panel Display using Projection within a Wedge-Shaped Waveguide." Technical Report, Department of Engineering, University of Cambridge, UK, 2002. Available from World Wide Web (http://ds.dial.pipex.com/cam3d/technology/technology01.html) 2002.

[190] A. C. Traub. "Stereoscopic Display Using Varifocal Mirror Oscillations." *Applied Optics* 6:6 (1967), 1085–1087.

[191] Trimension Systems Ltd. Available from World Wide Web (http://www.trimension-inc.com/), 2002.

[192] J. Underkoffler, B. Ullmer, and H. Ishii. "Emancipated Pixels: Real-World Graphics in the Luminous Room." In *Proceedings of SIGGRAPH 99, Computer Graphics Proceedings, Annual Conference Series*, edited by Alyn Rockwood, pp. 385–392. Reading, MA: Addison Wesley Longman, 1999.

[193] J. Vallino. The Augmented Reality Page. Available from World Wide Web (http://www.se.rit.edu/~jrv/research/ar/), 2004.

[194] D. Wagner and D. Schmalstieg. "First Steps towards Handheld Augmented Reality." In *Proceedings of International Conference on Wearable Computers*, pp. 127–137. Los Alamitos, CA: IEEE Press, 2003.

[195] Roy Want, Andy Hopper, Veronica Falc and Jonathan Gibbons. "The Active Badge Location System." *ACM Trans. Inf. Syst.* 10:1 (1992), 91–102.

[196] M. Walker. *Ghostmasters: A Look Back at America's Midnight Spook Shows.* Cool Hand Communications, 1994.

[197] R. Want, B. N. Schilit, N. I. Adams, R. Gold, K. Petersen, D. Goldberg, J. R. Ellis, and M. Weiser. "An Overview of the PARCTAB Ubiquitous Computing Experiment." In *IEEE Personal Communications*, pp. 28–43. Los Alamitos, CA: IEEE Press, 1995.

[198] Roy Want, B. L. Harrison, K.P Fishkin, and A. Gujar. "Bridging Physical and Virtual Worlds with Electronic Tags." In *ACM SIGCHI*, pp. 370–377. New York: ACM Press, 1999.

[199] C. Ware, K. Arthur, et al. "Fish Tank Virtual Reality." In *Proceedings of Inter CHI'93 Conference on Human Factors in Computing Systems*, pp. 37–42. New York: ACM Press, 1993.

[200] B. Watson and L. Hodges. "Using Texture Maps to Correct for Optical Distortion in Head-Mounted Displays." In *Proceedings of IEEE VRAIS'95*, pp. 172–178. Los Alamitos, CA: IEEE Press, 1995.

[201] Greg Welch and Gary Bishop. "SCAAT: Incremental Tracking with Incomplete Information." In *Proceedings of SIGGRAPH 97, Computer Graphics Proceedings, Annual Conference Series*, edited by Turner Whitted, pp. 333–344. Reading, MA: Addison Wesley, 1997.

[202] Pierre Wellner. "Interacting with Paper on the DigitalDesk." *Communications of the ACM* 36:7 (1993), 86–97.

[203] Y. Wexler and A. Shashua. "Q-Warping: Direct Computation of Quadratic Reference Surfaces." In *IEEE Conf. on Computer Vision and Pattern Recognition (CVPR), June, 1999*, pp. 1333–1338. Los Alamitos, IEEE Press, 1999.

[204] T. E. Wiegand, D. W. von Schloerb, and W. L. Sachtler. "Virtual Workbench: Near-Field Virtual Environment System with Applications." *Presence: Teleoperators and Virtual Environments* 8:5 (1999), 492–519.

[205] M. Wojceich and H. P. Pfister. "3D TV: A Scalable System for Real-Time Acquisition, Transmission, and Autostereoscopic Display of Dynamic Scenes." *Transactions on Graphics (Proc. SIGGRAPH 2004)* 23:3 (2004), 814–824.

[206] Xerox PARC. "Electronic Reusable Paper." Available from World Wide Web (http://www2.parc.com/dhl/projects/gyricon/), 2003.

[207] M. Yamaguchi, N. Ohyama, and T. Honda. "Holographic 3-D Printer." *Proceedings of SPIE V* 1212 (1990) 84.

[208] R. Yang, D. Gotz, J. Hensley, H. Towles, and M. S. Brown. "PixelFlex: A Reconfigurable Multi-Projector Display System." In *IEEE Visualization*, pp. 161–168. Los Alamitos, CA: IEEE Press, 2001.

[209] T. Yoshida, C. Horii, and K. Sato. "A Virtual Color Reconstruction System for Real Heritage with Light Projection." In *Proceedings of Virtual Systems and Multimedia*, pp. 158–164. San Francisco, CA: Virtual Heritage Network, Inc., 2003.

[210] Y. Yu, P. Debevec, J. Malik, and T. Hawkins. "Inverse Global Illumination: Recovering Reflectance Models of Real Scenes from Photographs." In *Proceedings of SIGGRAPH 99, Computer Graphics Proceedings, Annual Conference Series*, edited by Alyn Rockwood, pp. 215–224. Reading, MA: Addison Wesley Longman, 1999.

[211] Z. Zhang. "Flexible Camera Calibration by Viewing a Plane from Unknown Orientations." In *7th Int. Conference on Computer Vision*, pp. 666–673. Los Alamitos, CA: IEEE Press, 1999.

Index

aberration, 20
absolute optical system, 19
accelerated search techniques, 202
accelerometers, 5
accommodation, 32
accumulation buffers, 53
acquisition of depth information, 277
active
    shuttering, 38
    structured light technique, 127
adaptively refine, 193
adjustment values
    color, 263
    intensity, 263
alpha
    blending, 129
    mask, 129
ambient intelligent landscapes, 298
ambient term, 50
amplitude, 14, 244
anaglyphs, 37
angle
    of incidence, 18
    of reflection, 18
    of refraction, 18
antialiasing, 188
apex, 177
arbitrarily shaped surfaces, 267
arbitrary reflectance properties, 240
arbitrary, geometrically non-trivial, textured surfaces, 214
area of rejuvenation, 189
artistic filters, 313
attenuation factor, 49
Augmented Paintings, 12
augmented reality, 1
    screen-based, 84
authoring, 6
automatic keystone correction, 277

back buffer, 52
back-projection, 38
background image, 263
balanced load, 275
barycentric coordinates, 51
base mesh, 193
Being There project, 11
bi-linear
    interpolation, 191
    texture filtering, 176
binocular field of vision, 33
block-based simplification, 193
bodies of revolution, 203
boundary angles, 204
bounding sphere, 177, 198

calibration, 131
cathode ray tubes, 4
CAVE, 7
cell phones, 5, 79
centered on-axis, 158
ChromaDepth, 37
color
    bleeding, 231
    pigments, 255
color-blended, 188
colorimeters, 267
concave mirrors, 24

condition of Herschel, 26
cones, 31
consistent occlusion, 213
content
  generation, 301
  management, 301
continuous level-of detail transitions, 196
convergence, 32
convergent lenses, 29
convex mirrors, 22
coordinate system
  camera, 46
  Cartesian, 150
  clip, 46
  eye, 46
  normalized device, 46
  screen, 163
  window, 46
  world, 46
cornea, 31
correct occlusion effects, 220
critical angle, 18
cross-fading techniques, 276
crossfeathering, 227
cube mapping, 54, 243
curved optics, 175
curvilinear image warping, 151
cyberspace, 1

de-gamma mapping, 263
degree of curvature, 191
Delauny triangulation, 262
depth buffer, 52
desktop
  configurations, 38
  monitors, 38
desktop-window environments, 305
detector, 26
diffuse term, 50
digital
  holography, 247
  storytelling, 300
diplopia, 32
direct illumination effects, 240
discrete grid blocks, 193
displacement error, 192, 194
displacement map, 268
  pseudo pixel, 270
display
  air, 210
  area, 268
  autostereoscopic, 35
  BOOM-like, 36
  embedded screen, 39
  fog, 210
  goggle-bound, 35
  hand-held, 79
  head-attached, 36
  head-mounted, 4
    optical see-through, 74
    projective, 36, 76
    video see-through, 74
  holographic, 39
  information, 255
  large screen, 314
  near-field, 159
  parallax, 39
  parallax barrier, 42, 247
  portal, 96
  projection, 38
  pseudo 3D, 41
  re-imaging, 39
  real image, 40
  retinal, 36
  solid-state, 325
  spatial, 7, 36
    configurations, 5
    optical see-through, 149
  stereoscopic, 31
  surround screen, 39
    projection-based, 7
  technology, 5
  tiled screen, 274
  transmissive, 277
  ultimate, 2
  varifocal mirror, 41
  volumetric, 39
    multi-planar, 41
divergence, 32
divergent lenses, 30
double buffering, 52
drag-and-drop interactions, 309

dynamic
  correction functions, 231
  elimination of shadows, 277

edge collapse operations, 193
edge-free view, 174
electroholography, 44, 247
electromagnetic waves, 14
electronic paper, 325
electrons, 14
embedded
  information windows, 311
  multimedia content, 313
emission term, 50
energy state, 14
energy-flow, 235
environment light, 262
Everywhere Displays projector, 213
Extended Virtual Table, 11
extremal Fermat path, 202

film
  half-silvered, 152
  light directing, 152
filter times, 192
fine-grained retriangulation, 193
first rendering pass, 175
fish tank VR, 38
fixed focal length problem, 75
flat shading, 51
focal
  distance, 22
  length, 22
  point, 22
focus area, 312
force feedback, 305
form-factor, 235
fovea, 32
foveal field of vision, 33
frame buffer, 52
frame locking, 172
front buffer, 52
front surface mirrors, 158
front-projection, 38
full color anaglyph, 37
function pipeline, 44
  fixed, 45

gamma correction, 263, 273
GenLocking, 172
geometric optics, 66
ghost voxel problem, 325
global illumination, 5
  solution, 235
global or discrete LODs, 194
global positioning system, 5
Gouraud shading, 51, 240
gyroscopes, 5

hand-held, 71
hardware acceleration, 5
high black level, 274
higher light intensities, 274
hologram, 43
  color white-light, 245
  computer-generated, 44, 247
  copies, 307
  multiplex, 246
  optical, 43
  rainbow, 245
  reflection, 245
  white-light, 245
  white-light reflection, 245
holographic projection screens, 150
HoloStation, 11
homogeneous, 17
  media, 17
homographic relationships, 277

I/O bulbs, 213
iLamps, 11, 213
illumination, 213
image, 19
  composition technology, 169
  geometry, 178
  planes, 71
  space, 153
  space error, 198, 199
immersive, 7
implicit hierarchical LOD structure, 193
in-out refraction, 25, 159

in-refractions, 28
indexed triangle mesh, 178
inter-reflection, 5
interaction, 301
  devices, 6
  techniques, 6
interference
  fringe, 44
  pattern, 16
interocular distance, 32
interval techniques, 202
iris, 31
irradiance map, 238
irregular surface, 98
isotropic, 17

keystone deformations, 261

L-vertices, 195
Lambert's law, 271
Lambertian surfaces, 271
latency, 270
laws of refraction, 17
lectern-like setup, 309
lens, 31
  variable-focus liquid, 83
lenticular sheet display, 43, 247
light
  coherent, 15
  monochromatic, 15
  polarized, 16
  reflected, 16
  refracted, 16
  scattered, 16
  visible, 15
light emitting polymers, 69
light generators, 69
lighting
  direct, 231
  indirect, 231
  synthetic, 230
limited resolution, 273
linear texture interpolation, 201
liquid crystal display, 36
local contrast effects, 265
long-range tracking, 297
low dynamic range, 274
Luminous Room, 213

magnifying glass, 311
matrix
  intrinsic, 97
  model-view, 45
  perspective texture, 165
  projection, 45
  projection transformation, 46
  reflection, 154
  scene transformation, 46
  view transformation, 46
matrix stack, 56
micro-mobile, 284
mirror beam combiner, 152
  half-silvered, 149
  hand-held, 79
mirrors, 20
misregistration, 5
mixed convexity, 191
mobile AR applications, 5
monocular field of vision, 32
multi-focal projection techniques, 278
multi-plane beam combiner, 167
multi-texturing, 234
multidimensional optimization, 202
multiple viewers, 170
mutual occlusion, 221

negative parallax, 34
neutrons, 14
non-Lambertian specular reflections, 267
non-linear predistortion, 191
novel lamp designs, 278
nucleus, 14
numerical minimization, 203

object collision detection, 297
object registration process, 153
object wave, 44, 245
objects, 19
oblique
  projection, 112
  screen display, 39
observers' sweet-spot, 268

occlusion, 5
  shadows, 223
off-axis projection, 48
on-axis
  situation, 48
  viewing frustum, 177
opaque, 39
OpenGL, 45
optical
  combination, 71
  combiner, 69, 149
  path, 71
  see-through, 67
  system
    centered, 25
    on-axis, 25
  wavefronts, 244
optical axis, 25
organic light emitting diodes, 69
oriented convex hulls, 199
orthogonal projection, 112
osculation ellipsoid, 202
out-refraction, 26
overmodulated, 276

P-buffer, 53
painting style, 313
parabolic concave mirror, 189
parallax, 34
parameters
  extrinsic, 315
  intrinsic, 315
  vertex, 51
parametric functions, 180
paraxial
  analysis, 158
  rays, 160
Parks effect, 81
passive
  shuttering, 37
  stereoscopic projection, 151
PC clusters, 172
Pepper's ghost configurations, 40
per-pixel illumination, 214
per-vertex
  level, 175
  viewpoint, 175
personal digital assistant, 5, 79
perspective
  division, 46
  N-point problem, 223
phantoms, 242
phase, 14, 244
Phong shading, 51, 240
photometric emulsion, 244
photons, 14
photoreceptors, 31
pictorial artwork, 255
pixel-precise illumination, 299
planar lens, 25
plane of incidence, 17
planetary model, 14
polarization glasses, 37
polarized, 16
polygon order, 155
positive parallax, 34
presentation forms, 300
presentation tools and techniques, 310
principle of the optical path length, 17
programmable pipelines, 6
progressive meshes, 193
projective textures, 54
projector
  light, 221
  mobile, 278
projector-based illumination, 213, 221
protons, 14
Pulfrich effect, 37

quad buffers, 52
quadtree, 193
  data structure, 195
quantum leap, 14

radiance map, 231
radio frequency identification and geometry, 278
radiometric compensation, 255
radiosity, 5
  method, 235

ray-triangle intersection, 199
reach-in systems, 38
read-back, 53, 177
real image, 20
    pseudo, 157
real object, 20
rear-projection, 38
rectangular quad, 202
reference wave, 43, 244
refinement criteria, 194
reflectance, 226
    information, 230
    map, 233
reflected
    eye coordinates, 155
    screen coordinates, 164
    world coordinates, 155
refracted world coordinates, 162
registration, 2, 4
reillumination, 313
render-to-texture, 53
rendering
    level-of-detail, 193
    multi-pass, 54
    real-time, 5
        photo-realistic, 5
    stereo, 97
    stereoscopic, 71
    two-pass, 165
rendering pass, 54
rendering pipeline, 44
    fixed function, 6
    multi-channel, 172
    programmable, 64
response function, 232
retina, 31
retinal disparity, 32
retro-reflective surfaces, 36
rods, 31

scattering
    Rayleigh, 16
    secondary, 234
    sub-surface, 267
screen space error, 198, 200
secondary screen, 152
selectively refining, 194
self-correcting, 284
self-occlusion, 227
self-shadows, 267
semi-immersive, 7
Shader Lamps, 213, 279
shaders, 64
    pixel, 65
    vertex, 65
shading model, 49
shadow map, 54, 58, 241
shadow mask, 223
shadow-casting, 5
shape-adaptive projection, 277
shuttering, 36
sine-condition of Abbe, 26
Smart Projectors, 12, 276
Snell's law of refraction, 159
SoftGenLocking, 172
software frameworks, 7
solid-state devices, 41
spatial augmented reality, 1, 8
spatial resolution, 175
spatially aligned, 71
specular term, 50
spherical
    lens, 28
    mirrors, 189
spyglass, 152
stencil
    buffer, 52
    level, 53
    test, 53
stereo separation, 34
stereo-pair, 33
stereogram, 44
stereopsis, 32
stereoscopic
    viewer, 3
    vision, 31
stigmatic
    mapping, 159
    pair, 19
stigmatism, 19
stimulated emission, 15

structured light, 268
projection, 267
sub-pipelines, 167
sub-pixel precision, 270
swapped, 52
synchronized pointing, 309
system lag, 5

T-vertices, 195
table-like display devices, 292
Tablet PC, 79
target perspective, 268
temporal coherence, 193
tessellated grid, 178
texels, 54
texture mapping
perspective, 224
projective, 55
texture memory, 53
texture transfer, 192
time-multiplexed line strip scanning, 268
total internal reflection, 19
tracking, 4
electromagnetic, 4
infrared, 4
inside-out, 4
marker-based, 4
markerless, 4
mechanical, 4
optical, 4
outside-in, 4
transformation
curvilinear optical, 175
curvilinear refraction, 161
model, 175
model-view, 45
object-image, 204
reflection, 154
rigid-body, 154
scene, 45
screen, 163
view, 45
viewport, 46
transformation pipeline, 45
transmission, 245
transparent, 39
film screens, 150
projection screen, 150
screens, 149
semi-conductors, 150
tri-linear
interpolation, 191
texture filtering, 176
triangulated irregular networks, 193
truncated tri-pyramid shaped cells, 202
two-dimensional look-up table, 158
two-dimensional tessellated grid, 175

ultra-portable, 284
umbral hard-shadows, 226
underdrawings, 313
uniform illumination, 227

variational calculus, 202
vergence, 32
vertex split operations, 193
video mixing, 71
video see-through, 67
virtual
environment, 2
images, 20
objects, 33
points, 20
projection canvas, 268
virtual reality, 1
Virtual Showcase, 11
visual artifacts, 310
visual conflicts, 294

wall-like display screen, 39
wall-like projection systems, 292
warping, 271
wave optics, 66
wavelength, 244
white-balancing, 231
window on the world, 84
window violation, 86
windows mode, 306
workbench-like display screen, 39

X-vertices, 195

z-buffer, 52
zero parallax, 34